D0504092

The WATER ROAD

A Narrowboat Odyssey Through England

PAUL GOGARTY

ROBSON BOOKS

First published in Great Britain in 2002 by Robson Books, 64 Brewery Road, London
N7 9NT

A member of **Chrysalis** Books plc

'Pathology of Colours' appears in *White Coat, Purple Coat* published by Hutchinson,
1989, copyright © Dannie Abse. It is reprinted here by permission of the Peters, Fraser
and Dunlop group on behalf of Dannie Abse.

'The Red Wheelbarrow' appears in William Carlos Williams's *Collected Poems*
published by Carcanet Press Limited, 2000. Reprinted by permission of Carcanet Press
Ltd, copyright © the Estate of William Carlos Williams.

The extract from *Idle Women* by Susan Woolfitt (1995) is reprinted here by permission
of the publisher, M. & M. Baldwin, copyright © Harriet Graham and Adam Woolfitt.

'In a Little Wigan Garden' was written by Harry Gifford/Fred E. Cliffe, copyright ©
1934 Campbell Connelly & Company Limited, 8–9 Frith Street, London W1. Used by
permission of Music Sales Ltd. All rights reserved. International copyright secured. The
song was recorded by George Formby on 21 March 1934 and subsequently appeared in
the film *No Limit* in which he starred.

The extract from *The Road to Wigan Pier*, copyright © George Orwell 1937, is
reproduced by permission of A. M. Heath & Co Ltd on behalf of William Hamilton as
the Literacy Executor of the Estate of the late Sonia Brownell Orwell and Secker &
Warburg Ltd.

The excerpt from *FASTER*, copyright © 1999 by James Gleick, is used by permission
of Pantheon Books, a division of Random House, Inc.

The extracts from Joseph Campbell's *The Masks of God: Occidental Mythology*,
copyright © 1964 Joseph Campbell, are reproduced by permission of Souvenir Press, 43
Great Russell St, London WC1B 3PA.

The maps on pages 146–9 originally appeared in *The Canal Age* by Charles Hadfield,
published by David & Charles copyright © Charles Hadfield Estate. The maps were
prepared by Richard Dean in 1968 and are reproduced here with his permission. For
larger and more accurate mapping, see the Historical Canal Maps series compiled and
drawn by Richard Dean, details from 49 Grange Road, Biddulph, Stoke-on-Trent ST8
7RY, or at www.cartographics.co.uk

British Library Cataloguing in Publication Data
A catalogue record for this title is available from the British Library

ISBN 1 86105 515 3

Typeset by FiSH Books, London WC1
Printed and bound in Great Britain by Creative Print & Design (Wales), Ebbw Vale

Contents

Illustrations

18 Medmenham Abbey, once home to the debaucheries of the Mad
    Monks of Medmenham at the Hellfire Club.

All photographs are by Paul Gogarty unless otherwise stated on the
caption.

To Susanna, Larne and Max

Acknowledgements

A very big thanks to all those who feature in the book and to the dozens who don't – the friends who visited me; my niece, Karen Norman, for typing up notes when my laptop crashed; Mike Constable from Thinktank Science Museum in Birmingham; Kevin Maslin, canal photographer; Dr Carl Chinn, Birmingham University lecturer, radio presenter, author, you-name-it; Mike Taylor, Birmingham city planner; Andrew Fielding, curator at Lion Salt Works in Marston; Tom Brownrigg, principal engineer at Anderton Boat Lift; Amanda Nash and Emma Middleton at the Black Country Museum; Nick Fazeley of the Dudley Canal Trust; Gary Taylor of Brindleyplace; Dennis Fink, Docklands Manager; Robin Smithett, photographer and author; Neil Stanton at Bulbourne Workshops; Tim Colghan at Braunston Marina; Heather Duncan, British Waterways Special Projects Officer in Tamworth; John Redmond of the Environment Agency; Willie Wilson at Imray Laurie Norie & Wilson; Tony Conder at the National Waterways Museum, Gloucester; John Potter of the Arnold Bennett Society; Mike Clarke, canal historian and author; all at the Sobriety Centre; Gay Blake-Roberts at Wedgwood; Headley Terry at the Etruscan Museum at Etruria Junction; Nigel Richardson at the *Daily Telegraph*; and GeoProjects and Nicholson for their guides.

Particular thanks go to Peter Woodley and the rest of the team at Adelaide Marine; Ed Fox, Eugene Baston and the other B.W. staff; Waterways Holidays U.K.; and to Nick Crane, a man for whom going the extra inch is just a starting point. I had so much help, inevitably someone has been overlooked for which I can only apologise. Last but not least, the biggest cheer is reserved for my family who had to put up with so much and for whom my journey also became their own.

'Time is a gentle deity.' Sophocles

'Believe me, my young friend, there is nothing – absolutely nothing – half so much worth doing as simply messing about in boats.'
Kenneth Grahame, *The Wind in the Willows*

'Life is a bath. All paddle about in its great pool.' Seneca

'The great majority of people feel that reality is terrible.' Tschumi

Introduction

The year unfolded apocalyptically enough. Forty days and forty nights was a mere shower compared to the wettest year since records began in 1766. While rural England was still reeling from B.S.E. and salmonella, over the winter engorged rivers broke their banks, homes flooded and livestock drowned. Seemingly overnight England had become Florida Keys without the sunshine.

Then, sixteen days before I was about to set sail on my four-month circumnavigation of the greatest national concentration of waterways cut solely for navigation in the world, another disaster hit. Foot-and-mouth was discovered in Northumberland and spread across the country like tabloid gossip. Virtually the entire canal network was closed down for the first time in more than two hundred years.

God clearly had it in for us – pestilence, floods, plagues, hurricanes, hailstorms and train crashes (the Old Testament missed that one). They were all signs. A whale even washed up on a Sussex beach. What more proof was needed? The country was at the end of its rope. All that was missing was a great fire.

Funeral pyres were duly burning across the country as I eventually got under way six weeks behind schedule. Fields grazed by sheep for a thousand years were empty, tourism had fallen into the abyss and the stock market was about to crash. It was clearly the end of the world.

I slipped the moorings to the sinking island and sailed off into the parallel universe of the canals. If mainland England was currently damned to hell, the canal network was blessed to heaven. The inland waterways are the retreat of romantics and eccentrics, a water margin where the best of England is preserved and a new future is being dreamed.

There were many reasons for my 900-mile journey by traditional

1

narrowboat. The renaissance of the canals and attendant resurrection of inner cities suggested this was the moment for a waterborne English pilgrimage. The turn of the millennium and my own half-century also required marking in some way. But perhaps the greatest motivation of all was a desire to understand a conundrum that had puzzled me on my brief earlier visits to the canal. Why is it that on some anonymous high street just fifty yards from the waterfront people hurry past each other seemingly persecuted by life, and yet those same people walking the canal towpath smile and greet everybody they meet? Sailing through England's back door, over the next four months I hoped to discover the nature of that spell.

In September I would return home just in time to watch the Twin Towers crumble. The apocalypse was alive and kicking. There were lessons to be learned and I'd seen the writing in the trembling water.

A Four-month, 900-mile Journey through England

1

High Tide

The Tidal Thames

I'm slowly lowered into the dank bowels of the lock. The upper world with its bedlam of cars, trains and humans is lost to the tumult of rushing water. *Caroline* is straining at the leash, desperate to squeeze through the raised paddles and be off. I tighten my grip on the rope to the lock's holding bar. Eventually the great moving walls to the river swing back and I enter the vast belly of the Thames.

Unlike the empire builders turning left on the ebbing Thames with their unslakeable thirst for open water and new colonies, *Caroline*, my 50-foot hired narrowboat, is snatched like a twig by high tide and spun to the right. It feels like it's been an eternity coming, but I'm finally under way, barrelling up the ancient salty river to England.

'But why do you want to go?' That had been my twelve-year-old son Max's first question when I initially broke the news of the journey. His eyes were x-rays, boring through the muscle and fat looking for the cancer of betrayal in the bone. How do you convince a twelve-year-old that there's nothing suspicious about saying you love him on the one hand and that you're taking off for four months – 'That's for ever!' – on the other?

'I just have to do it,' I answered limply.

'Who's making you?' Max asked the unanswerable.

There are no grey areas in childhood. The world is flat: you either want to do something or you don't. Max cuts through bullshit like a diamond through paper. Throw in some philosophical waxing or middle-aged angst and his eyes simply turn up the heat.

I grabbed a lifeline. 'You'll come and visit some weekends and maybe you can stay for a week during the holidays.' Max lifted an eyebrow to signify – what? – distrust? Resignation? Acceptance? Who knows. What-

5

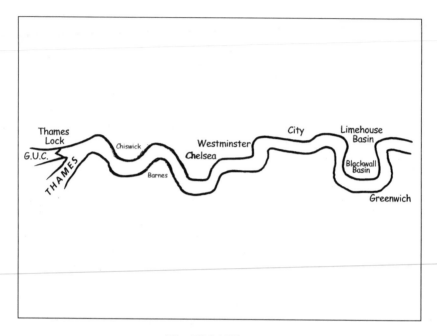

The Tidal Thames

ever it was, he turned on his heels and returned to the greater certainty of his PlayStation.

That first abrupt conversation took place in autumn 2000. Now, as I make a start of things out on the tidal Thames, it's early spring 2001 and the world feels as huge as it did when I was a small boy. An old lighter barge strains against the current making for the still-gaping jaws of the Blackwall Basin Lock and its journey's end. A tour boat swings wildly round my stern, returning upstream having completed its London story. *Caroline* rolls in its wash, engine growling, tiller reluctant to bring the boat to heel. The buffeting unceremoniously reminds me of the rules of water navigation: keep to the right and stay put. There is, however, another rule, one that's not in the manuals but which overrides the first: bigs rule the water just as they do the playground.

Right now *Caroline* feels like a leaf on the Amazon. Like countless sailors before me departing the capital, my stomach is in emotional knots. 'So why do it?' I hear my Max asking. The excitement, bubbling along with the tide, answers for me. I am ditching the virtual highway for the sensate world of escape. Some who measure the world in betrayals and deficiencies may well see my journey as just another midlife crisis, an indulgent and selfish abandonment of Max, his fifteen-year-old sister Larne, and my wife Susanna. Those of a more restless, more generous disposition may be gentler judging why I simply had to make the trip.

Riding the tide, London is a flick-book of images. In minutes the unloved £750 million Dome birthday cake with its lopsided never-lit candles gives way to exuberant screeching gulls piping me through the seaport of Greenwich. Down by the jetty are the diminutive *Gipsy Moth* and lumbering tar-black *Cutty Sark*. High above them, in the royal park the red time-ball of the Royal Observatory plummets as it has at 1 p.m. every day since 1833 as a reminder to shipmasters to adjust their chronometers. I look down at my own watch, vibrating on a wrist glued to the tiller. Six minutes slow. I leave it as it is.

At this same moment no doubt, a battery of camera-toting tourists are capturing loved ones straddling the Greenwich meridian, one foot in the western hemisphere and one in the east – irrefutable proof for empire builders that the world expands outwards from Greenwich. The line they straddle marks the zero meridian, the beginning of time. It also marks the beginning of my journey, the casting off from Greenwich mean time in search of English slow time.

Where I'm heading, into a spider's web of canals known collectively as 'the Cut', I'll have no need of longitude or latitude. On a couple of previous weekends pootling the margins of England by narrowboat, I've already discovered it to be a world no less magical than Alice's, a secret network as powerful as ley lines. The Cut's history, gouged in the British landscape deeper than the chariot tracks left by the Romans, is a hidden garden flashed with kingfishers and colourful traditional narrowboats; a parallel universe ringing with the laughter of water gypsies, the thin cries of bats, and the booming Stygian silence of tunnels.

Over the next few months, *Caroline* will carry me on a single continuous figure of eight from Greenwich up through the Midlands and across the Pennines before returning south. What would take fourteen hours to drive in a car, will take one-third of a year by narrowboat. The route I travel will be as much James Brindley's dream as my own for it was his eighteenth-century utopian vision of a *Grand Cross* of man-made waterways linking our major cities to the sea via the main navigable rivers of the realm – the Thames, Severn, Mersey and Trent – that gave wing to the industrial revolution and the current network.

The timing of my journey, it seems, could not be better. The current renaissance of our inland waterways is being heralded as the New Golden Age of the Canals: waterfront cities are being reborn and as many miles of canal are currently re-emerging annually as opened at the height of late eighteenth-century Canal Mania.

Just as I have no need of longitude or latitude to navigate my course through back-door England, neither do I need rocket science to operate *Caroline*. To hire a narrowboat, you don't even need a driving licence – you simply aim the pointy end the way you want to go and stop when the canal does for its arcane ritual of lockgating.

Yes, anyone can do it but anyone can just as easily make a complete arse of themselves too. Particularly if, like me, you're sailing solo. As a novice with just a handful of days boating under my belt I dread the crowds attracted to locks spotting me dropping a rope into the drink or smashing into a gate. Mooring has even greater potential for humiliation – attempting not to knock the flower boxes off roofs of parked boats nor scratch freshly painted hulls, while veteran boaters line the banks, arms folded, shaking their heads despairingly at every new incompetence. Then of course there's the possibility... probability... of falling in the canal itself.

What's making me anxious at this moment, however, is not locks or mooring. It's the fact I'm doing the toughest stretch – the tidal Thames – first, and according to my Nicholson guidebook *no* hire companies allow their narrowboats onto the river. Well, mine did. But what does that say about Adelaide Marine?

Along with the anxiety, and the emotional rollercoaster I'm riding leaving my family, I'm also on the crest of a tsunami of optimism and excitement. A lethal cocktail. And right now things are moving at a pretty hairy pace. In a couple of hours I may well be pootling at 3 m.p.h. along a tideless canal with plenty of time for swapping tea bags and chat, but for the time being steering demands constant adjustment, particularly each time *Caroline* is picked up by a powerful slipstream and partially turned.

From beyond the muddied banks of Deptford Creek comes the crunching sound of unwanted cars being pummelled in a breaker's yard. Beyond it stands the seventeenth-century Mayflower pub, named after the ship that sailed from its moorings to invent the modern world. A few hundred yards further on, the waterfront gallery of the Angel overhangs the river. This is where the *Mayflower*'s captain, Christopher Jones, bought supplies and a crew, and where Captain James Cook holed up while preparing for his antipodean adventure.

A police launch skips past, barely touching the water, creating an instant wash of guilt. A swarm of helicopters buzzes overhead. As I make another long serpentine loop with the river, Tower Bridge soars above me, an airy castle from a children's book across which hordes ebb and flow between work and home. After the Tower comes the architectural stew of the City and the gun-grey menace of H.M.S. *Belfast*. I sail on through sun gulleys with ominous rain clouds stacked on the horizon. On the left, the converted warehouse of Hay's Galleria, where tea clippers once unloaded from India and China, is filled with office workers munching on bagels and sipping lattes in trendy cafés beneath a glass-roof atrium. The pages of the flick-book keep turning.

The wash from *Caroline* funnels across to a small Mississippi mud beach where two boys, aged ten or eleven, are skimming stones from broken pallets beside a half-eaten sheep. Above them stands the preposterously small replica *Golden Hind* in its puddle at the corner of the Old Thameside Inn. Alongside it is a sign – 'The World Encompassed 1577–1580'. Nearby, a man is sitting astride the embankment wall, blowing on each individual crisp from a Walker's packet as if they're steaming chips, before popping them into his mouth. On the steps down to the mudflats from the Founders Arms, a black man and a white woman are passionately entwined, a Yin-Yang mandala. It feels wonderfully springlike and optimistic.

I'm growing in confidence, looking around more, taking in the sights, when suddenly the engine breaks back into consciousness. It's making funny – funny as in strange – noises. Over the next few months I'll learn to be more sensitive to *Caroline*'s moods and complaints. For the time being I try to ignore her belly grumbles.

I start picking up occasional snatches of music, and even loud conversation, from the bank. One voice reminds me of a teenage girl I'd heard a week earlier standing outside the Tate Modern screaming at her father, 'Where the hell were you? I was worried!' Dads do get lost.

Leading from the Tate Modern, the ethereal Millennium Bridge, the first new river crossing in central London for a century, floats above me, its aluminium vertebrae on the move, a giant sci-fi stingray leaping from the water.

The drizzle that has intermittently pattered on the deck becomes more insistent. Clouds have thickened, rain is now hammering down, and the enveloping darkness has transformed the river into steel corrugations. I think back to the illuminated skyline I strolled under that night a week ago outside the Tate Modern when a heavenly fire seemed to burn above the city. That was when I first started thinking about omens. What was the saying? Red sky at night, shepherd's delight? I don't know if we have shepherds any more.

I let go of the tiller and make a snatch for my waterproof. By the time the tiller's back in my hand I'm heading towards the northern bank. Fortunately nothing's coming.

I sail under Hungerford Bridge and Charing Cross Station. Later that evening I'd read in the *Evening Standard* that just a few hours before I passed it, a train ran amok hitting buffers and gouging a ten-foot hole into the concourse. The country is in the grip of Mad Train Disease after Mad Cow, Mad Weather, Swine Fever and Foot-and-Mouth.

As I approach Westminster Bridge a police launch does a show-off water plough to bring itself alongside *Caroline*. What have I done? 'Work's being carried out on Westminster Bridge,' a uniformed man shouts across. My neck muscles relax. 'Only one section's open. Follow the green, not the red, arrow.'

'But I'm colour-blind,' I shout back.

The uniform stares long and hard, assessing whether I'm just a loser or trying to take the piss. 'Follow the arrow and avoid the cross then. You're not cross-blind are you?' He smiles. If I *was* trying to be clever, he'd got me back with interest.

Just as I'm about to pass under the bridge I catch sight of my friend Nick Crane with his battered old Olympus shooting pictures of me and *Caroline*. Once through the far side, I slow and look up again. Nick's waving frantically between camera clicks. I wave back and suddenly my Conradian sense of destiny awaiting in the water margins of England is waylaid by the river police. An inflatable carrying ten frogmen slaloms alongside. I slow the boat. With the engine purring, I can now hear sirens seemingly coming from every direction, and above me helicopters are playing out scenes from *Apocalypse Now*.

I remember with a shock it's 1 May – May Day. And May Days are not what they once were. The city is heavy with paranoia. Eight hundred demonstrators have gathered on bikes at King's Cross. In other parts of town Anarchists, Greens, and common-or-garden anticapitalist foot soldiers have assembled. Six thousand police officers are on duty and another three thousand are waiting in the wings in case things turn nasty. Buildings have been boarded up and Westminster schools closed for the day. What has most bearing on my own present situation, however, are the hundreds of protesters milling about on the other side of the Houses of Parliament I am now idling outside.

The frogmen in the inflatable are clearly patrolling the building and, spotting an anarchist cyclist on the bridge – Nick – and his mate in the boat – me – exchanging signals beside the seat of British government, reasonably enough assume an attack is imminent.

I smile and raise a hand as greeting. The flat faces circled by sinister rubber hoods are expressionless. I smile some more. Not a flicker. They come in closer and escort me past the neogothic pinnacles of Parliament, the House of Lords and Westminster Abbey and don't quit – with another showy turn – until *Caroline* slips beneath Lambeth Bridge.

It's half an hour before I stop looking over my shoulder. The abandoned Battersea Power Station, known the world over to Pink Floyd fans, floats past, followed by a hundred-foot-tall Peace Pagoda with gilded Buddha and wind chimes. *Caroline* sprints between the towers of two medieval parish churches at Putney Bridge marking an ancient river crossing, and passes a string of rowing clubhouses. Ahead, a boat, crewed by six females, is powering towards me on the wrong side of the river with only their muscular backs to guide it. I move closer to the bank to accommodate them. Bad mistake.

It's then that I first hear fingernails being dragged across a blackboard. Thankfully the sound vanishes. Whatever it was trying to get in below decks has gone. Then it returns, just as in every schlock-horror movie. The next time it stops, so does *Caroline*. A swarm of butterflies takes off inside my stomach. The engine panics, straining for escape. But we're going nowhere. A wave hits the side of the boat and *Caroline* lists ominously.

The river is a bubbling cauldron and there are no other boats about (the rowing crew is probably back in its boathouse now enjoying a cup of tea). I try to accelerate into the centre of the river but we're as beached as the whale I recently read about washed up on a Sussex beach. I try reversing. It seems to work. I'm moving. In reverse, however, it's impossible to have any control over direction and within seconds the sand bar has us again.

I have only just started my great escape and already I've been snared. I put the boat into forward gear once more and turn the tiller from side to side. Not an inch. *Caroline* snarls. What would Harry Houdini do?

I try poling off the sand bar. The crowd on the bank that had been just two bored construction workers has grown to twelve or thirteen people who've interrupted their towpath journey for a bit of entertainment and to bear witness to an unfolding disaster in which a boat sinks after an epic struggle and its solo, sad occupant is drowned. As they listen to the item on the evening news, they will be able to say to whoever they share their lives with, 'I was there. I saw it.'

After maybe twenty minutes of deepening despair, I eventually manage, through a sequence of forward and reverse moves, and chance, to manoeuvre *Caroline* so she faces downstream. By putting all my weight on a pole at the stern, I slowly inch back the way we came. Free of the sand bar, I turn in a wide arc upstream and promise *Caroline* there and then never to leave the middle of a river even if the QE2 suddenly looms above us.

I have now almost been arrested and almost sunk. It's enough for one day. At Brentford, I pull off the river, following a sign welcoming me to the Grand Union Canal. It's taken three and a half hours to get here from Greenwich. I'm exhausted but still high as a kite. I'm at the end of river time and I won't see the Thames again until I'm on the final leg of the journey, travelling downstream from Oxford's dreaming spires.

I enter a short creek and sail straight into Thames Lock, disturbing a woman in her twenties who's attending flower boxes lining the wall. It's the last manned/womanned lockgate I'll encounter until I reach Wigan.

As *Caroline* and I rise from the maelstrom of the river up into the sanctuary of the Cut, I ask her where I can moor up. 'Anywhere you fancy. Pull up along on the near bank and pop back for a cuppa if you like. I've got another hour till I knock off.'

I can't believe my luck. I entered the English canal system two minutes ago and have already been invited to tea. I'll soon discover that hospitality and the Cut go together like love and marriage once did.

I moor up, stash my map, guidebook, spectacles and waterproofs, bolt the back cabin from inside and exit the fore-end, padlocking its doors as I go. In minutes I'm sitting in Anna Ward's skinny lockside cabin with a perfect view back to the Grand Union as it clambers out of the Thames. The canal is a millpond. James Brindley, the father of the inland waterways, likened water in a river 'to a furious giant running along and overturning everything'. He found, however, when he created England's first arterial canals that 'if you lay the giant flat upon his back, he loses all his force, and becomes completely passive, whatever his size may be'.

There is something of the retro-mod about Anna's short, parted hair. Her flat moon-shaped face, however, is interrupted by a ring through one eyebrow, she has a small heart tattooed on one hand garlanded with stars, and her neck is tracked by musical notes dancing above a stave. She is dressed in a British Waterways black sleeveless jacket and tapered green trousers. Anna moves easily and comfortably round the ten-foot by three-foot hut. As she connects the kettle to a plug socket on the floor, she tells me she has only been a lengthsperson – 'men call it lengthsman' – at Brentford a year. 'Before that I was based at Little Venice. Before that I was a coach and lorry driver. Before that a motorbike courier... There, you have my whole history now.'

Anna's clearly a well-rounded bloke, as happy shinning up ladders in a lockgate as she is tending her flower boxes. Having temporarily disappeared behind a cloud of steam, she reappears with two mugs of milky white tea and settles herself into a chair opposite me.

We quickly discover we have common roots in Chiswick: a bedsit I endured for a year in the seventies was located a couple of hundred yards from Strand-on-the-Green where Anna swam, fished and explored away her childhood.

She pulls out of her jacket pocket a crumpled packet of Cutty Sark tobacco and expertly conjures a roll-up which she then offers me. I decline and ask about the work. 'I open the gate two hours either side of high tide. Outside that punters have to book a passage by ringing ahead and giving an estimated time of arrival.'

Opposite us, no more than twenty feet away, is the red brick wall of Anna's two-bedroomed flat for which British Waterways (B.W.) docks £170 out of her £1,100 monthly earnings. When she gets a long weekend or week off, she usually joins friends on their boat on the Kennet & Avon, near Hungerford.

As we talk, my elbow accidentally knocks over a highly polished brass instrument that is leaning against my chair. I manage to catch it in time. 'Gauging rod,' Anna introduces it. 'Toll men used it for measuring loads 'cos boatmen would cheat and hide part of their load in concealed compartments. The rods would measure how low the boat was in the water and this'd be compared with the boat's charts which showed how low she was empty. The difference indicated how much they paid to use the canal. Simple really. All the best things are.'

I ask Anna about a cassette sticking out the top pocket of her jacket. She runs her fingers through her straight, greased hair and stares at it as if suddenly remembering a job she hasn't done. 'A demo some geezer sent me this morning. Wants to know if I want to join his band as lead guitar...I might.' Her freckled face breaks into a smile. 'Yeah, I might join. Make a fortune, buy a narrowboat of my own and do what you're doing. Four months...you jammy sod.'

'Do you get bored here then?'

'Nah. This is one of only two electronically operated manned lock-gates on the London Ring. I'm lucky. Sometimes, too, I get to help people beached on the mud. That's fun, having to get them off before the tide comes in because if they don't, get off I mean, it can rush in so fast, the boat sinks...People can be very careless.'

I quickly change the subject as water continues to rush through the cracks, finding every weakness in the lockgate. Although no one has entered it since we sat down, Anna claims hers is a particularly busy lock. 'Get loads through here and all sorts too. One thing though never changes...' Anna pauses. I wait. 'And it's the same with the manual locks too...' Again Anna waits, milking my expectation before letting the rope out. 'It's always the bloke steering and shouting at the back and the missus, for the sake of a quiet life, playing mate, doing the running about with the ropes and the heavy work.' I laugh. It's easy to laugh with Anna. She's comfortable with herself and with others. 'In your case, being solo, you've got no choice. You've got to be husband and wife.'

'Do you get many like me coming through on their own?'

'Not many. A few. Couple of weeks back we had a single-handed traditional Ovaltine boat pass through with its old livery and "Drink Ovaltine" written on the side. Well smart.' Her face breaks into a smile at the memory. 'Some of the boats that pass through can be a hundred years old you know.' I raise an eyebrow and nod my head to show I'm impressed. 'No, there ain't that many solos but there are a lot more younger people on the Cut than when I first started coming down. A lot of 'em sell up and live permanently afloat. And they're not all poor neither. There's wealthy yuppie types who come through on their all-singing, all-dancing floating palaces with driers, microwaves, stainless steel kitchens, Jacuzzis, the lot. They'll be having helipads next.'

We laugh. 'The flashy kitchens are usually down to the wives. The bloke comes home from work pissed off as usual and says all matey like "Let's go live on a boat." The missus groans and then agrees, "But I don't want no crappy work surfaces mind."' Escape is always preferable with a down landing. Anna pauses and then resumes her momentum. 'The other thing there are a lot more of is *noddy boats*.'

'Noddy boats?'

'What you're sailing. Hire boats. Less working boats, more pleasure boats.' Anna knows Adelaide Marine, the company I've hired from, well. 'They ain't the flashest but they're pretty reliable and they're the only company that lets its punters out on the Thames.'

Fearing the wrath of working boats, I ask Anna if she has any tips for me on my voyage. 'There's no proper freight carriers on the Cut, except up on the Aire and Calder which you'll do as you turn south after the Leeds & Liverpool. The only working boats on the rest of the network just carry a bit of domestic coal and diesel, flogging it to other boats and waterfront homes. Tips?' She pushes out her top lip with the bottom and rocks her head from side to side to indicate it's no big deal. 'Take your time. Don't do nothing you're not sure of – like don't moor up somewhere when you're not sure it's deep or wide enough to get in and out. You can easily get stuck on the bottom and it's a right bugger gettin' off. Familiarise yourself with the boat and, as you're solo, always have a long mid-line close to hand 'cos you won't be able to be at the back and the front at the same time.'

I'd already been given this last piece of advice by several other people when planning the trip. Today I'd first used the mid–line to hold fast in Blackwall Basin lock and I'd also used it when mooring just half an hour ago. I actually managed the job pretty neatly though I say so myself. I had the mid-rope handily placed at the stern so that once into the bank in neutral, I simply stepped off, nonchalantly, with the rope and fastened it to a metal mooring ring before securing the bow and stern ropes.

I'd expected Anna to be more impressed by my plan to negotiate single-handedly six hundred plus locks, plus lift and swing bridges, but she seems to think nothing of it. 'There's an old boy on the estate' – Anna points to anonymous apartment blocks looming above her lockside home – 'who natters to me about the war years on the Cut. Now that was tough. He'd get paid weekly in the boozer on Friday nights like everyone else and there'd be a right old knees-up. Real community it was then.' She looks wistful. 'The war years weren't bad all the time. Living on the Cut was fun when there was more freight like there was then; all the old boys mucking in together, making bacon sandwiches and tea on the boats. Don't do it no more though. And it weren't just men either. Have you heard about the Idle Women?'

I nod my head having read several books written by the Second World War waterway volunteers. Anna looks surprised. 'D'you know why they were called Idle Women?' she checks me out. 'Wasn't 'cos they were lazy.'

'Wasn't it the I.W. badge on their lapels – Inland Waterways – wasn't that it?'

Anna nods. 'They were the Land Army of the waterways. Kept the canals open during the war. Tough they were, too, carrying fifty tons of freight up and down to Birmingham – coal, armaments, all sorts.'

'Wasn't their depot somewhere near here?'

'Very close to where you're moored. Gone now though.'

I ask Anna if she's read *Troubled Waters* by Margaret Cornish, an Idle Woman I'd had the good fortune to meet who's promised to meet up with me during my trip despite the trouble she has with her 85-year-old legs. 'Ahh, that's such a good read,' Anna explodes, 'and have you read *Idle Women* by Susan Woolfitt? I couldn't believe how unfazed she was when her boat and the butty got mashed when that German bomb hit the Cut.' I nod my head. Together Anna and I reminisce about the night Susan, Kit and their new recruit Ruth had a hole blown through the side of their boat and the doors torn off.

A little later, back on *Caroline*, I pull out Woolfitt's book from the small library I've stashed on board. I find the incident on page 98 where the author talks of the sky being lit up like daylight and the boat tipping 'to a fearful angle' before her survival instinct kicked in. 'I remember thinking: Well if I *must*, I don't mind being blown to pieces, but to be sunk in the Cut and drowned struggling in the cabin is quite another thing... I raced my thoughts up the steps and out of the door.' Within minutes of viewing the devastation outside – 'the fire burning like a furnace over the water and fire-engine and ambulance bells shrilling urgency from all the streets round us' – Susan was padding back down the stairs; 'I think we all want tea and if the Primus is still with us I'm going to make some.' Tea. It's always tea with the British. If we have a cup of tea, everything will be all right. I settle down with a final cuppa. Tomorrow I head inland.

The cabin feels cold. It will take some time for us to get used to each other, to warm to each other. I feel a tingle of excitement. Tomorrow will be the start of slow time and a journey beyond my own earliest marker, Tommy Steele's 'Butterfingers'. Tomorrow, up at the Gauging Lock I'll sneak through England's back door into a 3,000-mile network of inter-connecting inland waterways leading to heartlands and lowlands, bucolic idylls and satanic mills, engineering marvels as brilliant and complex as the spinning jenny, and tunnels haunted by asphyxiated boatmen. It will be a journey across the face of England with all its exultations and darkness: traditional pubs and rave boats; glorious sunshine and sheeting rain; canals that have been resurrected and enjoying their new summer, and those still abandoned like shameful secrets.

I pull out the bed, switch off the light, and turn away from my family.

2

Moving but Barely

The Grand Union Canal

I wake to an overcast sky and the bleariness that's a sure sign of hangover, imminent illness, or the legacy of sleep broken by capricious giants violently bashing you against a canal bank all night because you left the mooring ropes too slack.

Like Susan Woolfitt, I put the kettle on and, while waiting for it to boil, unpack books from the last kitbag, beefing up my mini-library. Exploring my new home, I discover the living space can work as a continuous flow-through, or by pinning back the toilet doors at right angles, be transformed into three self-contained units. The boat's design is as simple as it's ingenious, as practical as it's magical. *Caroline* is the childhood den I created from a linen cupboard when we lived in Malaysia, the hideaway I made of a boiler room in an abandoned building in Shropshire, and the dilapidated gypsy caravan my sister and I requisitioned in a Wiltshire wood. As the son of a sergeant in the regular army I was used to upping sticks every two years and the dens were the security blanket I dragged round with me.

The bare facts concerning *Caroline* are that she was built in England in the 1960s and somehow made her way to France. Abandoned near Narbonne, on the Canal du Midi, she filled with rainwater and sank. Peter Woodley, who'd already bought two other English-built, French-sojourning narrowboats (all France's narrowboats continue to be British-made), bought the wreck off her English solicitor owner and brought it back to Adelaide Basin in Southall. As Peter's base, Adelaide, was named after a British Queen (the consort of William IV), he christened the new addition to the family after Queen Caroline of Anspach, wife of George

II, who acted as regent during her husband's absences abroad.

Caroline had had her continental fling and seemed more than happy to be pottering along England's backwaters, embarking on the real Grand Tour. After so little time together, we're still on formal terms: I have to check the manual to be reminded how to turn on the heating; and only remember that the gas supply switches automatically to a fresh bottle when an old one runs out, after wasting an inordinate amount of time fiddling around.

All I'm praying, as I warm up the engine and turn the ignition key, is that *Caroline* has forgiven yesterday's indignity of running her aground. I need not worry. Like every other day over the next four months, she starts first time with a gruff bark that rises several octaves when she gets into her stride. Within days we'll be singing a duet.

I slip the mooring and we're off under a whiteout of cloud. It's not the most auspicious spring weather, nevertheless my eyes and ears feel reborn. I can't quite believe my luck at escaping the spray-gun attack of modern urban existence. The day's planning is simple: follow the canal. Unlike driving on roads where you constantly fret about missing a turn, canal junctions are major events often separated by days or weeks rather than minutes.

As I set out on the water road, cherry blossom drops like confetti and a passing swan bristles its wings into an angelic arc. Beside the towpath in a marine yard, a fire is burning in a tin drum while a Danish flag flutters from a nearby houseboat and a windmill spins manically on a Dutch barge. Within minutes I'm at Brentford Gauging Lock, setting out to investigate, swinging my windlass as extravagantly as a Greek does his worry beads.

The lock, I soon discover, is operated electronically by my B.W. Yale key and I have no need of the windlass. I sail *Caroline* and close the lower gates but a tree trunk wedges between them like meat trapped in teeth. I reopen them, get back on board, reverse, and manage to manoeuvre the trunk out the gate. I feel rather proud of myself. The rest of the operation is a breeze.

Before resuming my journey, with no deck hands to carry out general duties, I nip down into the cabin to make coffee. While waiting for the kettle to boil, I open the *Shell Guide to the Inland Waterways* and on the first page of the section on the Grand Union Canal, read 'From the

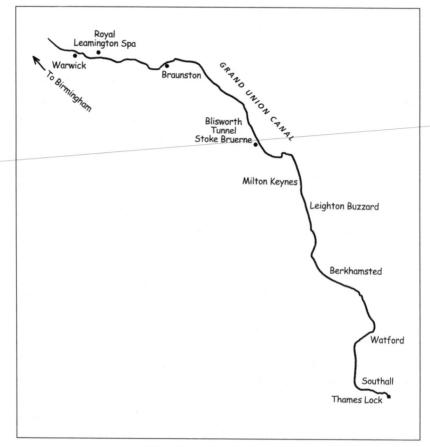

The Grand Union Canal

Thames at Brentford, Middlesex, to Birmingham, Salford Junction, 135m 2f [that's furlongs to those who still understand them] with 165 locks.' I freeze. More than one lock a mile. An experienced boater with crew does a lock in ten to fifteen minutes. I estimate it will take the rest of my life to reach Brum.

I skirt Brent River Park, 400 hectares of woodland, hedges, fields, ponds, streams and a succession of parks that stretch seven kilometres from Western Avenue at Hanger Lane all the way to the M4 at Osterley. A tube train hammers overhead one moment and the M4 the next. In between, despite being enmeshed in London's suburbs, I could well be in leafy Hampshire.

As I watch the tailgating cars and stacked planes awaiting their turn to come in to land at Heathrow, *Caroline*'s regular, dependable 3 m.p.h. doesn't seem so slow. There are no hold-ups, no queues, progress is absolutely dependable and even better, sailing the country's first nation-wide navigation network I'm rediscovering the joy of travel.

In the evolution of transportation, it seems to me, we've accelerated from pony and trap to the speed of sound and on to virtual travel, in what historically is the blink of an eye. Somewhere along the way, the journey got hijacked. Road arteries clog, and train commuters rage at every delay in getting them where they need to be next. Before long the only way to get everything done in 24 hours in super-accelerated lifestyles may be by living lives virtually – working, playing, having sex and relationships in cyberspace.

A butterfly cult of instant redundancy helps fuel the illusion of progress, serving short concentration spans and a fundamentalist faith in the notion that all things are hot and then they're not. Several careers are needed in a lifetime, and several husbands and wives. We travel through life like a dose of salts.

As *Caroline* sails through the snowstorm of blossom, I try to remember when the idea of my slow-time journey first formed. Singapore. Almost a year to the day. As I sat on the thirtieth floor of a hotel watching traffic streaming across a causeway into Singapore's forest of concrete and glass, ineluctably my only other visit to the steaming metropolis swam into consciousness.

Forty years earlier, aged nine, I'd been bundled, along with my brother and sister, into a trishaw by my father. It was our first time abroad and together we made a hallucinogenic journey through a night perfumed by the sweet sickly smell of durian ('tastes like heaven, smells like hell') and ringing with the guttural cries of hawkers yelling from stalls garlanded by bare electric bulbs. The night crackled, the trishaw kept its steady, human-powered pace, and we children in the back exploded with excitement.

Forty years on, alone in that hotel room, watching cars silently stream behind the plate-glass window to and from a city that never slept, it struck me something had changed in our relationship to travel. To those shuffling the causeway, journeys were simply inconvenient interludes between departures and arrivals; *dead time* which could only be filled *usefully* by catching up on business calls on the mobile. Movement and multitasking provides the illusion one is getting on, getting through. But getting through what? The reality is, those commuters are like Sisyphus daily pushing their rock to the mountain top, only to find it waiting for them again at the bottom the next morning. It was that evening in Singapore that I started actively seeking a more humane navigation through life.

Twelve months later, that navigation is taking a breather at a second lock. I stare at the ingenious beauty and simplicity of the mitred gate, an invention that belongs in the upper echelons of history-changing advances. Europe's first pair of horizontal swinging gates, very similar to these, was operating on a Milanese canal in 1497, invented and installed by one Leonardo da Vinci, when employed as engineer to the Duke of Milan.

The water steps by which the canal climbs the contours of the country have a noble pedigree then. If you can't go through or round a hill (James Brindley's preferred method), then you take giant's steps up it. If the obstacle is a river, you simply create a metal trough, or aqueduct, and sail over it (known in the eighteenth century by an incredulous population as 'Brindley's castles in the air'). When it's a gentle incline, you may have just one lock gate. When there's a larger rise to negotiate, the locks come thick and fast and are known as a flight.

The gate I'm standing in front of is the bottom lock of the Hanwell Flight, a sequence of six that will lift me 53 feet 2 inches in less than half a mile. The black and white staircase climbing the hillside, designated a

Grade I scheduled ancient monument, is as impressive as it is daunting to the novice. British Waterways cares for around 3,000 listed structures and 130 similarly scheduled monuments. Indeed there are moves afoot to get the entire network listed as a UNESCO World Heritage Site.

The locks may be pretty, but all bar the bottom one are against me (i.e. the water inside the gate is at the upper-pound level and in need of lowering before *Caroline* can enter).

Having swung into the first lock, I whistle triumphantly. I close the lower gate I've just sailed through and, with the windlass, raise the paddles of the upper gates. After twenty minutes, the lock has still not filled but I notice a narrowboat that's moored beside a garden in the pound above is now resting on mud instead of water. Shit. I look at the water in the lock again. It still isn't rising. I look at the upper pound. I look at the heavens, hoping a deceased lengthsman might take pity on me and descend to act as ghostly crew member and advisor.

When the mist clears, I notice the last person through the lock has left the lower-gate paddles up and so water is just rushing through, draining the upper pound. Basic stuff you should check. With my face flashing like a beacon, I chase back to close the paddles. I then run and slip my way up to the next lock while rain – which has appeared, instead of the phantom lockman, out of nowhere – hammers down. I half-open all four paddles on the next lock to let water through to refloat the beached narrowboat. Thank God no one's on board. Thank God no one's about full stop.

By the time I quit the final lock in the flight, I have spent two hours getting the water levels right to climb the hillside. It's hard to imagine *Caroline* has ever had such an incompetent helmsman. I go gentle on myself and call it a learning curve.

Already today I have discovered coots run on water, swans rule the Cut and that there is truth in the saying that the quickest way to work a lock is slowly. I get into the slow groove. Relax. After all, I escaped unspotted. Despite the hammering rain, despite my running aground, despite my Hanwell cockup, the feeling of elation is unquenchable. Without wishing to sound too religious or poncey, I'm feeling blessed. When our children were born, I got my head down and brought in the bacon. As the children became self-propelling, and my wife retrained as a psychotherapist, I told anyone willing to listen that, at some point on

the hazy horizon, my turn would come for some time out. The hazy
horizon, now clear of its rain clouds, is here.

I moor at Tesco's at Bull's Bridge and stock up on provisions. The
supermarket has replaced a major British Waterways depot where
Margaret Cornish, Susan Woolfitt and the other Idle Women used to
await orders during the Second World War.

After Hayes I squeeze past an old working boat breasted up with its
sister butty (a butty is a motorless water trailer that evolved from old
horse boats). From the hold of the latter, an attractive woman spinning
yarn greets me. I blink a couple of times for visual reassurance and throw
a greeting back. Leaning on the tiller in a Kangol beret, staring into
space, is what I presume to be her male partner. I moor beyond the boats
and wander back.

Kim Russell confirms that the couple are indeed as much a pair as
their two boats. In fact they've been together long enough, notwith-
standing their youthful good looks, to see two daughters and a son depart
for equally nomadic lives in the merchant navy.

Kim is dressed in battered boots, a blue turtleneck sweater and khaki-
and-oil two-tone Pepe jeans. I suspect the long sideburns emerging from
the black-peaked Kangol cap ('second-hand shop') have been worn that
way since they were fashionable last time round. His day has been spent
lending a hand to a gang of B.W. workmen ferrying giant reels of fibreoptic cable to points along the Cut.

Hearing Kim and me chatting up on deck, Sue's head emerges from
the butty, inches from a dipper and washbowl decorated in bright
traditional roses-and-castles narrowboat designs. Close up, she's even
more attractive than down in the hold. Dressed in tight black jeans with
a shirt tied at the waist, and her long hair held up by a brightly coloured
scarf, she is the picture of alternative good health.

Sue invites me on board for tea with an accent that still drags her
Dutch roots with her. We descend the steep stairs into the back cabin of
the butty *Naples* and sit on a narrow upholstered bench that converts
into a bed. Opposite us, in touching distance, is the range Sue keeps
going all day for warmth, bread-making and tea.

During the day Kim occupies the motorised boat and the butty
becomes Sue's personal estate. Kim notices me looking from wall to wall,

mentally measuring dimensions. 'Small isn't it? Eight-foot-six by six-foot-six total living space. That there,' he points to a small cupboard to the right of the range, 'is the soap hole.' He moves his finger. 'The cupboard door there that drops into a table, is known as the bed hole. The engine room's the engine 'ole, bridges are bridge 'oles and sewerage stations – you guessed it – stink 'oles.' There's a pause, a swig of tea and a shaking of the head for effect as Kim sums up, 'Yup, there's an awful lot of 'oles on the Cut.'

Above the stove are lace plates and handspun wool drying after Sue has washed the oil out. 'Sue spins inside when it rains and out when it's sunny. Then she plies the wool on the wheel, dyes the skeins different colours and lays them out to dry on the trees along the towpath. It looks like an art gallery out there sometimes.'

Before they bought *Naples*, and bow-hauled it (pulled it by hand) 32 miles through 29 locks over three days from Tipton to Fradley, Kim and Sue had already spent a decade sailing the more challenging waters of the Atlantic. When they finally jumped ship, they returned to England and joined an alternative circus in the mid-eighties where they sold hash cakes and brandy coffee instead of popcorn in the interval.

'Then we set up our own family circus with our kids because Kim's never been very good in larger groups,' Sue continues the saga while filling the copper kettle for another brew. Having finished my mug, I excuse myself, nip back to *Caroline*, fetch a bottle of Sauvignon Blanc and then shoehorn myself back into the cabin for more story spinning. It reminds me of primary school when we used to sit on the carpet and Miss Firth would tell us stories of King Arthur, Guinevere and Sir Gawain.

When they travelled with their own circus, Sue did stilt walking, juggling and a fire show. Kim also did fire breathing until he burnt down a jewellery shop in Spain. They then bought a showman's trailer they pulled behind a fire engine and moved on to Ireland.

'When I first went to look the trailer over at the Hoffmann Circus, I found a bear staring at me inside.' Kim grins and refills our glasses. 'We had a coconut shy, a skittle alley and swing boats all stored in the back and still managed to fit the five of us in.'

What strikes me most listening to their stories is that Kim and Sue have chosen a life rather than stumbled into one. And according to Sue, of all the lives they've led so far, their present one is the most interesting. 'At sea, there's a more hectic rhythm to the day. You live with the weather, feeling for the changes in anything. You're on watch twenty-four hours a day. The Cut's a much easier, far less stressful life but it's more interesting and unpredictable too. You can't get lost, you can't sink unless you're stupid, so you have time to watch, and follow whatever comes up. It's a perfect life for us because we don't need much. We cut wood, help out if anyone needs help, I make a few hats and tea cosies we sell at Camden Market or festivals.'

Rays of sunshine start raking down through the hatch so we decide to take our wine and tea up onto the deck so Sue can continue spinning her yarn and yarns. She settles onto a stool and starts turning her wheel. Next to her is a sign, 'Chopped wood 50p a bag'. An Asian man on a bike stops on the towpath beside us, cuts some bluebells with a pair of scissors and cycles off, one hand holding the handlebar, and the other his blue bouquet.

Kim settles himself against a pile of logs. 'This is the only way in the U.K. we can live the travellers' lifestyle with integrity without running foul of the establishment or police. We stay within the law but outside society's constraints. I enjoy cutting wood and the wood feeds the fire, which in turn feeds us. Seems to me other people often do jobs they hate just to feed themselves. By reducing our needs we can live on very little and not impact on the environment. We're lucky. It's a different attitude of mind. Being a traveller is a way of life for us and company for a traveller is our luxury.'

These water gypsies never stay put anywhere more than a couple of days. 'We like to keep moving. It's good to see the seasons in different parts of the country. Down on the Kennet and Avon one early summer morning last year – it must have been around five-thirty – I looked out the hatch and within an eighty-yard stretch there was a swan, moorhen, duck, hare and a deer. All we needed was a badger to make the set. People watch nature programmes and often don't realise that a couple of hundred yards from their back door down on the Cut they've got one of the best nature walks going. Look at it here now. You'd never know that we're right by an industrial estate halfway between ugly Hayes and West Drayton. It's like deepest country.'

I look up. Pink cherry blossom explodes along the towpath. Each time

I see it, I hear my dad singing 'Cherry Pink and Apple Blossom White' in the kitchen as he prepares dinner. With the bluebells and occasional bloom of daffs I could indeed be in Hampshire instead of on the outskirts of the capital.

From beneath the nearby bridge, a pneumatic drill bursts into life, tearing up the peace and the towpath, so that the fibreoptic cables Kim has ferried can be laid. When England's first itinerant navvies (navigators) dug the canal network by hand, the Cut would have been a sea of mud unlike the minor disturbance that's experienced today as yellow-vested B.W. workers courteously warn those passing to watch out for rogue cables.

The fact that it's the towpath rather than the water that's carrying freight in the twenty-first century and that the cargo being carried is mythic – underground dancing pictures invisible to the naked eye – makes this an even more fantastical future than that imagined by those witnessing, open-jawed, the miraculous opening, two centuries earlier, of what was then known as the Grand Junction navigation.

The Grand *Union* was created through a number of river and canal amalgamations between 1929 and 1932. At the time, there were still as many unpowered vessels like *Naples* being pulled by horses as there were motorised boats.

I decide, with the first really clear sky I've had since departure, to spend a couple more quiet late-afternoon hours moving on before calling it a day. I leave Sue spinning and Kim wandering off for a chat with the B.W. workmen.

It's already becoming apparent that the journey will be as much about stopping as going. I pass a stretch of permanent moorings and slow down as canal etiquette demands. Some of the narrowboats are spick and span, belonging to the brass-polishing brigade, others are garden centres of pot plants and flowers. There's an abandoned, half-submerged 35-footer; and an old working boat laden with pots of paint, tarpaulins, and oil drums. Some have cats on board, others dogs, one a parrot, and at the stern of the last boat, a woman is releasing five white doves from a cage. The names of the boats – *L'Escargot*, 'Not so fast in the morning, Slow in the afternoon' – tell their story. Waterside concrete and brick homes with their dinghies, ceramic ducks and wooden decks have turned themselves inside out to the canal. The water itself is the colour of pickled olives and boiled socks.

A necklace of lakes and shimmying reeds slow-mo by. Three swans make a fly-past. Fields run up hills and long raking sun-shadows down. I become mesmerised by the subtle gradation of colours on a moorhen as it seemingly propels itself across the water by means of rhythmic thrusts of its yellow-tipped red bill, its red frontal shield following in due course along with brilliant white undertail coverts. I have identified the coots as the adolescents of the Cut, indifferent to size, demeanour and seniority. A pearl-grey male mallard cruises by with its wedding ring round its neck and its delicately flecked female honey by its side. A yellow wagtail hops along the towpath. An egg floats by in a nest, like Moses in his basket. By day's end my reborn eyes ache with seeing.

Over the next few days a rhythm establishes itself. Mornings start with birdsong and boat checks (topping up water and oil, greasing the tube, freeing debris from the propeller through the weedhatch), followed by a few hours pootle, moving but barely. Days often end with a pint at a canalside pub followed by a stir-fry in a kitchen where I need only take one step in order to reach everything I need.

Temple Thurston once compared canal-boat travel to 'motion asleep'. It's a sleep that's ineluctably taking over my life. My practical dexterity is also improving. I discover a way to steer without the use of hands, freeing them to make notes or consult maps. This is achieved by placing the tiller in the small of the back and swaying my body in the opposite direction I want to go, balancing on one leg like Ahab. By now I am leaving the toilet seat up, the surest sign that my journey is defiantly solo.

Just before Springwell Lock, near Rickmansworth, moored amid a flotilla of craft, I spot an old Woolwich working boat named *Hyperion*. Internal bells ring. Excitedly I realise it's the boat Margaret Cornish ran during the Second World War and that she's told me to keep an eye out for.

I moor, cross a bridge and walk past a farm to reach it. On the towpath I meet a man with tousled hair and spectacles who looks to be in his late twenties. I ask if he's the owner of the *Hyperion* and he nods the suspicious nod of all renegades before his natural enthusiasm for his boat takes over.

'One half of a pair, but *Hyperion*'s engine's being fixed and the other half, a butty by the name of *Serpens* is up on the Aylesbury Arm at the minute.' He pauses momentarily. '*Hyperion* was the first small Woolwich to be built and launched by Harland and Wolff in 1935.'

I have a suspicion I have met my first anorak but as soon as I mention Margaret Cornish, Kevin Jackson – like Anna and Sue before him – immediately invites me for tea.

I descend into the black hole of a matchbox-sized back cabin. *Hyperion*'s living space appears to be about half the size of Kim and Sue's and as far removed from orderly elegance as coal is from fibreoptic cable. I step off the stairs almost onto Kevin's bed. An amorous cat and the smell of recently fried bacon greet me, along with a free living space roughly eighteen inches by two feet.

The yellow and red of the cabin gave way to grime decades ago, a crumpled sleeping bag lies across the bed, and spilled rice is covering the floor like the memory of a wedding. 'Accident last night,' Kevin explains.

Kevin is 39 and single. Looking around it's not hard to see why. For the past seven years he has scraped a living as a contract worker carrying out towpath maintenance and anything else that's going on the Cut. During a stint as a soldier, the only regular job he's ever known, he managed to save enough to buy the boats. 'First off I lived in a flat for a year when I came out of the forces. Never again. That's why I got the boats. Keep moving, that's what I say. If you don't know who your neighbours are, it's best to get on with everybody. First rule of the Cut. I like that. I like it too that those who don't obey it, get bad names fast – like rubbish dumpers, 'cos the Cut's our back garden isn't it? Or lock thieves.'

'Lock thieves?'

'Boaters that nick your lock even when it's against them. That's doubly bad 'cos they're wasting water as well as pissing you off. Don't jump the line, if there is a line that is, and usually there isn't.'

Decorating the walls are a few old-fashioned pin-ups, some photographs of his family and a few nick-nacks he's collected. Kevin drops a hatch down to rest our mugs on – Princess Di for me, Elvis for him – restricting the floor space further.

When Kevin bought *Hyperion*, he knew nothing of its past. 'One day in a lock, someone asked if the boat was *the Hyperion*. I stared a blank. "The *Hyperion* Margaret Cornish drove." They tried again. I still hadn't a clue what they were on about so the woman hares off and comes back with a book called *Troubled Waters* about Margaret's wartime experiences. Eventually I bought a copy and then last year, out the blue, I get a phone call from B.W. – I was dredging on the Basingstoke Canal at the time – and they ask me to escort Margaret, Olga and Virginia up the Grand Union for a celebration of the Idle Women at the Crick Boat Show.'

'How did Margaret cope?' I ask, remembering she's 85.

'She took over the steering and wouldn't let go. Loved it. When we got to Crick, B.W. had a few big cheeses for her to meet but Margaret preferred the proper boaters, not the Fancy Dans, like when we bumped into Vicky bow-hauling a butty on the Aylesbury Arm.'

Could it be the same Vicky that Margaret had requested I pass on a colourful blanket she'd knitted to? It is. Unfortunately the beneficiary of Margaret's kindness took off yesterday to visit her mother. Kevin offers to stow the blanket for her. God knows where.

My host bends awkwardly in the cramped space to refill my mug. 'Long-term back trouble,' he explains. 'It was a bit better yesterday. Vicky did some reflexology on my feet before she left. That eased it.'

A voice calls Kevin from the bank. I follow him up out of our warren into daylight. Daniel, tall, bespectacled and wearing a Second World War flying hat, wants to know when Vicky's due back.

Daniel lives on the *Argo* and *Tadworth*, another pair of working boats, moored behind *Hyperion*. Kevin introduces us. Daniel in turn introduces his wife Debra to me by pointing to the rear of the *Argo*. It is now Debra's turn to point, this time with raised paint brush as greeting, before continuing transforming the magnolia-coloured hatch at the stern into brilliant red. 'Debra paints for a living like Vicky. She specialises in roses and castles,' Daniel explains.

For seven years Daniel and Debra delivered coal and diesel along the Grand Union on the *Argo* and *Tadworth*. Then they decided to get married but couldn't afford to pay for the wedding so Daniel took a job in an office in Hemel Hempstead working as a computer analyst for the Dixons stores group ('Got the experience for the job operating our fuel business on a Psion organiser'). He's still there.

Kevin starts the outboard motor on his inflatable and whips me back to the far bank to pick up the blanket. It's not quite the royal launch but the speed comes as a shock after *Caroline*. Down below decks I find the bright blanket made of brown and yellow squares. I put it in a Tesco bag and hand it to Kevin. We say our goodbyes, he whips back to *Argo* for more tea and I get moving again.

The horses may have been pensioned off but the spirit and the boats remain. The Cut in the landscape I'm already discovering is not just the legacy of an ingenious but outmoded navigation, it's a slice through society, and rural and urban England. But above all it is a national memory chest. The canals provide a route to another England, one we've travelled so far and fast from: a refuge of distinctiveness and regionality against the onslaught of the hypermarket, TV lives, virtual highways, and the creeping paralysis of suburbia. Tiptoeing into the inland-waterway network, I've entered a time capsule where people smile and greet each other from passing boats instead of snarling across congested roads; where courtesy is shown, assistance given without asking, and the speed of movement – a sedate walking pace – always allows for the exchange of stories.

At Cassiobury Park, I unlock my bike from the roof of *Caroline* and spin across the River Grove where a labrador is lifesaving a small branch. I zip down an avenue of 300-year-old lime trees and past football fields to paddling pond where several small children are playing. I stop to watch them splashing each other. Slowly it dawns on me I'm being watched too. The toddlers' mothers are tracking my every move and it's not because they fancy me but rather because I am probably a dirty old man. I look down at my grubby jeans and sweatshirt, feel my four-day bristle and understand. I cycle like the wind back to *Caroline* and take my first shower for five days.

With three inches of scummy water slopping about at the bottom of the shower, I discover the water pump has come away from its tubing. I ram the head on and hold it together underwater while I pump. I'm a little apprehensive about the electric wires sharing the slimy water with my hand and wonder how long I'd lie decomposing on the floor before being discovered if I'm suddenly fried.

Twenty yards ahead of *Caroline* is a fisherman by the name of Kenneth Ball, who I discover is a flautist and bandmaster with the Royal Air Force Central Band when he isn't fishing. So far this morning he's caught two bream, 'but I threw them both back in the water so you can't see them.

It's etiquette, you always throw them back to keep the numbers up.'
Madness. If none are eaten surely a time will come when boats won't be
able to move for wading through fish.

Ken is clearly the man to get advice on what kit I'll need for my son
Max – apart from the £10 rod and reel I've already purchased – so that
he can fish when he visits me. Ken needs little coaxing. 'Split shot which
you pinch on the line. And you'll need a float – you just have to
experiment how far you place the float from the hook. It has a rating on
it so you'll know how to cock the flight.' As someone who doesn't know
his cock from his rating, I'm out of my depth already but the kindly Ken
still has one final tip for me. 'Remember in winter fish are lower in the
water than in summer.' He nods twice sagely. I flash the same fishy
Masonic sign back and make a bolt for the next lock.

Up ahead thunders the M25 – the road that leads nowhere. But then
if you're tailgating – just as the North Circular, A40 and M4 were when
I passed them – then you could say all roads lead nowhere.

I am completely at ease with lockgates now. Sometimes I operate them
alone, sometimes I share a lock and stories with another boat. On other
occasions someone walking a dog or simply hanging about lends a hand.
However we manage it, unfailingly the gates open, and out I pop each
time like a newborn child. Along with the stories and help, from time to
time there are more tips too. At one gate a woman painting a picket fence
advises, 'If you want a good night's kip, make sure you sleep with your
head facing the fore-end, it sits higher in the water and will stop your
blood swishing about in your head.'

The route, now tickling the toes of the Chiltern Hills, is as English as
dandelions and daisies. Children wave from bridges, and cyclists wobble
along the towpath avoiding fishermen munching on their doorstops.

The canal tiptoes into Berkhamsted, passing Bridgewater Boats where
I hired my first narrowboat a decade ago along with a cautionary tale
from the owner Mike Foster. Like all the best morality plays no lives
were lost and humour is as important an ingredient as hubris. A rather
cocky hirer phoned on his second day afloat to complain his boat had
been stolen and to inquire why Mike had not fitted better security.
'Where was it last?' Mike asked.

'We left it tied up in a lock while we went and did some shopping.
When we got back it was gone.'

Bingo. 'Go look in the bottom of the lock,' Mike sighed.

The biggest danger in locks is if the boat gets caught on the cill. 'You get lifted askew, water runs in, and it sinks in minutes.' Just in case I hadn't got the message, Mike underscored it, 'No one leaves a boat in a lock.'

And so instead of mooring inside a lock, I do so beside the Boat riverside pub and have a pint on the terrace. It's a sunny spring evening and there are more imbibers outside than in. Nearby is the all-white narrowboat *Zaporzive* plastered with fliers advertising yoga and jazz dance sessions in Berkhamsted Town Hall. I knock on the double-padlocked door but its owner has flown, possibly literally, if his claim, 'Miracles by appointment', painted in the shape of a lifebuoy, is to be believed.

That night at the Canal Side Curry Pub, over badami gosht, I think about the marginals I've already met who find refuge on our inland waterways. Someone has left a *Daily Mirror* on a bar stool next to me. On page eight I find a small item suggesting some members of the Real I.R.A. are also hiding out on England's canal network. The feature highlights the fact that the Cut is both a perfect place to vanish as well as an easy navigation into the heart of all our major cities.

At the Tring Summit I follow a delightfully sinuous cutting, gently making my way, with the herons and swans, across forever-England countryside. There is a loud rustling in the undergrowth and a russet-coloured muntjac deer rushes past.

Why narrowboating has the image of aquatic knitting beats me. As far as I'm concerned anyone preferring to hole up in their London homes rather than slinking through the countryside on a narrowboat is certifiable. A bottle of wine and a few cold beers taste a whole lot better on deck, than trapped inside four walls and a roof.

In the garden of the Grand Junction Arms, a hog is being roasted and a swelling crowd listening to Wild Willy Barratt and the Sex Beasts. Boaters lie in the sun and sit at tables, swapping navigation tips and rounds of drinks. On the opposite bank, the handsome 200-year-old brick water tower of the British Waterways Bulbourne Workshops reminds me that canal architecture ages far more gracefully and organically than railways and roads. Beneath it is a submerged 80-ton hopper in which a pair of new lockgates are luxuriating in the water. Unlike humans who do everything to avoid the wet, new oak gates need to get in the canal as soon as possible to prevent cracking in the sun. Because of the appalling winter, there's a long backlog of essential repairs, and the two gates are holidaying in the hopper while they wait their turn to be installed.

I am now at my highest point so far. The rain that seeps into the Chilterns here has 500 feet to fall with the canal to London. On the other side of the hill I drop 42 feet through the seven-lock Marsworth flight. Bordering the Cut are four reservoirs that have been designated nature reserves. On the banks, amid the walkers, twitchers, and fishermen seeking to better the 24lb-record pike caught in 1913, is Gladys Phelps. She appears to have her binoculars trained on the sedge warblers and reed bunting flickering in and out of the reed fringe on the far side, home-building after their trip up from southern Africa. In fact she's concentrating on something closer to hand: two great crested grebe in the heat of their exotic mating ritual. 'Marvellous sight,' Gladys sighs appreciatively.

'Do you get many here?'

'More these days. In the mid-nineteenth century, when Rothschild owned the lake, they were hunted almost to extinction because women wanted their feathers for hats. The reservore's played a major part in their reintroduction to our shores.'

Her odd pronunciation of reservoir doesn't diminish the fact that Gladys knows her onions, her local knowledge gathered in a childhood spent living in a lock cottage a couple of hundred yards away from where we're now standing. 'My father was the lengthsman on the Marsworth flight. When I was a small girl we'd get an average of twenty pairs of working boats a day through with their coal, cereal and tar. That was in the forties when there were still plenty of boatpeople about.' Gladys raises her binoculars again and then drops them. 'False alarm,' she smiles. 'You know when I say "boatpeople" it doesn't mean just any Tom, Dick or Harry living on a boat, don't you?' she asks, pausing only momentarily before continuing. 'Boatpeople are the old-timers. The families that ran cargo for generations and lived on their boats. When the winters were bad – and they were in those days – and the canal froze over, sometimes they'd be holed up for weeks here and my father would give their children lifts to school in Marsworth with us. They might pick up a few days schooling here and there at other bases along the Cut but that was the only education they ever got. After school, when the ice was thick enough, we'd skate on it. If there was snow on the banks we'd sledge down onto the ice.'

'Sounds idyllic.'

'It was,' Gladys replies.

On a whim the following morning, I cycle a short way along the underused Aylesbury Arm. Beneath the Wendover Woods nestling on the edge of the Chiltern escarpment, I come upon what appears to be a boat graveyard. A man in his early thirties, with stubble for hair, is throwing a trunk rather than a branch for a playful Rottweiler. Surrounding him is the debris left after a hurricane has torn through the place uprooting trees and scattering boat carcasses.

The scene of seeming dereliction is in fact Jem Bates' field of dreams, a dry dock where old tugs and traditional narrowboats come for resurrection. The uprooted trees are oaks – wind-blown or collected from road clearances – that Jem planks up and leaves for a couple of years to season before using them to restore traditional boats with the same patience and care that other men reserve for Sundays and classic cars. Jem's passion is for the shapeliness, character and warmth of wood. My *Caroline*, with her steel shell, would not find a home here.

In the era of quick turnrounds and speed-is-of-the-essence, I discover Jem takes a day just to winch a boat from the water. 'Every boat that comes here is a veteran – the last wooden narrowboat was built at Nurser's Boatyard in Braunston in the late forties – so they all need lifting carefully.'

The most unique feature of the dock, however, is not the fact that it is the exclusive domain of wooden boats, but rather that it's shared. Men are bent in two caulking and blacking, sanding or sawing, and tiptoeing across planks. Jem offers the occasional tip or shows how something should be done, assisting others in becoming self-sufficient. The yard has become their field of dreams too.

Jem is 33 and 'sort-of-single'. Chrissie, his partner, has created her own wonderland garden which Jem proudly shows me. He sits down by the pond and indicates a chair against the wall. I'm reluctant to move it as it seems so perfectly placed, part of the aesthetic grand plan that aligns a silver Buddha beside old canal signs, and potted coal scuttles beneath wind chimes fanned by field pansies, forget-me-nots, lavender, woad, spinach, artichokes and, finally, the chair. 'Don't worry, there's no feng shui here,' he laughs, reading my mind.

Jem's stubbled head looks more down to genes than style, his jeans and Fruit of the Loom T-shirt sure signs of a healthy disregard for fashion. While studying at Warwick University, he paid his way by working in a boatyard and 'that,' according to Jem 'was that'. His life was fixed. He managed to get a 2.1 in English and American Literature but for the past eleven years, boats have been his life. Chrissie and Jem instead of burrowing, head down, through urban careers, turned a little bit of wasteland into their own eco-friendly slice of paradise.

Encounters on the Cut are as beguiling as they are fleeting: the very nature of the trip – always passing through – means that only *Caroline* and I have anything more fixed than cameo roles. I sail in and out of locks as I do lives. It is a logical progression from a military childhood that saw us up sticks and move every couple of years. As the son of a soldier, you learn to make friends knowing you'll lose them, you learn fast that all things pass.

The abandonment of my own family is perhaps a logical pathological extension of this. By sailing away I'm leaving them before they leave me, proving I can survive our children's departure. The truth is I'm absolutely dreading it.

At the tiller, a week into the journey, I'm already pondering the possibility of following the water road for a good deal longer than my present journey. Perhaps if Susanna and I bought a home that moved round the country, daily new encounters like the ones I've enjoyed so far would go some way in filling the void left when Max and Larne leave. One thing's for sure: inhabiting vacuum-sealed lives in London won't.

As my mind skips across the furrows life's ploughed me, the canal becomes increasingly more remote and impulsive. The green tunnel I followed out of London has long gone. Running to distant hills are open patchwork fields grazed by dairy cows, sheep and horses.

Below Slapton Lock I moor and walk half a mile to the Carpenter's

Arms. Eighteen months earlier, I visited the pub on another narrowboat Chilterns jaunt with my family. Then, as we walked up to the village from the Cut, the heavens opened and we arrived at the pub looking like drowned rats, fearful that a grim child-hating, walker-hating, rain-hating ogre would bar us admission. The kids were famished and I was dying for a pint but the thatched roof suggested tidiness and cutesy order, and a frosty welcome inside.

As I opened the door, a large woman made a beeline for us. I prepared myself for a fight. 'Look at the state of you!' she launched into her tirade. But the haranguing didn't come. 'Let me get you some towels and see what clothes I've got you can change into.' As she popped our wet bits in the dryer, we slipped into our new, decidedly odd but dry outfits, and settled down to hot soup.

The weather this new spring day is altogether different and I'm alone as I enter the bar to find Ann Vogler still ensconced there. Ann remembers my last visit because I'd written about the episode in the travel section of the *Daily Telegraph*. 'We had it up on the wall for a while. I got ribbed mercilessly by everyone for being a debagger.'

With a pint of Haddenham in hand, I browse the books Ann leaves scattered round the pub along with the playing cards and Scrabble. 'We've got another twenty thousand second-hand books out in the malt house with a pretty good section on canals if you'd like to look.' Good food, beer and books – can life get better?

I power through the mixed fish mornay – the best meal I've had since leaving home – and take my beer into the converted library/book store where I find around 150 tomes on England's canals. Hadfield, Rolt, Smiles, Paget-Tomlinson – they're all here collecting dust. I start browsing. An hour slips by before I remember Ann had warned me that an American woman lost track of time and got locked in the book shop a while back. I take three volumes: *Tom Rolt and the Cressy Years* by Ian Mackersey, *Slow Boat through Pennine Waters* by Frederic Doerflinger, and *5000 Miles and 3000 Locks* by John Gagg. When I pay, Ann throws in *The Worst Journey in the Midlands* by Sam Llewellyn, a rain-sodden epic of Noah proportions.

It strikes me *Caroline* looks a bit like Noah's Ark as I walk back along the bank in the penumbra. A couple of swans, the homeless of the canals, idle up to pay respects, or rather beg. They stare forlornly through the window with their begging bowls as I make tea. My home feels like home tonight.

Living on a narrowboat reminds me of when Susanna and I inhabited an eight-tatami-mat-sized flat in Japan's Kita Kyushu, the world's biggest steel town. After a while you wonder what on earth you ever needed all that extra space for. *Caroline* and I are children of the sixties. Not old, not new. In between things like the canals. She doesn't breathe lightly like some narrowboats that whisper their way along the Cut; she has the wheeze of a smoker.

The next morning I wake to midges dancing in sun motes, and birds trilling. I am unmoored from London. The magic has hit: the trains and roads have been erased from the landscape and villages can't even be bothered to make it down to the water's edge.

In the afternoon, the Grand Union takes a piggy back along the River Ouse as it slips between the twin towns of Leighton Buzzard and Linslade past ceramic herons and assorted garden statuary, church steeples and an apple orchard ablaze with spring blossom. I slide into the centre of town: no traffic lights, no queues. In Market Place, the Leighton Conservative Club, Going Places and the Willen Hospice shop mark out the values, aspirations, and fates of the populace. Off to the right is the sandblasted old post office and All Saints Parish Church where a most effeminate Jesus is being crucified beside a blooming magnolia. It's 6 p.m. and Leighton Buzzard is dead. I just manage to squeeze into Waitrose before they pull the shutters down.

At Stoke Hammond Lock, a foot-long pike throws itself out of the water and at my feet. If I was religious I would have knelt down and prayed. Instead I throw it back in the water.

The days keep getting better. Often, for hours at a time, I'm the only boat moving. I pass a man with a cocked rifle, I watch a farmer moving paving stones in a wheelbarrow, and slip past a large Dutch barge named *De Witt Seep*. There are no straight banks any more but protrusions and cavities where wildlife hides and hunts. I watch the vicious courtship of mallard ducks and a child swaying on her garden rope-swing. Sentinel crows, high in beech trees, watch my movement from their castle eyries. I pass Willowbridge Marina where a sign boasts that any boat can be craned out of the water in ten minutes. It's all whistle-clean and business-like and light years from Jem's boatyard.

For the best part of a day I stare through the window of Milton Keynes. The town is a suburban pastoral of nymphs undressing in rock pools, the kind of town encyclopaedia salesmen like – nuclear, smart, self improving.

A statuesque heron acts as Flint McCullough scouting ahead of me. A woman is asleep in a comfy armchair in her sweaty glass conservatory, kids flown, husband, if there is one, at work. In a Barratt-land close, a newspaper boy hands a *Daily Mail* to a man hosing his Peugeot. A small girl waves from one bedroom window, a woman in another puts on make-up.

This is a town with a penchant for bird perches, rock pools, double glazing, ornamental external wall lights, and kitchen extensions. I swing past a moored boat suffused with the heady aroma of a wood-burning stove, and detour briefly into Great Linford, a traditional village marooned in the Milton Keynes agglomeration. Inside the 1,000-year old church, a seventeenth-century rector, Richard (Sandy) Napier, transcribed medicinal recipes dictated to him by the Angel Raphael and regularly held communion with the Angel Gabriel. Amid the tottering, blind gravestones in the churchyard, I read 'Until we are together forever'. The best story of all.

I continue snaking my way along the secret green alley to the Midlands. My back is improving its steering technique and I've started talking to swans. It commenced with a 'Hello swan,' and soon we were discussing the upcoming election.

Graffiti on bridges is becoming more interesting; alongside spray-can tagging and timeless classics such as 'Trish was 'ere', is the intriguing 'Gone but not forgoting' (Bridge 71A) and the acid-fuelled 'I can't see what I read but I know there's no eye.' Beneath one of the bridges an attractive sixteen-year-old girl sits, legs folded, dressed in Levi jacket with hippy bag grasped tightly to her belly. She sits under the modern concrete bridge doing nothing, staring into space. I smile at her and ask if she's OK. She smiles back and nods. Maybe she's heartbroken. More likely she's swallowed something that in turn has swallowed her.

On the Great Ouse Aqueduct, a six-inch lip is all that's between me and a sixty-foot drop to the river. On the far side an elderly American walking with his wife and two friends, notices the Adelaide sign on *Caroline*'s side.

'Did you bring that all the way over from Australia?' he asks, proudly displaying his geographical literacy to his friends.

I assume he's joking and knock the ball back. 'Yeah, the Pacific was a little choppy though.' The man turns to his wife and lets out an admiring 'Phew', before turning back to me with new admiration, 'I bet it was, I bet it was.' They wander off deep in conversation. The man in the stars-and-stripes baseball cap has met a hero.

Beyond them is a line of moored boats, three of which are draped in the red and white flag of St George. Maybe it's not rampant patriotism but an acceptance that, Scotland and Wales having sailed off into semi-autonomy (and wishful thinking that Northern Ireland would do the same), it's time to map out a new identity. The Grand Union is dissolving.

Following a seven-lock climb, I arrive at Stoke Bruerne, a village that is to the canals what the Cotswolds is to land. At the final lock I'm greeted warmly by the same group of four Americans I'd met earlier, perching on the upper lock balance beam, and asking me to take a group photo of them. I duly oblige.

The senior in the stars-and-stripes baseball hat then gets into a gallop with his local knowledge. 'You know working boats used to come through here in ancient times and the lock keeper probably sat right here.' He indicates the balance beam beneath him. 'They didn't get paid in those days but you'd see the old guy in the pub in the evening and . . .'

'I'm sorry, but could I get by while we talk?' I interrupt his dissertation as gently as possible. 'I need to get across the beam to lift the other paddle.'

The American is taken aback. 'Oh sure.' He steps off the beam before incredulously asking, 'You have to do it yourself? Geez. Elaine, you hear this, the guy has to do the locks himself.'

Having slinked silkily with *Caroline* through the willows, blossom and the Americans' admiring gaze into the lock, I pose for a picture with Elaine. Beyond the lock are two lines of moored boats flanked by a couple of restaurants, a pub, some pretty cottages and the Stoke Bruerne Canal Museum. A sign on the bridge declares Northampton to be six and a half miles away and Milton Keynes ten.

Inside the converted warehouse that is the Stoke Bruerne Canal Museum the most memorable exhibits are the moments captured in real lives in black-and-white. One is of a girl walking a shire horse that's pulling two fully laden boats. With no health care, no school, and no pocket money, she's responsible for fifty tons of freight and has walked all day. From the photograph, taken in the 1940s, I'd guess her age to be five.

Alongside the photograph is one of Sister Mary, the Mother Theresa of Stoke Bruerne, who brought boat children into the world between 1930 and 1960 and is still spoken of reverentially by all who knew her. A third photo shows Brigadier Fred Fielding and his wife Ivy, operating their Salvation Army school boat around the same time. In the photograph, the brigadier looks as fierce as an American evangelist, but his reputation was considerably softer: reliable friend to the boat people first, Christian missionary second.

I ask the woman behind the desk at the museum entrance about an oil painting on the wall depicting an old boatman with his van horse and butty. 'That's Jo Skinner, the last of the long-distance horse-drawn freight operators. Everyone else after World War Two turned to diesel. Not Jo. He stuck to the horse and only retired in the mid-fifties, when it fell in the canal. Those canals that weren't already closed were really dangerous because no maintenance work had been carried out for years. When the horse tumbled into the water from the collapsing towpath, it was the final straw for Jo. He threw in the towel.'

The man who'd captured Jo for posterity, with consummate timing appears at that very moment in the doorway from the shop. Brian Collings is the manager of the museum as well as the windlass of the inland waterways, the keeper of its secrets. The curator of tales and water wisdom, artist and raconteur is just returning from lunch.

Before he hares off for his next meeting, Brian whisks me up to the first floor and swings back the wooden doors that corn sacks once disappeared through, hoisted down to waiting boats. Beneath us a 40-foot and a 70-foot narrowboat are entering the lock and gongoozlers (those who gather to watch boaters) have gathered on the overhead bridge to observe. I presume the overview of all the activity going on below is what he's brought me here to see, but Brian is pointing to a small white hut beyond the lock.

'For sixty years during the nineteenth century, that's where the leggers kicked their heels waiting for work. When a boat arrived bound for the Blisworth Tunnel, a few hundred yards north of here, two leggers would hop on board. Once they got to the tunnel, they'd lie out on two planks projecting like wings from either side of the boat and inch it – walking their feet along the walls – slowly through 3,057 yards of darkness.'

'What happened to the horse while the boat was being legged through?' I ask.

'It was walked across the hillside by the junior crew member and hitched up again on the other side.'

Brian pauses and shifts his pointing finger forty degrees to the thatched pub opposite. 'And that,' he stabs his index finger three times, 'is the most important place in the entire village.' The finger leads to the Boat Inn. 'If you fancy meeting up for a pint and a chat that's where I'll be this evening.'

I get to the Boat Inn just before eight and sit at the bar chatting with Nick Woodward, who runs the pub with his brother Andrew. I guess him to be in his early forties. He's dressed in a blue polo shirt that partially conceals a thin gold chain. Nick is the latest in a family line that has run the pub since the nineteenth century. 'My dad, Jack, ran it before me – he still has a flat in the pub and will be in later for his usual pint – but the family have been here 123 years of the 140 it's operated as a pub.'

The public bar where I'm sitting is pretty much as it was when the first Woodward moved in: the original undulating flagstoned floor, the brick fireplace, the oak benches. The only incursion from the twentieth century is a large mural of the Cut painted by Brian Collings in 1968.

The flow of our talk is interrupted by a steady stream of regulars to the bar and the crash of a wooden 'cheese' smashing into a diamond of nine skittles on the leather-covered alley next door. 'Northamptonshire skittles – the only pub game alive today I reckon that's still free!' I recognise Brian's voice. I turn to see him approaching the bar, already pulling out a pouch of tobacco for his greedy pipe. He's wearing the same broad boater's black leather belt over black trousers he was in at lunch time. He has, however, changed his shirt and, by the look of his wet slicked-back hair, has just jumped out of the shower. Despite his raking cough he lights his pipe and sucks in hard on already sunken cheeks.

We adjourn to a large leather-covered round table whose gnarled oak trunk is festooned with nails. 'They're there to stop the hobnailed boatmen's boots wearing away the wood – the leather on top's there for the same reason.' The table is 160 years old and as fabled as King Arthur's. In one book I'd read that the two corpses of asphyxiated boatmen had been laid out on it after a nineteenth-century accident in the tunnel. Brian puts me right. 'Romance. Poetic licence. They'd never fit – look at it for God's sake. For some people the truth's just not big enough... The men were actually laid out on the floor along with the two who survived first-degree burns and were waiting for the doctor to arrive by horse... I like accuracy. The stories handed down need to be as close to reality as possible.'

On the wall above us is a black-and-white photograph of Sister Mary with Eamonn Andrews on *This is Your Life* in 1959. There's also one of Jack Jones, the lengthsman who started the museum, sitting in the pub with Frank Smith. 'Frank was known locally as Craney and he made rope fenders. When he retired, he stopped his nightly visits to the Boat and six months later he was dead. See what happens if you stop going to the pub?' Brian sums up.

The history of the Boat Inn and that of the village and the Cut, are as interwoven as the strands of a rope. 'When the Grand Junction Canal came through towards the end of the eighteenth century, it sliced the village in two and brought about a massive physical, social and economic revolution. An army of navvies descended on the place and services had to be provided for them – food, drink, spades, boots, lime, mortar, ropes, fenders, horseshoes. They all had to be made locally or fetched here.

'All the businesses in Stoke Bruerne became canal-related. Ropes were laid out in one alley – it's still called Rope Walk – stables were created where the pub restaurant is now and the pub also had its own butcher's shop where the skittles are. Where the museum is, that was a corn mill. And you know what's interesting too?' Brian rhetorically asks. 'When I keep saying "where the restaurant is now", "where the museum is now", it's obvious we've had a second revolution too. It happened in the sixties as freight died and leisure cruising took over. Now the businesses are shops, day-trip boats, the museum, and restaurants.'

Julia from the museum appears at the door with a photo album under her arm. 'Thought you might be interested in these,' she smiles and exchanges the album of her life afloat for a pint of Morland. The party is growing at our round table. The first pictures in the album span the years she spent running camping boats between 1976 and 1982. 'Camping boats for kids started when the old working boats were pensioned off. They were basically floating tents. Tarpaulins covered the butty and the kids bedded down beneath them.'

The tattoo on Julia's muscled bicep ripples as she takes the weight of the album to turn the page to a winter day and an iced-over canal. 'Fish would flap on top of the ice and gulls swoop down on them. We used the motorised boat to break the ice ahead and pulled the butty behind. Unfortunately the ice bust a hole in the fore-end.'

The next photograph is of a perfect summer day and a boy throwing himself extravagantly from the camping boat into the Cut. 'Wouldn't do that now would they? Wouldn't be allowed. Parents would warn them about ear infections and the like even though the canal's cleaner than it's ever been. Who has the best time though – the boy in the picture or modern boys?'

At 10.30 p.m. on the dot Jack Woodward pops in for his customary pint in the bar he has spent all 76 years of his life in. He joins the party and immediately makes it clear to me he has no intention of leaving either. 'When I die I'm going to be buried under the top-floor bar.' Jack is a small, frail man with a Chaplinesque white moustache and a voice as faint as a whisper. I lean forward to catch words made more elusive coated as they are in their thick Northamptonshire burr.

Jack was born in 1925, a period when tourists were first starting to find Stoke Bruerne. 'A cycling magazine mentioned you could get teas at the Boat Inn but it were a trickle of visitors compared to today. Right up to the sixties it were coal carrying what mattered. Then in 1963 the museum opened and pleasure boats first started appearing. Now the whole town's leisure and tourism.'

Jack's earliest memories are of the boat people arriving late at night after a twelve-hour day. 'They'd feed and stable their horses here, then go back to the boats for their own tea – whatever they'd caught, rabbit or hare but more often than not it were just potatoes. Then they'd pop back to the pub for a pint. And they usually sat at this table – it were their favourite. Me Dad let them keep a tab and they always paid either in 't'morning or on next trip. Honest as the day is long. Mind, at that time the table were on a swivel and if their pint were low and yours weren't, it soon would be.'

The skittlers have now adjourned and two dogs are wrestling on the floor beside us. Like every other boy in the village, Jack often got into scrapes with the boat children. 'As they wound the paddles in the locks, we'd creep up and call them water gypsies – they 'ated that and they'd chase us. Sometimes they'd catch us and give us a hiding. Sister Mary would sort them out. She were their favourite. Worshipped by 'em she were. I used to get her to syringe me ears when they blocked. She lived in the house where Bruerne Lock Restaurant is now – you can still see her surgery hatch. Before that it were the village store. She'd be turning in her grave if she could see the prices the restaurant charges now. When we had those hard winters and got froze in – that's when we used to get proper frosts – Sister Mary would look after the boat people, give 'em food and medicine. Sixty-three were the worst. You could take a horse and cart out on t'cut the ice were that thick.'

I walk back along the towpath around 11.30 p.m. A mist has engulfed the canal. I find my way to *Caroline*. There's a feeling of an animal about her, a strong reliable swimmer I ride the back of. We've started saying good night to each other.

The 3,057-yard Blisworth Tunnel is just five minutes away. Outside it, a moorhen with seven chicks is turning circles beneath a diffuse early sun, as if uncertain whether to enter or not. Maybe they think the tunnel is populated by mink, their most dangerous predator.

As I enter the tunnel, the temperature plummets and bird song is replaced by the lonely sound of dripping water. The initial penumbra reveals a grotto coated in three-inch-thick calcified limestone. Then the darkness swallows us. I am inside the longest publicly navigable tunnel in Britain (a month later, it will drop to second place with the reopening, after 57 years, of the 3.2 mile Standedge Tunnel on the Huddersfield Narrow Canal).

At any moment I half expect *Caroline*'s headlight to pick out Charon rowing towards us. The ferryman of the dead was beaten to a pulp with a boathook when he refused Heracles admission to the Underworld. I have my baseball bat – the equivalent of the shotgun to the stagecoach and the boathook to Heracles – handy in case he tries any funny business with me.

I stare up the first of seven brick ventilation shafts installed after the accident that Brian had talked about in the pub. The day of the fatal crash in 1861 was particularly cold and the tunnel was filled with choking smoke as all the boats being pulled through by steam tugs had their kitchen ranges on for warmth. The smoke was so thick, visibility was down to a few feet when the tug collided with an oncoming boat and the two men were asphyxiated. I imagine the horror of the scene: the choking smoke, the pandemonium, boats sinking snared on each other and their crews seemingly with no means of escape.

Before steam power, it was the leggers who puffed their way through the hellhole, soaked and exhausted. And before them it was the navvies slowly inching their way to Hades by hand. It seems a miracle, at a time when life was as cheap as labour, that again only two men lost their lives during the notoriously difficult three-year construction (the fatalities occurred when a kibble, being hoisted to the surface with the workmen inside, fell 80 feet).

As I take another shower from water seeping through the hillside and cracks in the arching brick tunnel, I estimate *Caroline* is roughly halfway through and 120 feet underground. It's another quarter of an hour, and another mile, before the pinprick of light in the distance leads me beyond Hades to Elysium. The sun flares through a steep green canopy in which an orchestra of song birds are rehearsing their evening concert, the setting as naturally perfect as that of any river. This must be how Lazarus felt when he rose from the dead.

I slow the throb of the engine to nature's heartbeat. After a mile or so, the embankment flattens and I'm in open country. Horses flick tails and crack jokes. Several miles further, however, near Norton Junction, the rural idyll is shattered by a Babel of traffic. I tether *Caroline* and clamber up onto Bridge 18 to check what all the fuss is.

Beneath me another narrowboat, without a care in the world, slips slowly along the 200-year-old canal. One hundred yards to my right the 50-year-old M1 is suffering chronic angina. Fifty yards to my left, a Virgin train hammers through en route to Birmingham. Beyond it, I can see more cars streaming along the arrow-straight Roman road known as both Watling Street and the A5. In a single sweep of the eyes, I take in

2,000 years of navigation. As those travelling the motorway and railroad do the same, I have absolutely no doubt whose journey is being most envied this sunny May afternoon.

It strikes me, as another train rattles the world, that evolution is perverse rather than progressive. Just as the computer should by rights have come before the pencil, so too planes, cars and trains are primitive experiments leading to the perfection of a narrowboat gliding swanlike along a canal.

I walk under the railway up to the Roman road to the Heart of Shires Centre, a crafty, foody collection of shops selling local products around an open square and bandstand. I order a turkey with hot sage and cranberry baguette from Emily's Tea Room and munch on it in the sunshine. I then stock up on local honey, local apple juice, local eggs, and local apricot-and-ginger pork sausages from a shop two doors down. The portly owner, Richard Wincott, is having a tough time of it.

'Thank God we've had the shop to keep us going. We've only just sold the first batch of pigs off the farm. Couldn't move them because we were in a foot-and-mouth contaminated area. We got £27 for each pig – the factory didn't want to take them because they were so big they couldn't fit on the racks. We bought them eighteen weeks earlier for £32 and they were three times the size when we sold them! And I'm still stuck with my heifers because I haven't been allowed to move them and now they're over thirty months old. Too old. The only way I can get paid is to have them slaughtered and burnt but I can't do that, can I?'

I have to confess that what shocks me most listening to Richard, is not so much his terrible plight but his sentimentality in a business not noted for it.

'No farmer can stomach their animals suffering. We all care about our animals too much, that's the trouble. You become very close. I talk to mine every day. If you're kind to them, they're good to you.'

Richard's nephew in Darlington has had it much worse. One hundred and ten sows and 700 of his bacon pigs were recently slaughtered. 'On the day it happened, my nephew cried. It was all done in eleven hours. By six-thirty they'd killed the herds, taken them away, and the farm was empty. He just sat down and cried. A grown man.' Richard looked close to tears himself.

When I first left London on my inland voyage, six weeks later than

LEEDS and LIVERPOOL CANAL

NOTICE

GARGRAVE LOCK FLIGHT

Due to Foot and Mouth Disease, Gargrave Lock Flight from Holme Bridge Lock No.30 to Stegneck Lock No. 35 are in an exclusion zone and must be operated by British Waterways Staff with boaters remaining on their boats

All the above locks will be operated between the hours of 8:30am and 5pm, seven days per week until further notice.

To ensure boats complete the passage through the locks by 5pm, the last boat into the flight at either Holme Bridge Lock or Stegneck Lock will be at 2:30pm.

No mooring will be allowed within the flight. Temporary mooring will be allowed at Bank Newton Locks where access to the facilities in Gargrave can be gained.

For further information contact the Waterway office on 01274 611303

Dated 23rd May 2001

Foot and Mouth briefly closed the entire network and continued to affect stretches of the Grand Union and Leeds & Liverpool during my journey planned because of the foot-and-mouth epidemic, I'd expected to find a ravaged, diseased countryside. Instead I have found one as glorious and unblemished as childhood memories. True there have been stretches where boaters have been requested not to stop, or to use disinfectant trays stepping on and off their boats, but my encounter with Richard is the first time I've got some measure of the real impact of the latest disaster to hit our shores.

In Braunston the clamminess that's been hanging around like a wet blanket all day becomes a deluge. I walk from the nondescript village through the thunder, torrent and lightning back to the boat. *Caroline* is as dry and welcoming as any house.

When the rain stops I stroll to Braunston Marina, visiting the dry dock with its steeply pitching roof and red brick walls, more redolent of a Methodist chapel than a boat workshop. If I'd been walking in from the eighteenth instead of the twenty-first century I doubt I could spot a jot of difference – boats are still floated in before the sluices are dropped and the water scurries out leaving the dock dry and ready for work.

I move on to the old forge that is now home to a fender maker where I buy a new twelve-metre stern rope made from natural fibre. I then stick my washing in the launderette located in a building built by the Idle Women. With time to kill before I pick it up, I wander the basin covetously eyeing up a newly built £180,000 Dutch barge built on site, and more modest second-hand narrowboats. The *Lucy*, one of the last butties built in 1953, is going for a song. A 57-foot five-berth narrowboat built in 1980 is on sale at £19,950. A 38-footer with a Lister engine is priced £12,950. I start wondering how much it would take to make my relationship with *Caroline* more permanent – my guess is it would cost around £16,000 to make the ultimate downsizing move.

In a corner of the basin is the *Water Lily*. The 40-foot narrowboat is home to Jim and Dorothy Collins who are reputedly England's last pukka boatpeople still living afloat. Dorothy, one year Jim's elder at 79, was born on a butty and began work at the age of five, leading the family horse as it pulled the coal-laden butty from Marston on top of the Ashby Canal all the way to Oxford in a week. As an adult she had her own babies afloat and there is little doubt that both she and Jim will also die afloat.

Dorothy and Jim's last working boat had been the *Raymond*,

constructed in this same yard in 1958. It was the last working butty to be built in Western Europe and, like Dorothy and Jim, found its final resting place in the marina. When the couple stopped running freight, the rest of the Braunston boating community made a collection to restore *Raymond* as a Basin attraction and gave Dorothy and Jim the *Water Lily* to live on in return. For the first time in their lives the pair have hot water and a shower. They've used neither: the hot tap is probably locked solid now as it's never been turned, and the shower is used as the coal shed.

Back in the launderette I transfer my wet clothes into a dryer and move on to the chandlery shop next door. Beside the windlasses, canal books, Grand Union Canal tea towels and sweatshirts, I find a scrapbook of newspaper cuttings on Braunston and the marina. One item is an obituary to Rose Whitlock who died on 17 May 1999 at the age of 81. When her father died in 1937, Rose, aged 20, became captain of an all-female crew aboard a pair of working boats. Her mother, sister and a ten-year-old orphaned niece set the all-time coal-carrying record that same year on one of Samuel Barlow's boats, reloading at Longford north of Coventry a week after departing with a 54-ton load which they had unloaded by hand themselves in London. They had made a negotiated round trip of 206 miles and negotiated 186 locks. In my books that's a more impressive feat than Everest without oxygen.

The following morning, having topped up the gear and engine oil, I'm filling up with water at a B.W. tap, when I get a call on the mobile from Susanna. My father-in-law, Dannie Abse, is in intensive care, critically ill after a heart attack. My own heart sinks. The doctors have given him little chance of surviving the next 24 hours. Susanna asks if I'll return to London and visit him in hospital. She shouldn't need to ask.

The journey from Rugby (six miles from Braunston) takes less than an hour by train. At Euston I put my head down and scurry through the crowds spilling like marbles across the concourse. I treat the return to the sprawling metropolis as a dream, that way I may be able to protect the gossamer spell of the Cut. I descend to the Northern Line underground, cocoon myself in a bubble, take the tube to Belsize Park, and walk down the hill to the Royal Free Hospital where both my children were born.

Outside the hospital, sick people lean against walls smoking. Inside, trolleys are being pushed, tears are falling, people are holding other people up, and one or two are even laughing. Up on the twelfth floor, life is on hold beneath a funereal shroud. Opposite the nurses' desk, as directed, I find Dannie. He has what look like fifty corkboard pins and blue Post-it notes attached to every part of his upper body. The bedclothes are rolled down to the thinning forest of his pubic hair. His face is an imposter, a lifeless mask from Greek tragedy.

Dannie cannot move his head but manages to welcome me, telling me I shouldn't have left the boat. I point out it only took 55 minutes by train to cover what had taken two weeks afloat. He almost manages to laugh.

I touch his arm and tell him how sorry I am. 'Not as sorry as me.' His kindness and generosity make me feel guilty at my churlish reluctance to leave the Cut.

Dannie's wife Joan is shrunken in a corner. Susanna is waiting outside with her brother and sister, trying to be there but to add no weight to the delicate balance of life. They know that Dannie needs to keep all his energy inside, healing, not wasting it on others – 'not letting it into the air' as Susanna puts it.

Dannie is familiar with hospitals having spent a lifetime wearing the white coat of the doctor. But he has also spent another lifetime donning the purple cloak of the poet. As I stare down at the dartboard of life-supports and monitors that his chest has become, one of his best-known poems – 'Pathology of Colours' – echoes through the wards.

> I know the colour rose, and it is lovely,
> but not when it ripens in a tumour;
> and healing greens, leaves and grass, so springlike,
> in limbs that fester are not springlike.
>
> I have seen red-blue tinged with hirsute mauve
> in the plum-skin face of a suicide.
> I have seen white, china white almost, stare
> from behind the smashed windscreen of a car.
>
> And the criminal, multi-coloured flash
> of an H–bomb is no more beautiful
> than an autopsy when the belly's opened –
> to show cathedral windows never opened.

So in the simple blessing of a rainbow,
in the bevelled edge of a sunlit mirror,
I have seen, visible, Death's artifact
like a soldier's ribbon on a tunic tacked.

Dannie is hanging on. It will be several weeks before there can be any meaningful prognosis of the damage done to his heart. Susanna generously tells me I should go back to the boat, that coming home for a day or two will just mean another painful departure. And so I go.

Back on *Caroline*, a black cloud descends. I phone home. The other home, where my parents live.

'I've only just got back in. How do you already know?' My mother confuses me.

'Know what?'

'Oh, you don't. Your father's back in hospital.'

Mum has just that minute got back from Macclesfield General. In the night Dad's emphysema got so bad, he couldn't breathe and she'd had to call an ambulance. A fortnight earlier Dad had just completed the last of 37 daily radiotherapy treatments for prostate cancer.

Mum tells me there's nothing I can do and not to come up. I ring the hospital and speak to my father who's been on oxygen all night and is feeling a little better. I do what all Englishmen and Englishwomen do in a crisis: go to make tea. I turn on the tap to fill the kettle. No water. I try the other taps on board. Zilch. I phone Adelaide Marine and Paul, one of the mechanics, tells me to leave one tap open as it's probably an airlock. Ten minutes later I ring back, 'Nothing doing.' Paul gives me the number of a local mechanic. The local man instructs me to back into Braunston Marina and turn *Caroline* round so I can travel a couple of hundred yards back the way I came to their dock. I am halfway through the tricky reversing and turning manoeuvre when I notice water gushing out of the overflow into the canal. I put *Caroline* into neutral, nip downstairs and turn the tap off. Panic over.

I resume my forward journey. Soon my attention switches to the oil warning light which is flashing on and off. I know the oil level is fine as I'd checked it that morning. This time it's Dave at Adelaide who picks up the phone. He decides there must be a loose wire and tells me to ignore the flashing light. I take off again.

A half-hour later, a high-pitched drill sounds as if it's ripping through the boat's innards. Again I put the boat into neutral and ring Adelaide. This time it's John.

'*Caroline* sounds as if she's in terminal pain.'

'Where's the noise coming from?'

'Sounds like from behind the control panel where all the switches are.'

'Impossible. There's nothing but wires there. Check again.'

I do as I'm told and find that my transistor radio has somehow switched itself on and is screaming above an electrical party. I promise John he won't hear from me again until I return to Adelaide Marine in fourteen weeks time.

I finally admit to myself that the problem is not *Caroline* but me. I moor up at Long Itchington and sleep fitfully for an hour through the afternoon.

When I wake and look out the window a dead carp is floating by. I was hoping the mood would have lifted. Outside, the day is a pointillist canvas of drizzle. The low cloud lends a soft ethereal quality to the landscape. There is a lusciousness to the countryside but I feel bone dry inside. I am not sure whether it is worry about my father and father-in-law, or the return to London that has unsettled me.

I decide finally it's because I feel guilty at abandoning my family again at a particularly trying time. I try to make sense of why I leave them so often. It's my job, I tell myself, I'm a travel writer. That's what travel writers do. But who chose the job and why? Why leave when I find the leavings so difficult?

The truth is that when I'm leaving, all I can see is loss. But it's also true a new vista of possibilities opens as soon as I have left. Freud in *Civilization and its Discontents* identified many human foibles but one is often overlooked. 'We are so made that we can derive intense enjoyment only from a contrast and very little from a state of things.' Continuing to travel, despite the initial devastation of loss, is an affirmation of life and hope; a basic trust in the fecundity of life. If each departure is a small death, each new journey is a resurrection. And life on *Caroline* is a daily procession of openings and possibilities.

In sombre, reflective mood I travel on to Leamington Spa, a dazzle of white Regency architecture and blazing municipal gardens complete with sprinkling fountains and a weir cascading beneath a blue and white cast-iron bridge. As I pass a statue of Victoria a 'Keep The Pound' Conservative Party election bus swings through. 'Chuck it,' I encourage Victoria to lob the mitre she's holding in her hand. It's not the pound I've got anything against: it's the stoking of fears I object to. Don't take chances, conserve, keep the pound, keep foreigners out, don't welcome, believe in the bogeyman and keep the doors locked. It is the antithesis of canal life which turns itself outward to embrace life. Those that can't embrace their time, or others, cannot embrace themselves.

I notice a flyer for Steve Harley and Cockney Rebel who'd played the Leamington Spa Centre last Sunday. So that's where he went. Next week it's Rupert Bear (who seems to have aged better). A two-bedroomed home in Warwick, on the other side of the canal, is advertised in an estate agent's window for £89,000.

Back on the Cut, I'm greeted by a man on a narrowboat moored next to *Caroline*. I know I'm now properly in the Midlands because the head saying ''Allo' sticking out the rear hatch is wearing a Noddy Holder top hat festooned with badges. On Bridge 42 I read 'Self judgement is self destruction.' They're weird in the Midlands.

A little further on, I moor beside the Cape of Good Hope pub on the edge of Warwick and spend an evening watching Liverpool winning the UEFA Cup and, during the adverts, listening to a room full of adult males talking tosh.

Tomorrow it's the Hatton 21, my first endurance test, and the start of the climb up into Birmingham.

3

Regeneration

The Birmingham Canal Navigations

The 47 locks I count in the Nicholson guide leave me in little doubt Birmingham sits atop a mountain. Having only ever previously approached it by car, I'd always assumed it was as flat as the accent. The first rung is the Hatton 21, known locally as the Stairway to Heaven – its soubriquet surely ironic rather than a measured suggestion that Birmingham and heaven have anything in common.

Before setting off, I make two quick calls. The first finds no change in Dannie's condition. At this stage, no news is good news. My second call to the Macclesfield General finds Dad much improved after a night on oxygen. He hopes to be home in a few days.

I put the mobile on the charger, check the weed hatch, and slip the moorings.

A mile later, approaching Hatton's distinctive paddle-gear rising like candles up the tiered layers of the flight, I'm torn between aesthetic appreciation and cold fear. Through the drizzle, just ahead of me, I notice another boat entering the bottom lock. I put on a spurt, hopeful the occupants will see me and wait.

On board the 70-foot narrowboat *Dover* is John, a retired solicitor from the Surrey–Kent borders, and his hosts – bizarrely – from Hawaii. Ann and Dougal Crowe, I quickly learn, are old hands who have holidayed on English canals every second year for the past fourteen. As the two male members of the *Dover*'s crew work their windlasses, Ann leans across conspiratorially from the tiller of the *Dover* and whispers, 'John's recovering from a stroke that's affected his balance but he's very proud and you just have to let him get on with it.'

'OK,' I reply, a little flummoxed.

'He's used to being active – we first met trekking up at Chomolhari,

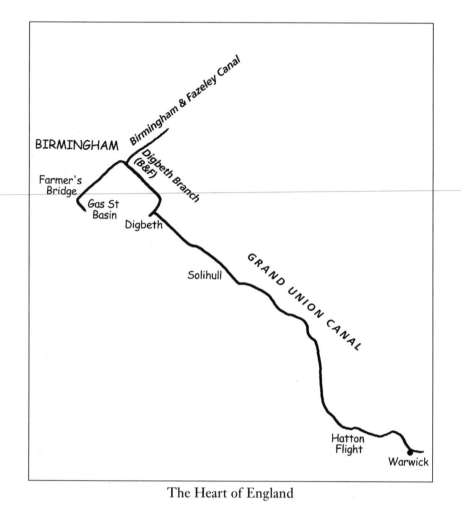

The Heart of England

in Bhutan, thirteen years back. He likes to pretend the stroke never happened.'

I reassure Ann that I will in no way try to interfere if John wants to do the locking, and try not to sound too gleeful.

'We always tend to do active holidays together,' she continues. 'And the weather over here helps.' I laugh. 'No, really. We get too much heat in Hawaii. We come for the rain.' Are they insane? Probably. Their penchant for sunless skies certainly explains why Ann has a smile as wide as the Golden Gate Bridge as we stand in the drizzle at the stern of our boats waiting for the water to lift us. 'It's not the only reason of course – the weather. The Cut itself is the big draw. We've done French canals and they're just not a patch – the landscape's too monotonous. We've made so many friends over the years here that we keep bumping into again and again on different stretches. It's such a closely knit community, like a village. Everywhere leads back to everywhere else.'

The marketing departments of British Waterways and the English Tourist Board should sign the couple up.

At the next lock, as John and I work the paddles together, it becomes obvious very quickly he's dotty about Ann. 'You know she was the Times Square billboard model longer than any other girl in history?'

I look across at Ann still smiling at the tiller. Last week she celebrated her sixty-fifth birthday with the flight of 30 locks at Tardebigge. She reminds me of Barbara Stanwyck and every big-hearted, handsome cowboy mum that wagon-trained through my childhood.

John walks on to prepare the next lock and I lead our mini-flotilla towards him. Following behind me, Dougal and Ann look like snowbirders sailing the Florida Keys on their winter vacation. Only they're doing the opposite: swapping the Maui sun for British rain.

Attempting a little mental arithmetic, I become increasingly grateful they made that switch. Without my new friends, working on the basis that half the locks would be against me, I calculate I'd have to turn my windlass 2,760 times and open or close 74 heavy oak gates to get to the top of the hill.

At the fifth lock heavy rain gusts across the pound in clouds, and steam billows from the deck vent of the water heater. The encroaching vegetation is playing at rainforest: lush, succulent, riotous. At gate six, we dip over a hillock and get a real perspective of the giant's staircase we're climbing to heaven. As a distraction Ann suggests we swap boats. The engine on the 70-foot *Dover* is a whisper compared to *Caroline*'s crotchety growl. It doesn't, however, have the low speed tick-over and manoeuvrability of *Caroline*. Anyway, *Caroline*'s family.

At lock nine we decide to breast the boats up by lashing them together so that only one of us needs to steer both into the locks. The rain is lifting and we get a clearish view back down the flight, through the black and white candlestick paddle gear to St Mary's Church in Warwick. By the time we quit the last of the 21 locks, it's 11.45 a.m. It's taken three hours to climb 146 feet 6 inches over a mile and a half.

At the top of the flight Neil Ratcliffe is behind the counter of the Hatton Canal Shop. I buy a book on the north-eastern navigations and pass on the fudge, Tetley tea bags, embroidered lock scenes, canal jigsaw, and the selection of china mugs ('I'm a canal mug', 'Windlass Loser').

Neil, who's had the shop nine years, and lives five miles away on a 70-foot Fellows Morton & Clayton ex-working boat, fills me in on the latest news – of the punch-ups that followed the UEFA Cup final, and John Prescott punched someone who threw an egg at him on the election campaign trail. The Cut is known as the Towpath Telegraph for good reason.

'What's the record anyone's done the Hatton Flight in?' Dougal asks Neil, after he informs us someone in a hurry fell in the top lock the day before.

'Dunno,' Neil replies. 'A record seems to run contrary to the nature of the canals, don't you think?'

Back aboard *Caroline* I think about this. Can the canals be said to have a nature? After three weeks afloat, unhesitatingly I'd say yes. Far removed from the busyness of rivers, the Cut possesses a placid, mature and meditative nature...a bit like an old sheepdog.

I lunch on the *Dover*. Ann makes Welsh rarebit, a fruit salad and provides a steady stream of beers. The boat is a palace with individual cabins (each with its own TV), a proper dining room and even carpets on the walls. Ann and Dougal have hired it, as they have on their last three narrowboat holidays, from Black Prince. Amazingly they haven't been impressed. 'It's in nowhere near as good condition as the others we've hired previously,' Ann announces.

'Next time we're going to try Napton Boats,' Dougal adds.

As Ann clears away the dishes, she generously invites me to be their guest that evening at the Tom o' the Wood pub at Rowington. 'It's by Bridge 63. We've eaten there before and it's pretty good. Come aboard for a drink first. Any time after six is good.'

Back on *Caroline* I watch a coot sailing across the canal followed by its fanning trail and three chicks. I burrow through the 433-yard Shrewley Tunnel. As *Caroline*'s light scrapes the calcified arches of limestone, I feel like Jonah disappearing down the ribcage of the whale. At the far end I enter the rain-soaked green grotto of a single heaving organism.

It's probably no more than five miles to Rowington and there are no locks. I moor opposite a house whose entire upper canal-facing wall is glass. A woman sits in her armchair staring out over her pretty garden and the Cut. It's 4 p.m. At 10 p.m. when I return from the pub, after beers, wine, steak and tales of Hawaii, she's still there.

The *Dover* sounds a farewell blast on its horn around 8.30 a.m. and my new friends pass out of my life en route to a couple of nights theatre down the Stratford-upon-Avon arm. Although I enjoyed our day together, I relish the solitude. As Adam Phillips wrote in *Houdini's Box*, 'People often feel most alive while they are escaping.'

Birmingham is getting ever closer and yet, for the time being, the Cut remains resolutely rural. Buttercups and dandelion clocks cradle an abandoned plough as beautiful as a Giacometti sculpture. Lambs, no more than a few days old, cavort. Foals skit, a heron pounces on a fish. A pheasant, rushing through undergrowth, brings a shower of elder blossom down on itself.

And every passing stranger continues to greet me. It doesn't even surprise me any more. The warm feelings everyone seems to have towards *Caroline* and other narrowboats are something I now take for granted – even car passengers wave as they hurtle by.

After the Knowle flight of five locks, a sign declares London to be 124 miles away, and Gas Street, Birmingham just 13. Close to the sign, a fox stands staring at *Caroline* from the scrub. After the exotically named, prosaically minded Catherine de Barnes, the landscape finally puts on its suburban coat. Through the window of a waterside cottage I watch a cartoon on TV. If I didn't already have enough reading material I could almost lean through the window and nick a book from the shelf. A phone rings from a second home as loudly as if it's on board; a man building a patio curses, wipes his hands on his trousers, and goes inside to answer it.

The pages in the Nicholson guide that up to now have been predominantly white are overtaken by a dense and indivisible concentration of streets and buildings indicating England's dark heart: it will stretch from Solihull to Digbeth to Birmingham city centre to Dudley and all the way to Wolverhampton. I phone home, as a plane comes in to land two miles away, ruffling my hair and drowning out *Caroline*'s feeble growls. Dannie's own dark heart is still holding up.

Trees arch over the Cut; the canal hunkers down. Time to burrow.

A tin of bluebells has been poured down the embankment, but so have bags of rubbish that now bob along the water until snagged by uncleared trees and branches. A couple sit on a step together in suburbia, the woman dressed in a yellow dressing gown, the man in jeans and T-shirt. Their hands hug steaming coffee mugs for warmth, their voices are silent, eyes staring into emptiness. They are locked in one of those domestic cataclysms familiar to all who cohabit. This morning the future looks bleak beyond their neat lawn.

The embankment rises up and becomes more precipitous as if blinkering the canal from the awfulness beyond it. Brookside crescents appear in gaps, a hive of weekend car-washing, DIYing and grass-cutting. A swan nests beside a garden gate. A Breezer bottle, a Big Mac box, and a very old black plastic dustbin with the imperative 'Keep Britain Tidy', bob by. The bright beaded necklace of moored narrowboats I've slipped past every few miles on my journey up the Grand Union has vanished. I feel like *Caroline* and I are the only mariners to have ventured into the heart of darkness in a hundred years.

I slow the boat, attempting to weave between the rubbish. Sometimes I fail, something wraps itself round the propeller and I reverse to try to free it. When unsuccessful, I have to stop the boat, remove the weed hatch and thrust my arm down through the murky water until I can free the offending item from the prop. Sometimes, too, *Caroline* scrapes her belly over unwanted objects lobbed off a bridge – a supermarket trolley, a bicycle, a car door. On the bridges themselves graffiti declares 'Kishore Afghan was 'ere' and 'Babagar Sucks Dick'.

I pass a line of fishermen at a winding hole (a point wide enough to turn a boat round in) just as a robin hitches a lift on *Caroline*'s roof. Why is it the vast majority of fishermen choose Shakespeare tackle boxes?

Does the brand come with a copy of the collected works inside just in case these most patient of men do finally get bored watching a piece of barely visible fishing line?

Amid a profusion of bankside red poppies, a line of old brick warehouses and small factories rise up requisitioned by Double Glazing Supplies, Rentokil, and Ballantyne Electrics. I dip under roads and railway bridges, burrow through tunnels and slip through a modern industrial park. I am now on the Birmingham Canal Navigations (B.C.N.), the Big Bang of English waterways from which the nationwide constellation spread out. Birmingham was the world's first industrial city, boasting more canals than Amsterdam or Venice (in its 1865 heyday the network consisted of 160 water miles, and even in the grimmest hours of the 1970s it still managed a more than respectable 100).

Locks on the B.C.N. are a breeze because they're all single which means less walking, less heaving and less winding. The minor downside is there's no sharing the work load with another boat.

After the second lock in the flight of six known as Camp Hill, a cyclist in his early twenties who's run over a pigeon, yells across to me with a boyish grin and without breaking stride, 'They usually move faster than that.'

The pigeon is in the canal, flapping frantically. A white man with his veiled Asian wife and daughter watch me reversing, trying to reach the bird which is now attempting to get out of the water on the opposite bank. Gingerly, I come in close, lean over and scoop the pigeon into a bucket. I place the exhausted creature in the hedgerow. It has lost half a wing, will never fly again and is very weak. I know the kindest thing is to kill it but I'm too squeamish and leave it there. The Muslim wife claps her hands from the bridge and says 'Well done' in a broad Brummie accent. The little girl asks if it will live. 'If Allah wills it,' mother cops out.

A little further up the flight, having opened another lock, I head back towards *Caroline*, ducking my head to avoid the low roof of the tunnel. Unfortunately I forget there is a second lower tunnel inside the first. My forehead crashes hard into its concrete lip, the force knocking me to the floor. My head is numb, stars are out. I stagger back to *Caroline* and, looking in the mirror, see that blood has run into my left eye. I toy with the idea of leaving it there to look hard and scare off any hoolies I might encounter. Instead I wash it thoroughly and smear antiseptic on. Cuts easily get infected if canal water gets into open wounds.

At Warwick Bar, I set off on the bike to explore Digbeth. In the early eighteenth century the area was as quiet as the proverbial dormouse, a sleepy village with open fields, orchards and no canal. By the end of that century it was a maze of docks, aqueducts, bridges, tunnels, locks and wharves that were home to Typhoo Tea, H.P. Sauce, and *canal* carriers like Pickfords.

Today the scrapyards and cement-batching plant I cycle past are the crows picking at the scraps on the post-industrial landscape. In Fazeley Street, a Pickford truck swings wildly round a corner and onto my side of the road. Beside Small Heath Boxing Club is an old sign of Fellows Morton & Clayton Canal Carriers and above it another for Clifton Steel Limited, the late-twentieth-century occupants.

On Andover Street, I finally meet the twenty-first century and Digbeth's rosier future. The Abattoir, UB40's recording studio, is expected to become part of a performing arts centre – just one component of a grand design to transform the area into Birmingham's education and arts zone known as Eastside. Nearby, Bond Wharf has already been converted into a café, arts centre and office complex housing designers and architects. Moored alongside is the *Thinktank* narrowboat belonging to the Birmingham Museum of Science and Industry which is also relocating here. Other university departments and a major library are due to follow, the whole area on the cusp of rebirth.

For the time being, however, the battle rages between the goodies and the baddies over which future awaits Digbeth. Planners introduce tree-planting programmes which are then trashed by local vandals; warehouses are expensively converted and then the hoolies flood the basements by leaving all the paddles up on the locks.

According to the city map, Brindleyplace and Gas Street Basin are no more than a twenty-minute walk from Bond Wharf. By canal, according to Nicholson's, it looks to be around three miles, seventeen locks and four hours. Fortunately, at the start of the next flight I find a willing helper. Raymond, a Pakistani-Anglo in his early twenties, pushes the gate back so that I can enter without mooring up outside first. Raymond is kicking his heels, waiting for a friend he regularly meets at the bottom of the Ashted flight 'To escape me mom and me 'ome in Moseley'. He has the same nasal Brummie accent immortalised by the female panellist on *Juke Box Jury* who pronounced 'Oill give it foive' to just about every new release she heard in the sixties.

As we work the lock together, I learn Raymond has only been to Pakistan twice: the first time for three weeks and the second time for three years when his grandfather was dying and there was nobody to care for him. 'Me uncle was supposed to be looking after him, but he didn't, he just abandoned him. Let him rot.' Raymond took over his grandfather's care and didn't leave until his grandfather died three years later. He is still bitter about the experience. 'I've never spoken to me uncle since. If he walks in a room, I walk out.'

Raymond is a quality control inspector at a factory making plastic parts for cars. He says he's always filled with dread on his daily drive into work. 'But once I'm there, it's sort of OK.' He assists me through the next two locks before apologising. He can't risk going any further in case his friend has turned up. He too vanishes from my life.

At Belmont Row Bridge a large woman and three children are out on their roof garden throwing bread to swans below. The garden is located on the tin roof of a timber yard, and their home is a carbuncle that's grown out of the canal bridge. The flowers they've grown cascade over the roof to a miniature mill they've built for the swans to nest in on the water. Two are waiting below, gobbling up the raining bread. I take a picture of the woman. She smiles and points to the swans. 'They're regular visitors.'

The final skinny flight up to the city centre is a set of thirteen locks known as Farmer's Bridge. When the stretch opened in 1789, the price of coal immediately dropped from 15s (75p) to 4s (20p) a ton. It was already the busiest flight in the country when it got gas lights in the nineteenth century and began operating 24 hours a day. Locking up through 200 years of history, my feet stand in hollows worn by a million other shuffling feet. I slip through the Jewellery Quarter where two-thirds of Britain's rings, necklaces and bracelets are still made; and where Washington Irving wrote 'Rip Van Winkle'. *Caroline* pants her way past the walled-up entrances to factories and wharves into which Joey boats once slid like envelopes through a letterbox delivering goods. As I ratchet up, the sun ratchets down. I time myself in and out of a lockgate. Six minutes flat. A tramp grasping a can of Tennants reckons the average for the flight is an hour and a half but that he'd once done it in 25 minutes. I do some speedy mental division: 13 locks in 25 minutes. 'But that's one every two minutes!' I protest. 'They don't even fill that quickly.'

'My locks do,' the tramp deals peremptorily with my cynicism.

It is a typically chummy flight. The sight of a narrowboat, as usual, loosens everyone's tongue. A woman with a foot in a cast sits on a balance beam with crutches beside her. 'I broke it last month skiing in Courcheval,' she explains without prompting. 'It's the first time I've been out the flat since.' At the next lock, a man opens the gate for me and warns, 'The rest of the flight is all against you, I'm afraid.' Further up, a group of twelve young men and women descend the towpath carrying canoes. With nowhere to paddle in the vicinity, I wonder if they're a troupe of performance artists. One of the party catches my puzzled expression and offers illumination. 'We're walking to the Grand Union. Our leader, Adolph, still reckons you have to work for your exercise.' At the second-to-last lock a bespectacled student in a green Shackleton College rugby shirt and cycling helmet tells me about the time he spent two weeks on a narrowboat travelling the Leeds & Liverpool Canal with his parents. He also recommends a good pub and curry house for the evening.

Eventually I reach the top of the flight and the B.W. water point. Two drunks offer assistance and attempt to attach *Caroline*'s hosepipe to the tap. The result is vintage slapstick with them soaking and cursing each other by turns.

I moor for the night beneath the National Indoor Arena (N.I.A.) opposite Brindleyplace. No parking meters. No warden. And it's free. You're meant to stay just 48 hours but who's checking? Once again, I set off to explore.

Along the waterfront and up through the colonnaded squares of Brindleyplace, terraces are already heaving despite the fact it's only 7 p.m. Espresso culture and the new fad for eating alfresco brought back from foreign holidays has taken the city by storm. For those, like me, visiting the place for the first time in ten years, Birmingham city centre is unrecognisable. Where once there were crumbling warehouses and general urban dereliction, now there are bars, restaurants, offices, penthouse lofts, an art gallery, Sea Life Centre and a theatre. The mixed-use brownfield regeneration is the most successful turnaround in the country. In the last five years property in Brindleyplace has doubled and trebled in value as Brummies return in droves to live where they once endured.

Birmingham may once have been the first city to trip off the tongue when conversation turned to urban architectural barbarism but now Britain's second city is serving as the blueprint for urban regeneration. It is also responsible for two new wisdoms that have spread round the country fanned by its success: a) any property redeveloped overlooking water will enjoy a 20 per cent financial premium over property redeveloped nearby; b) where waterfronts are reborn, so are inner cities.

I briefly detour into the Fiddle & Bone, a live music venue located in a former primary school and transhipment warehouse, where a Brummie Irish band are playing to a burgeoning crowd aged 20–60 singing the wrong words to 'The Leaving of Liverpool'.

At the nearby Celebrity Balti House on Broad Street, a waiter complains to me about his increasing workload while taking my order. 'Every night it gets busier. In the seventies Broad Street was so run down and shabby only drunks and thieves spent time here. Now you can't move at weekends.' The Celebrity Balti was the regeneration pioneer, moving in eight years ago while the major redevelopment of Brindleyplace was still under way. I stare out the window down Broad Street at the tide of human flesh – and there's plenty on show – eddying between the back-to-back bars. The place is booming.

The Aussie at the next table seems as shell-shocked as I am but for different reasons. On his way to meet his brother – who's studying at Birmingham University – at the restaurant, he passed a bar advertising 'women in thongs'. 'They make out it's a big deal. I mean if that's their idea of big-night entertainment I'm better off at home in Perth.' He shakes his head, nonplussed. His brother cracks up before enlightening him. 'Thongs aren't sandals, mate. They're dental floss women wear round their privates.'

I head back onto the street. A hen party cruises by in a stretch limo, windows down, calling suggestively to me and every other male to come inside. They overtake a Ford Transit out of which two girls, no older than sixteen, are bursting out of the rear doors and their boob tubes. They stand on the pavement thrusting their bits back and thanking their dad for the lift. A group of four girls greet them on the corner. They are dressed for their tattoos: one cropped top reveals a dolphin above a breast, another girl has a rose on a shoulder, a third has Chinese lettering striding across the lower back, and the fourth displays a Gibson guitar across a tightly strung stomach. A group of young men approaching in the opposite direction steer themselves as close as they can to the bevy of girls, legs wide, muscles accentuated by tapered short-sleeved shirts. As they pass, they discuss which of the enticing cheap offers advertised outside the bars to take up first on their quest to get ripped as quickly and cheaply as possible – 'Fosters £1.50 a pint', 'Buy two large glasses of wine – get the rest of the bottle free', 'Two Bacardi Breezers £5'. The Quakers who once ran the town and had it tucked up in bed by 9 p.m. would be turning in their graves at the ten pints and step-on-someone's-face-on-the-way-home culture that's gripped both sexes in the city.

I head back to my oasis. No late tube, no exorbitant taxi, no long wait for a night bus. I simply stroll a couple of hundred yards, draw the curtains and hit the sack. I sink into a deep sleep. Several hours later I'm dragged to the surface by the beating of a drum and raucous female laughter. I drift off again. The banging resumes. This time it's more persistent and more familiar too. *Caroline* is bashing against a wall. But there can be no earthly boats moving at this time of night creating a wash. Another bang. I pull back the curtain. Everything's wrong. There is now water between me and the National Indoor Arena and the only other boat that had been moored alongside me (complete with wedding bouquet pom-poms) is across on the opposite bank. I look out the port window. I've been bricked in. *Caroline* is on the wrong side of the Cut, banging against the rear wall of the Malt House pub as if trying to get in after hours.

Someone – I have no doubt it was the drum-majorettes – has slipped my ropes. I put on shorts, a sweatshirt and trainers, open the hatch, and emerge on deck. Modern cities never see true darkness and I quickly get my bearings. Shuffling along the gunwale to the middle of the boat, I give it a good shove off the wall. I then grab a pole and punt through thick mud. Eventually I reach the other bank.

Very carefully I refasten the ropes in such a manner that someone would need a saw or blowtorch to remove them. I return below deck, too wide awake for sleep, and try to read but there's only enough power left in the battery to give out a dull glow to the cabin lights. I turn the fridge right down to conserve energy and pray there's enough juice left to start the engine in the morning.

I take out a torch and read the same paragraph in Philip Roth's *The Human Stain* five times. My mind can't sit still. There's something unsettling about being set adrift at night. I'm reminded of just how public and vulnerable a narrowboat is. I think about Dannie and how adrift he must have felt in the dark with daggers thrusting into his heart. I think too about how Dad felt when he grabbed for, but could not find, breath. Unmoored. Floating to oblivion.

I think of my own children growing too fast. In a few years they too will sail off. Fathers and children departing, me adrift midstream. In the dark it's difficult to throw off the blanket of maudlin. I put down the book and check there are no bogeymen still about. The ground and water rumble as if gripped in an earthquake as another train leaves New Street Station and passes underground.

Munching toast and sipping tea at the fore-end the following morning, I field greetings from office workers and stare into the shimmering reflection of the Malt House. The image is not quite a mirror, colours are slightly muted, its bulk greater than its earthly twin. The sky is an eternity of blue broken by four horizontal white bars, an outsized Greek flag.

In Brindleyplace those walking to work display a relaxed waterside gait, unlike the persecuted scurry you witness in streets. Others relax over lattes and sunshine. There is a traffic-free riverlike flow of contentment between and through the squares, beneath the warm brick colonnades and along the terraces.

Following the towpath under a low tunnel that supports a pub on Broad Street (formerly a school), I emerge into the quiet of Gas Street Basin. A wiry, fair-haired man is building kitchen shelves in a butty. He notices me observing him, smiles and wipes a bead of sweat from his brow. 'Beats work, doesn't it?'

'There's definitely worse ways to spend a morning,' I concur. The man's name is Graham Wrigley. *Ash* and the other six traditional boats moored at the pontoon are the working fleet of the Birmingham Canal Boat Services – formerly Birmingham & Midland Canal Carrying Company – of which Graham confesses to being 'general factotum'. The all-purpose, all-dancing new title suggests a willingness to turn a hand to anything to survive. The former title tells of a more clearly defined role.

'The last proper delivery I made was in 1971 – Dawes cycles to Ellesmere Port for export. Now I do the occasional run carrying spoil from new waterside developments, but mostly the *Ash* and the *Collingwood* are used as camping boats, and the other boats in the fleet we hire out for private cruises – stag parties and the like.'

Although Graham concedes Birmingham is in much better shape than twenty or thirty years back, he misses the old working boat community. An even greater loss, however, is the peace of the Basin. 'At night we're trapped inside a disco down here with all the flashing lights and thumping music.' Like Irving's Rip Van Winkle, Birmingham has stirred from its long sleep. The problem now is no one else can get any kip.

Graham disappears back into the cabin and I return briefly to *Caroline* to pick up the bike and explore a short stretch of the Worcester & Birmingham Canal down to Edgbaston.

A decade back, when the towpath was a virtual no-go area, I'd have been considered nuts to venture here. Then its banks were inhabited by muggers, dealers and other flotsam. Today it's occupied by joggers, mums and toddlers, and workers on their way to and from work.

At Bridge 82 students from the university are picnicking on the grassy banks, laughing and wrestling amorously. I leave the towpath and cycle up through the campus and botanic gardens and then, as usual, get lost. No one knows the way back to the Cut despite the fact it can be no more than half a mile away. One elderly man I ask even argues that there's no canal in his neighbourhood.

I decide to cycle back into town by road. A belching lorry called 'Bazza' attempts to force me into the gutter. Car steering wheels are gripped as if lives depended on it, eyes are glazed. I seriously doubt anyone is having a good time. My eyes water from exhaust fumes and a headache lurks. I feel like a rabbit caught in headlights.

It's difficult not to wax poetically about the pacific effects of water. There is something inside us that is only stilled when we're beside sea, river or canal. Birmingham's waterfront regeneration shows an awareness – unconscious or otherwise – of this fact and a rejection of its earlier post-Second World War slavish city design around the motor car, in favour of the pedestrian. Now the revival needs to travel further along the water road, seeping into other suburban pores, soft-landscaping its hard-edged neighbourhoods, creating a network of water villages linked by water buses.

4

Pioneers, Rural Idylls and Raining Maggots

The Wolverhampton Level, Staffordshire & Worcestershire, Trent & Mersey, Caldon and Bridgewater

Early the next morning I quit the city, slipping with the Birmingham Level Main Line past Winson Green Prison and Soho. At Smethwick Junction the canal splits and the landscape becomes an open book on the engineering predilections of the two greatest canal pioneers. Brindley's Old Main Line, which opened in 1769, meanders off to the right, doggedly hugging the contours of the land wherever it leads. Meanwhile Telford's canal, which opened 68 years later and shaved seven miles off the Birmingham–Wolverhampton journey, maintains an arrow-straight route utilising massive cuttings and embankments (it was the philosophy of the second phase, nineteenth-century canal engineers that 'every valley shall be exalted and every mountain and hill made low'). Without a moment's hesitation I plump for Brindley.

For a while the two canals and the railway line keep semi-rural company, before the M5 soars overhead on its slim concrete columns and *Caroline* and I scamper off with our new playmate along the 473-foot contour.

The coal mines of the Black Country are now sealed, healed and green but old transhipment points, such as Monmore Green, are still visible where rail and canal meet and shake hands like old friends. Unlike in the rest of the country where the two were bitter rivals, in the Birmingham area, railroad and canal continued to work together profitably until after the Second World War. It was only when army lorries returning from the front were sold off to private contractors that freight moved, literally lock, stock and barrel, to the roads.

Once off the Wolverhampton express line, the Cut grows increasingly unkempt – canals in poorer areas tend to be shunned by narrowboaters as well as by B.W. maintenance staff. Spring does its best to rise above the floating detritus in her white pinny of blackthorn blossom. Two women in overalls, enjoying a fag break on the towpath outside their factory door, give me a wave. I wave back and ask them what they make inside.

'Nuts and bolts,' one replies.

'Come in for a screw if you like,' the other offers to an explosion of laughter from her colleague.

A little further up, I moor alongside the Black Country Living Museum in Dudley, a 26-acre reconstructed canalside village anchored in the early twentieth century. Opposite the hardware shop with its metal baths and trivets, the thick-paned windows of the Providence Methodist Chapel sparkle in the sun. Next door in the general store, an actor, revelling in her role as the illiterate storekeeper Mrs Gregory, offers me jugs of vinegar, paraffin and milk, as well as two Woodbine and a match for one farthing.

Like all the other employees manning the shops, fairground, school and village cinema, the woman is paid to pretend the village belongs to the same Black Country 'black by day and red by night, that cannot be matched for vast and varied production, by any other space of equal radius on the surface of the globe', (the American Consul visiting Birmingham, in 1868).

Dudley today, however, bereft of industry like South Wales and the North East, is singing from the heritage hymn book. Lacking any meaningful manufacturing or mining to speak of they've fashioned a nostalgia industry from the industrial ashes. The reality in 2001 is that the chainmaker makes chains and nails that are never purchased, the mangle and tin bath in the hardware shop are stage props, the Edwardian schoolhouse has an unqualified teacher and no pupils, the fairground echoes emptily, trams are stationary, and the village cinema is playing to an empty house.

The chippy is one of the few places selling anything more ethereal than memories. I carry my cod and chips over to the Bottle and Glass Inn and buy a pint of Holden's Mild (brewed just a mile away). After lunch I join a minibus party from the Stourport Retirement Home filing through the chainmaker's cottage. There are only a couple of men in the group ('The men die quicker than we do'). A couple of the women squeal with delight at the sight of ripped up newspaper hanging on string in the

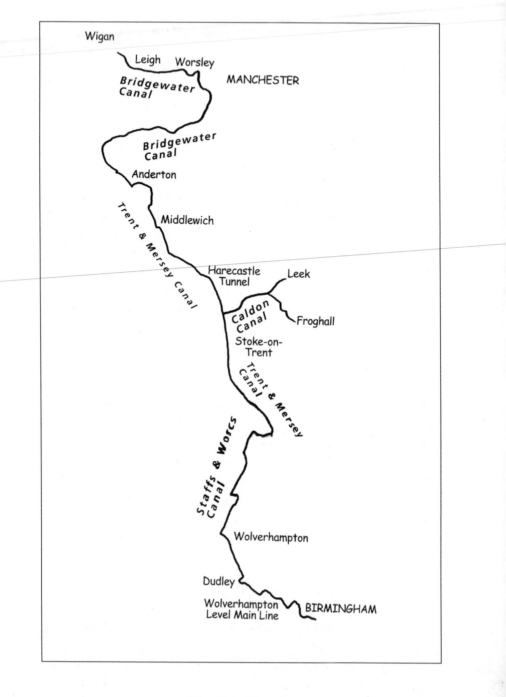

The Four Counties

toilet. In the kitchen, another woman champions the advantages of a flat iron warming its bottom on the range. Her name is Mary. 'Never a crease with a flat iron unlike the Rowenta rubbish I use now.'

I ask her why she feels such a strong yearning for the past, 'After all, everyone's cleaner and better off now aren't they?'

Three women shout me down at once.

'More community.'

'Simpler days.'

'You knew where you were then.'

The women clearly wish they were still blacking their old lead grates rather than listening to radiator pipes gurgling through the rooms of their retirement home. Their preference for a backside blacked with newspaper print rather than smoothed by Delsey, I suspect, has more to do with a yearning for lost youth than a lament for the passing of an era when the life expectancy of those working the mines was 34. Whatever, the Black Country Museum has unblocked a landslide of memories.

As the women continue on their heritage-fest in the pharmacy to a Greek chorus of 'oohs' and 'ahhs', I stroll down to the mouth of the Dudley Tunnel where a narrowboat tour of Lord Dudley's underground mines is about to depart.

Behind the cavelike entrance we sail down the exhausted seams of coal, iron ore and limestone that fired the industrial revolution. The entire hillside had been pickpocketed by joey boats and today the Earl of Dudley's mines are as quiet as the grave. A couple of women raise their umbrellas to protect themselves from the dripping water, taking care at the same time not to snap off calcite straws hanging from the roof like church tapers.

Briefly we emerge into a sunlit open quarry, the vertiginous walls ringing with the song of blackbirds. A curtain of ivy leads us into another limestone grotto and another. The heritage industry has transformed an ugly hellhole into a thing of beauty. The horrors of the past – the long hours, the clamour, the choking fumes – have all been erased as has the fear of sickness, accident, the sack, and early death. Butterflies flicker through sun motes between tunnels.

Half an hour into the trip our guide kills the engine. In the silence he asks for a volunteer to assist the deckhand in legging the boat through the final section of tunnel. It's my opportunity to join the cast of actors and play at miners. I lie on a plank, place my feet on the tunnel wall and lean my shoulders intimately into those of the deckhand who's doing the same on the starboard side. Together we inch the boat forward, walking our feet along the brick wall. After twenty yards my legs already ache. I conclude I'd have been dead in a week if I'd been born in the wrong place at the wrong time. I'm rather sceptical therefore when the guide tells us that the leggers had the longest life expectancy of any mine employee. 'And it wasn't just down to the regular exercise and exposure to semi-fresh air they got walking back over Castle Hill.' Equally important apparently was the fact their alcoholic consumption was only a fraction of that of workers digging the seams. At a time when ale rather than water was used to slake miners' thirst, a 12–16 hour shift in dry, choking conditions required an equal number of pints to keep going. The leggers, however, had to carry their own drink with them and therefore drank considerably less.

With perfect symmetry we reach sunlight as the guide concludes the Dudley Tunnel story. By the 1960s the leggers and the miners had vanished, the tunnel was closed, and the canal so filthy 'You could walk across it in a pair of wellington boots and you wouldn't get your feet wet.' Abandoned and dangerous as it was, only rats were happy until the Dudley Canal Tunnel Preservation Society came to the rescue in 1964, reversing a council decision to transform the tunnel into a sewer. The canal and tunnel reopened in 1973.

Back aboard *Caroline*, I just have time to shower and change before walking up to the King Arthur pub, to meet a friend, Dave Walton, who has caught a train over from his Derbyshire home to visit his son at Birmingham University. The King Arthur is awash in a sea of hallucinogenic wallpaper on which prints of French musicians act as visual life rafts.

The music that's playing, however, is pure Arthurian – Wayne Fontana and the Mindbenders' 'Game of Love' (was he a Brummie?), Freddie and the Dreamers' 'You Were Made For Me', and the Beatles' 'Love Me Do'.

Five tattooed *mutilattos* stagger in under their chain-mail of rings, studs and chains. Dave and I follow them into the carvery and devour a mountain of food and two bottles of wine for a grand total of £22. We then retire back to the main bar, as mellow as we are bilious amid the gory battle scenes of patterned fabrics.

An extended family of twelve sit at the next table, already large and still expanding. The two youngest members, a huge girl and a very small girl, who I guess to be eight or nine, play pat-a-cake loudly, their hands and words dancing faster than thought.

> My boyfriend gave me an apple,
> My boyfriend gave me a pear,
> My boyfriend gave me a kick-up-the-bum
> And he kicked me down the stairs.

The rest of the family ignore them. The children repeat and repeat like a scratched record, singing as they did 200 years ago when they lived on a butty and walked the horses daily between Dudley's mines and the city centre.

Dave and I lean back reminiscing over our own halcyon days, teaching in a large comprehensive school in West London in the mid-seventies. In the intervening years, miscreants are transformed into angels. I'm closing in on Stourport Retirement Home territory. The fart-and-cabbage corridors are my sooty fires and flat irons. Faraday High School is the hearth where our hearts are, our own heritage industry.

Together we conjure up our heroes and heroines: Bimala who couldn't swim but leapt straight into the water the first time she saw a swimming pool. Daniel who always refused to participate in lessons until he was first granted permission to dance in front of the class. Most poignant, however, are the memories of the extra-curricular activities: the weekly five-a-side football against the fifth-formers; ferrying kids in my ex-post-office van to Bob Marley concerts; and Dave and I getting locked out of the Arundel Youth Hostel when returning from the pub after lights-out (the children had to sneak us in through the dormitory window). The violence, the nutters, and everyday deprivation of the inner-city comprehensive is erased as easily as miners' premature deaths.

At 11.30 Dave gets a cab back to his son's digs and I head back to *Caroline*. On the dust on a shed window someone has used a finger to write, 'Sex wanted'. It would be nice.

The next morning I slip under a bridge that declares 'We hate Wolves' and 'WBA are wankers'. The bearded half of the honeymoon couple I'd moored next to in Birmingham approaches me at the top lock of the Wolverhampton 21. He's forgotten his B.W. handcuff key (the name comes from the handcuff-like device that was first introduced on the Wigan flight of the Leeds & Liverpool Canal to reduce vandalism). Without it he won't be able to remove the chains to access the paddles. 'Bloody stupid idea anyway,' the honeymooner complains. 'The morning after the locks were fitted, the kids were making handcuff keys in metalwork at school.'

We decide I'll travel ahead down the flight and leave the chains unlocked for him. I notice, for the first time, that his narrowboat is called *Brindley*. The flight of 21 locks takes me three and a half hours. The groom reckons he once did it with a group of friends in half the time.

At the bottom of the flight I join the Staffordshire & Worcestershire Canal. The traffic, though objectively a trickle, seems like a veritable deluge. Four boats pass me coming from the opposite direction in the first twenty minutes. This is the equivalent of canal tailgating compared to what I'm used to.

It's a gorgeous day. Giant electricity pylons stride out across undulating fields for the horizon. I slip into Calf Heath Wood where a thin light rakes a carpet of bluebells. Up ahead I see two boys and a girl throwing stones at something in the canal. My heart skips with each skimming projectile. What luscious murder are they executing and will they turn on me in their blood frenzy?

As I get closer, they duck down behind a mound of stones that have been left there to complete the paving of walkways. I brace myself for the hailstorm and rehearse the Ali Shuffle. Astonishingly the assault doesn't come; in fact the children only emerge from their bunker when I've passed. For some inexplicable reason they're afraid of me. We are shadow boxers dancing with our demons.

The canal drifts into one of its river moods, taking three miles to cover the crow's one. At Gailey Wharf, while *Caroline* is having a sewerage pump-out and a bodge job done on the nonoperational bilge pump, Susanna and Max arrive by car from London. Larne has relayed her abject apologies but she could hardly be expected to waste tickets to two gigs and a party invite just to hook up with her old man. Max has replaced her with one of his best friends, Michael Addo, whose family migrated to North London from Ghana in 1981, eight years before he was born.

With the boys on board I know adventures will not be far behind them. Michael steers and the boat careers. Max takes over and rather imperiously brings us into the next lock perfectly. 'See, I'm better than you,' he dismisses me as a serious male rival to the head of the family. I wonder if he'll let me back in the door when I return after four months.

Standing beside the lock a boy proudly holds up the 3lb roach he's just landed and I take a picture of him. Max and Michael now want to fish. At Penkridge we're directed by a dog-walker to a sprawling industrial estate where we're assured we'll find Ian's Tackle Shop. We find it hidden in a lonely corner of the estate. Sheepishly we seek advice amid beefy, tattooed fishermen who have stood like sturdy oaks on canal and river banks all their lives. I pay £12.50 for weights, three floats, line and a box of pulsing pink maggots (we have a choice of red, yellow or pink but plump for the last as we're assured 'pink live longest'). The boys liberally sprinkle their modern argot of disgust over the brain in spasm in the Tupperware bowl.

'Erhhh, butters, rank, that's well dirtee,' Max pronounces his urban condemnation of unclingfilmed nature.

'Them are gross... butt-ugly, man,' Michael concurs. We snap the lid shut and create an outer confinement wall by lodging them outside the cabin at the fore-end.

We sail on past a line of neat post-war prefabs. A woman with steel hair and an orange floral-patterned housecoat waves before continuing to rewind the garden hose onto a large plastic donkey as her dog chases the hose's tail. Michael waves back. Three more couples sitting in gardens, enjoying the sunshine wave. Michael again enthusiastically waves back and then asks me how come I know everybody.

At Acton Moat Bridge, I encourage a younger boy with a water gun to attack the new crew members who are now relaxing on the roof with their eyes closed and ears insulated from nature by Walkman headphones. The boy can't quite believe his luck and attacks with gusto. Michael and Max are outraged and – unaware of my collusion in the guerrilla attack – demand to be put ashore so that they can seek revenge. Meanwhile the M6 rushes by.

At Acton Trussell, we moor beside the Moat House pub and the fishing rod is duly prepared. Once operational, the boys manage to tangle the line with every cast, drop twenty maggots for every one they successfully spear, and don't get a nibble. Half an hour later they give up and we walk across an apron of lawn to the front of the pub where there's a large pond and a further stretch of grass just about wide enough to play football on. T-shirts are shed for goalposts and we chase the ball as if our lives depended on it, dribbling, feinting, tackling like bulldozers. The score is five–three to the boys, and we're into the eleventh minute of the match when the ball lands in the lake.

Michael's plan for its retrieval involves me leaning out over the water like a masthead as both boys hold me. It very nearly comes to a logical conclusion. Across the pond, Asian bridesmaids aged around seven troop by in clumpy high-heels as a swan watches from its nest. Susanna calls across to tell us supper is ready just as Michael finally manages to retrieve the ball.

That night I sleep like an angel and the next morning I sit in the fore-end double bed reading *James Brindley* by Samuel Smiles. Susanna meanwhile continues to slumber. Suddenly something drops on my shoulder. My body involuntarily shudders. The something is moving and it's pink. I then notice there are several more pink cousins making a trans-cabin crossing across the roof. Time stops. My mind treads water, trying not to panic, desperate to hold chaos at bay. It's only a small step across the threshold. It's taken by another pinky crawling across my wife's face. Horror engulfs me. I know a tornado will blow through the cabin as soon as Susanna, now stirring and feeling for whatever is tiptoeing over her, discovers her home is infested with maggots.

For the next ten minutes bloodcurdling screams can be heard from *Caroline*. In a brief lull, we discover that Michael had left the lid off the Tupperware bowl out on deck the night before, allowing the pinkies to make their D-Day assault under the cover of darkness. More eruptions occur over the next two hours as Susanna screams and blitzkriegs the place with bleach.

Domestic life. Heaven and hell. It's a fine edge. Max can be unreasonable. I can be preoccupied or grouchy. Susanna can rage with P.M.T. or raining maggots. That's the reality of living with people and squeezing past them on a narrowboat. That's why Mohammed, Jesus and

Buddha were so perfect – they never shared their narrowboat with anyone. Their loss and their gain.

The countryside we are now sailing through would be sensational if bathed in yesterday's sunshine; instead our hoods are up against the cold and drizzle as we slip past five swans with their dawn begging bowls. Cows' udders sway uncomfortably as if about to explode. The world seems pregnant with something. After the maggot attack, I am apprehensive about what pestilence God or Michael might be planning. We duck under a series of handsome stone bridges, the canal narrowing and fattening like a python digesting breakfast.

At Milford we slip above the River Sow on an original Brindley aqueduct whose stone arches seem to be sinking into a swamp inhabited by geese. I jump ship, tiptoe round the geese shit and take a photo of *Caroline* sailing across the iron trough. When I make it up to the aqueduct, I discover I've been abandoned. Susanna has sailed on and moored on the opposite bank fifty yards ahead and has no experience of the tricky art of reversing. I'm buggered. Maybe this is her revenge for the maggot attack.

Fortunately, just as I'm considering swimming, another narrowboat sails onto the single-lane aqueduct and grants me permission to use it as a bridge. I walk onto the starboard side and step off on the port side on firm ground – the boat doesn't even have to slow down.

The next lock, Tixall, is the quintessential rural lockgate with its bank of primroses, sheepdog, and a retired boat slumbering on a lawn alongside the seven dwarves. We enter a sublime valley and the lake-like Tixall Wide, whose reputation for kingfishers is second to none. None of the shy, retiring creatures, however, appear prepared to put up with Max and Michael's curses as the boys try their hand at fishing once more with the final twenty orphaned maggots inhabiting their Tupperware mansion. They give up in five minutes flat and feed the remaining maggots to the ducklings.

The boys make ham sandwiches which they serve Susanna and me up on deck just as we join the Trent & Mersey Canal. At the first lock, Hoo Mill, Max slips as he hops back on board and is only saved from a ducking by Michael's quick lunge. Sixty seconds later, walking along the gunwale to join me at the stern, he slips again and, this time, there's no one there to catch him. All that's left of Max is the grey hood on his jumper floating limply on the surface.

Eventually his shocked face appears, eyes on stalks. I stretch down a hand and yank him on board as *Caroline* sails past. He slithers on the deck like an eel.

The family finally jump ship at Barlaston, beside a terrace of old white cottages fringed with yellow laburnum. I'm left with a great emptiness which I attempt to fill by washing up and clearing away the assorted debris of the 48-hour hurricane. A cyclist dressed in puce yellow, with wrap-round Oakleys, knocks on the window and asks to borrow an adjustable spanner. It's a welcome distraction. Prayer flags flutter alongside a tinkling Tibetan bell on a moored boat. That distracts me too. At Trentham Lock someone tells me a story about his father living on the Cut after returning from the front after the First World War. Wind ripples the water, a stream creates an eddy, a fish leaps and shatters the glass. I'm back under the spell. I catch myself singing again and decide to take an early afternoon stroll.

God has chosen instead to spend the afternoon landscape painting. On the water surface he has painstakingly copied each leaf, flower and grazing cow from the bank. I'm back in the landscape too.

Another group – six friends in shirts and shorts – are enjoying a barby on the towpath beside their hired narrowboat. They sit in canvas chairs listening to jazz, drinking cold beers and blowing on sausages that burn their fingers. Close by, two of their children are unsuccessfully trying to scoop minnows in a runoff. 'They're too quick, look,' one in a Stoke City F.C. shirt complains as another shoal vanishes like quicksilver.

Beyond them, a line of fishermen sit on boxes smoking Samson roll-ups, their forty-foot erections stretching from one bank to the other. One man is firing bait in a catapult, as a pike leaps six feet away. 'Bloody taking the piss intit?' His voice is a slack elastic band twanging halfway between the Midlands and Lancashire.

The path eventually leads to the Wedgwood Factory, outside which a bronze statue of a sombre Josiah Wedgwood in buckled shoes, wig, dress coat and breeches stands. At the base are the dates 1730–1795, and in his hands he carefully cradles his best-known creation, the Portland Vase.

The sculptor might just as easily have depicted him swinging a windlass, for it was Wedgwood, along with his friend James Brindley, who accessed the landlocked Midlands to ports on the west and east coasts with the opening of their Grand Trunk canal in 1777 (now known as the Trent & Mersey). In nearby Stone they were so excited when their stretch was completed (canals were the dot.com of the eighteenth century, the water road that was believed to lead to overnight riches), they fired a cannon to celebrate and promptly demolished the lockgates.

The building of the canal meant not only safer and cheaper export of Wedgwood's pottery (packages don't survive well on packhorses, however nimble of foot) but the speedier and cheaper import of clay from the Southwest too. Other cargo carried by the canal – coal from the Staffordshire coalfields, salt from Cheshire mines, and beer from the Burton-on-Trent breweries – resulted in a share bought for £200 in 1784 being valued at £2,400 by 1824.

In India travellers talk of bowel movements and brag about getting through the day on half a shoestring. On the Cut, conversation and thoughts tend to be preoccupied with locks: how many, how big, how long they take, and what surprises they may throw your way.

Ahead of me are the final three steps up to Stoke-on-Trent. With a collective 50-foot lift they are the deepest locks on the Trent & Mersey, and with a single ancient wall ladder inconveniently located too close to the fore-end, they're the trickiest too. Fortunately, as usual, at the second and third locks I find willing hands offering assistance, and willing tongues bearing stories.

The first, a retired miner, reminisces about his molelike existence at Cannock Chase. 'Pit ponies went blind in the dark it were that bad,' he tells me, making bad sound good. The second helper, another solo narrowboater, was made redundant from an engineering post at a power plant seven years earlier when aged 52. He confides that he had a complete breakdown and even seriously considered suicide over the following six months. 'Then I bought a narrowboat! Sounds soft I know, but it's God's truth it were the best thing that ever happened to me, losing that job.' He now spends his time drifting through England, meeting up with other narrowboat members of his new extended family.

At the top lock, I fork right off the Trent & Mersey and moor at the start of the Caldon Canal, which I intend exploring over the next few days. The conglomeration of towns surrounding me, collectively known to all who don't inhabit the region as Stoke-on-Trent, spreads in all directions. On my left is a life-size bronze statue of James Brindley standing on a small island in front of sheltered accommodation (the Caldon was the last canal he ever surveyed). To my right is Jesse Shirley's Etruscan Flint Mill which opened in 1857 to grind bones and flint for the pottery industry. It closed in 1972 but a new factory, still owned by the Shirley family, is at this moment grinding away like a dentist's drill at the rear of the old mill.

I unlock my own toothshaker from the roof and head off to explore, cycling first northwards along the towpath of the Trent & Mersey. A short way up, I spot a beehive brick roundhouse that is the only remains of Wedgwood's utopian model factory and village, Etruria. Once there was a second roundhouse and the pair stood like ornamental bookends either end of the pottery façade. High on the opposite bank, what remains of Wedgwood's elegant Georgian home has been incorporated into the Moat House Motel.

I continue up the hill into Burslem, one of six towns – the others are Tunstall, Hanley, Stoke, Fenton and Longton – that are the knots on the rope of Stoke-on-Trent. The streets and buildings of the mother town of the Potteries are immortalised – albeit under different names – in Arnold Bennett's 'Five Towns' novels (the author rightly felt five sounded better than six):

> In front, on a little hill in the vast valley, was spread out the Indian-red architecture of Bursley – tall chimneys and rounded ovens, schools, the new scarlet market, the high spire of the evangelical church...the crimson chapels, and rows of little red houses with amber chimney pots, and the gold angel of the blackened Town Hall topping the whole. The sedate reddish browns and reds of the composition all netted in flowing scarves of smoke, harmonised exquisitely with the chill blues of the chequered sky. Beauty was achieved, and none saw it. (*Clayhanger*, 1910).

As I cycle past 205 Waterloo Road, a large red-brick building in a short terrace that was one of the author's childhood homes, I try to imagine the nineteenth-century scene that seared itself into Bennett's memory:

the pall of choking smoke, the honeycomb of fire, the rattling trams, the grinding flint mills, the towering pot banks enclosed behind blackened walls, and the tarred workers quitting their workplace, heading for the nearest pub which was never more than a hundred steps away.

Beside a statue of Sir Henry Doulton (1820–1897) in the bifurcated Market Place, I find the old coaching inn I've been looking for. The Leopard features as the Tiger Inn in *The Old Wives' Tale* and *The Death of Simon Fuge*. It's also the backdrop for virtually the whole of *The Tiger and The Baby*. For boaters, however, the pub has a far greater claim to fame than mere literary backdrop for it was here that Brindley first met Wedgwood to plan their canal adventure together.

If Bennett was writing today, it's unlikely the Leopard's magnolia pebbledash and saccharine music would merit even a sentence. The inn, for its own part, appears largely indifferent to its historic associations. There is a plaque on the wall put up by the city tourism division to commemorate the author who put them on the map, but there's nothing at all to mark the meeting of Brindley and Wedgwood.

James Brindley was already familiar with the town, having built a flint mill here for one of Wedgwood's uncles in the 1740s. Wedgwood himself, like Bennett, was born in Burslem and served his apprenticeship here. As for Brindley and Bennett, they shared considerably more than a predilection for the same boozer. In 1759 Brindley had an affair with local girl Mary Bennett who later had an illegitimate son, John Bennett. It is commonly accepted Brindley was the father. The line leads directly to Arnold Bennett, his great-great-great grandson.

Stoke-on-Trent has few other famous sons to lay claim to – Enoch Wood (another Burslem potter and the owner of Europe's largest pottery), Sir Stanley Matthews, Robbie Williams and, moving down the B list, TV presenter Nick Hancock. None of those in the pub today looks likely to be joining them. A few are munching sandwiches, one's attempting *The Times* crossword, and two bearded men are discussing whether to holiday in Skiathos.

Back out in Market Square, with the town now bathed in the soft sodium glow of street lamps, Burslem's Georgian heyday becomes discernible as long as I keep my eyes trained on the upper stories of buildings. Drop them below architectural waist level for a second, and they collide with mid-twentieth-century featureless glass frontages of gas showrooms and bookmakers' offices.

Beyond the Atlantis lap-dancing club is the steeple of St John's Church staking out the burial plots of virtually the entire potter hagiology (with the exception of Josiah Wedgwood who's buried in Stoke Cemetery). On a corner, Enoch Wood's Fountain Works has been converted into Fountain Court flats. On another stands Vale Dry Cleaners which in *The Card* (later made into a film starring Alec Guinness) was a dancing academy outside which Ruth Earp, doing a moonlight flit, saw her removal van careering down the hill, plunging into the Burslem Spur Canal. The canal's still there somewhere, derelict, hidden like a shameful secret. Like most boarded-up stretches in the country, however, there are moves afoot to reopen it.

I leave Burslem by Queen Street and the ornate elevations of the Grade-II listed Wedgwood Institute, with its terracotta frieze of local industries. It was erected in 1863 as Britain's second free library to commemorate Josiah Wedgwood's Brickhouse Works that had formerly occupied the site. For a while it became a school which Arnold Bennett attended. Today it totters, its upper floors in a parlous state, its lower ones sinking ever deeper into the underground mines below.

Sloping floors are a way of life in Burslem, so is bankruptcy and there are few signs of better days ahead. J.B. Priestley in his *English Journey* wrote, after a visit to the Potteries, 'Out of these dingy houses come people with magic in their hands. People who daily put gold onto pots. May the ovens of Stoke-on-Trent never grow cold.' Unfortunately they're stone cold. The few Staffordshire potteries that are operating, diminish daily because of cheap Far Eastern competition. The city teeters at the edge of an economic abyss and one day soon it will sink completely into the Emmental of abandoned underground mines.

Today it's hard to see further than the discount stores and charity shops such as Mission Possible – 'Jesus is Alive'. Where once there was industry and community, today an economy and culture of asset-stripping prevails. On Nile Street – Bennett's Aboukir Street – stands the Royal Doulton factory shop (now rumoured to be importing its china from the Far East), one of a dozen discounted pottery outlets within a ten-minute walk of each other.

I look for Clayhanger's printing works but find the home of Darius Clayhanger, the first steam printer in Bursley, occupied by the Kismet

Restaurant and Best Wishes card shop. I swing past the George Hotel, which in the guise of the Dragon Inn, was where Edwin Clayhanger (son of Darius) on his 'free and easy' night out, could barely control his excitement when watching clog dancing for the first time.

Hanley, where I pedal to next, has fared even worse than Burslem but if there is a modern equivalent of clog dancing, this is where I'm most likely to find it. Designated the new cultural quarter, it has unfortunately succumbed, however, to cheap booze, cheap drugs and clubbing. Outside Ye Olde French Horn, in a chevron-paved square, a rather beautiful female statue stares down on drunks who have lost the power of speech and movement. To their right, if they could see, next to the British Heart Foundation charity shop, is a store called Nuke. Hanley looks as if it already has been.

My wife often accuses me of not being critical enough as a travel writer, of tending always to focus on the good. But in Stoke-on-Trent, I've met my match. The conglomeration is one large clearance sale – Poundstretcher, Bargain Deals, charity second-hand shops. Even the entertainment is hand-me-downs. A banner across a street proclaims that last week Stoke celebrated 'Britney Weekend' with lookalike, singalike and dancealike contests.

In the pedestrianised Piccadilly, I lean the bike against a 'No Entry' sign and don't bother locking it. I buy a beer at the Bar La-De-Dah ('4 pints Ice Cold lager £7.50') and then sit twelve feet from the bike at a pavement table. Sixty seconds later, a thirteen-year-old boy stops and asks the young man at a neighbouring table whether the bike is his. On learning it isn't, he starts furtively looking around, planning his getaway. I call across, trying to look as hard as nails, to let him know the bike's mine. He doesn't even deign to reply but instead turns back to the first man demanding cigarettes. The two head off together. I'm still trying to make sense of it when another man with very large breasts waddles by with a look as downtrodden as Stoke-on-Trent's.

Ten minutes later, the first man returns without the boy, asks me if I want a drink and joins me at my table. I am immediately suspicious. There is a sniff of drains washing up the street. We quickly discover, however, we have something in common – we both hate Stoke-on-Trent. Only Craig Chamber, a trainee chef, reckons no one hates it as much as he does. 'People have stopped living here. There's no work, no fun; just paranoia and doing people down.' My companion is slight of build, unshaven and is wearing a red ski jacket, jeans and leather trainers. 'That kid was going to sell your bike for cigarettes, you know?' I didn't know but guessed my bike was destined to fuel one craving or another.

It's Craig's day off and he has gravitated to town, even though he knows 'there are few deader places on earth'. Having quickly exhausted the charms of the city – 'There aren't any' – we move on to Craig's work. 'It's better than play, that's for sure.' Craig is planning to move down to London in a year to look for work in one of the posh hotels. Meanwhile, in his spare time, he's taken up writing – 'Poetry mostly and songs but I might try something longer.' He says he discovered 'the knack' while at college. When he learns I'm also a writer and that I even manage to scrape a living out of it, he treats me like a lottery winner.

'So what are you writing about now?'

'A four-month narrowboat journey I'm making.' Craig is temporarily stilled. His jaw drops. 'You mean you're getting paid for messing about for four months?' It's as if I've shared a prison escape map with him. When he rediscovers his tongue, a torrent of questions follow: How am I funding the trip? – The book advance, help from B.W. and Waterway Holidays U.K., and getting in debt. How long can you spend doing the whole network? – For ever. How much does a hire boat cost? – Depends on the time of year and how long you take it for. Is my wife all right about me being away? – No. Do I get lonely? – Sometimes. Why didn't I do a book on the Caribbean or somewhere exotic instead of England? – Because I wouldn't swap this trip for anything. Honest.

By the time the questions dry up, we have sunk two pints of Hoegaarden. Craig heads for the bus station. I head back to *Caroline*, passing a bridge en route on which someone has written 'Welcome to Heaven'.

Back on board *Caroline*, I open a bottle of red and turn on Radio Four. The voice of Robin Lustig eases my concerns that I have somehow wandered into hell instead of heaven. Robin lives just a couple of streets away from me in Muswell Hill and our daughters are in the same class at school. His comforting voice on *The World Tonight* brings a bit of home to my present itinerant life. Tonight he also confirms that the rest of Britain is still in place, breathing if not thinking. An I.C.M. poll he quotes estimates that in the 18 to 25 age range, 10 million votes will soon be cast in the second *Big Brother* series and less than 1.5 million in the upcoming general election on 7 June. The poll also requested the newest members of the electorate to nominate their ideal prime minister: Ali G got 24 per cent of the vote, Robbie Williams 16 per cent, Prince William 6 per cent, Madonna 5 per cent, Alex Ferguson 4 per cent, and the hermaphrodite Posh and Becks 2 per cent.

The next morning *Caroline* and I sneak past abandoned factories caked in age like old wine bottles in a forgotten cellar. The only beauty to be found in the stark post-industrial wasteland comes from sailing under the ornate wrought-iron bridges of Hanley Park, and spotting an occasional thirty-foot brick bottle kiln that managed to dodge the mass cull of the last century. In 1956 when the Clean Air Act was passed, there were around three hundred bottle kilns in the Stoke-on-Trent area. Now there are less than forty. These giant genie lamps stand beside the Cut waiting for men without eyelashes or eyebrows to return and stoke their bellies. They never will; the furnace operators are as extinct as dinosaurs.

I slip downstairs to make a coffee and, waiting for the kettle to boil, put on the radio once more. A DJ is playing 'Desolation Row' as a tribute to Dylan on his sixtieth birthday. I sail past a goat munching contentedly in an allotment, and pigeons having a panic attack in a loft. Suddenly before me stands my first lift bridge and it's here that I very quickly find out just why so many discouraged me from attempting the Caldon solo.

The Nicholson Navigational Notes give no clue to the difficulty of negotiating lift or swing bridges alone. '1. Some of the bridges on this section are very low. 2. Your B.W. key will be required for Ivy House Lift Bridge 11. 3. You will need a windlass and a B.W. key to operate Bridge 21, and a windlass for Bridge 23.' What it doesn't tell me is that the bridge-lifting mechanisms are located on the opposite bank to the moorings. How the hell am I meant to get from one to the other with the bridge up? I try to think my way through the operation and conclude the only solution is to clone myself or grow wings.

Eventually I settle on a course of action. I nose *Caroline* almost to the bridge and tie her mid-line loosely to the nearside railing. Once I've walked across the bridge, I insert my B.W. key in the control panel, drop the traffic barrier (pissing off car drivers in the process), and raise the lift bridge a foot or two. I then clamber up the inclined bridge and slip the rope through the crack before retying it to the upstream railing as close to the operating mechanism as possible. I complete the raising of the bridge and haul *Caroline* through with the mid-line. Finally I drop the bridge, lift the barriers and, as cars hare across trying to catch up lost time, I pull *Caroline* into a position where I can board her.

It sounds complicated but it's worse than that. As the guide produced by the Caldon Canal Society points out, 'Ivy House Bridge will not lift unless the road barriers are properly down, and the barriers will not lift unless the bridge is properly down, and the operator cannot retrieve the B.W. key unless the bridge is down, the barriers are up and the lid to the box is closed. Easy is it not?'

With each new bridge successfully navigated, and with each bend in the canal passed, there is a promise that open country is just one more bridge or turn away. But homes doggedly cling to the city's coat-tails. A Walls ice-cream van creeps down a suburban street, a sign on its rear advising all who follow to 'slow'. I couldn't go slower unless I crawled on my hands and knees.

A reed fringe gives a river feel to the Cut, encouraged further by its twists and turns. I pass water meadows filled with buttercups, a swan nesting on a bed of bulrushes, and a crow marching through a field of corn marigolds. Eventually I manoeuvre past a bizarre aquatic rounda-bout with seven feet clearance either side, dip under a bridge and moor up beside a large winding hole outside the Stoke Boat Club at Endon.

Beyond the red and white barrier barring entrance to all except members, I notice a man fiddling about with his narrowboat engine. The white hair that once resided on his head has slipped to his chin, and he's sporting a broad check lumberjack's shirt. I ask him if it's OK to moor overnight where I am.

'Should be fine. We don't get much trouble round here these days. And it's our club night tonight so if you'd like to pop in for a pint, you'd be most welcome. It's my turn on bar rota so I can come and let you in through the gate here,' the man points to the razor-wire garland to the padlocked entranceway. 'By the way, my name's Ray...Ray Dawson.'

The clubhouse is a bungalow with brutal strip lighting that members have attempted to soften with boat-rally pennants, memorabilia, trophies, newspaper cuttings, an honorarium of past presidents, and a heaving shelf of almanacs charting the interwoven histories of the club and the canal.

In 1961, ten years after the last commercial load was carried along the Caldon, one of the club members, Ben Fradley (now deceased), ripped down the 'Closed' sign at Etruria Junction and cruised the forbidden waters. His aim was to draw attention to the canal's criminal dereliction and to show it could become a much-needed green linear park for the Potteries. The Boat Club, with the Inland Waterways Association and the local council, formed a committee and in 1974 the canal finally reopened.

The members who've gathered in the clubhouse tonight are an eclectic bunch, full of kindly advice such as 'Don't leave your boat alone at Leek,' and 'Hug the towpath side at Cheddleton where the canal joins the river or you'll run aground.'

As I prop myself at the bar, I listen to their stories as they ferry drinks. One farming couple decided recently to diversify and so visited Cheshire's largest market at Chalford to check out what was on offer. They came away with a llama, 'But we found it confused the horses and kept wandering along the railway line so we had to get rid of it.' Peggy Young used to run horse-drawn boat trips from Froghall Basin until the horse died four years ago.

For most members, however, there is only one topic of conversation: boats. Dave Salt, the Caldon archivist and compiler of the Society's guide to the canal, is currently refitting his 42-foot Loftus Bennett cruiser and confesses to having spent the past three days trying to fix hardboard to the cabin roof to take carpet. 'I'm a car mechanic by trade and should keep to engines really.'

Fred, at 86 the Boat Club's oldest member, recently finished building his dream 40-foot narrowboat and immediately started planning a big trip with his wife. Two months later she died. He doesn't sail now but still has the boat moored in the marina.

I continue trawling as they continue drinking. In between I chat with Ray, behind the bar, and wander the clubhouse. Alongside the bar rota, club rules, and B.W.'s Foot-and-Mouth News, I note a number of sister canal-club publications – the North Cheshire Cruising Club's 'Ditch Crawler' and the Trent & Mersey's 'Grand Trunk' among them. The most interesting material, however, is pressed between the black covers of the club's archive annuals – each album another ring on the trunk of the canal and its society.

Inside the first, I thumb through black-and-white photographs showing the Leek arm filled in and used as a tip, the protest boat rally to Froghall Basin and the volunteer work digging out locks. In another, are the midnight barbecues in woods, glamorous black-tie dinner dances, and fancy-dress Christmas parties. By the time I reach the 1999 album, the club's photographer has finally switched to colour and parties have degenerated to a 'Conker Final Autumn Cruise'. The Stoke Boat Club members' glamour has been shelled with the ball gowns and youth.

It's not surprising. Most of the hundred or so current members are closer to their century than their teens. The problem the club shares with other boat clubs is in attracting new members. 'Most youngsters can't afford narrowboats. They'd start out with a second-hand cruiser but British Waterways wants to be rid of them.'

'What, cruisers or youngsters?' I seek clarification.

'Cruisers. They must do.' Dave Salt gets into his stride. 'Licence fees are weighted against us, we're charged full wallop for our petrol while narrowboats get subsidised farmers' diesel, and because of the fire hazard we're not allowed to carry cans of petrol on board – that means at around six miles per gallon you're not going to get very far are you?' Dave grimaces, rippling his beaver-lamb beard. 'Maybe B.W. think cruisers aren't historical enough. Or maybe they hate fibreglass. There's got to be some reason why they pick on us.'

I leave Dave to his rant at B.W.'s apartheid policy and return to Ray, who's just finishing his shift. I recognise the check shirt as the same one he's wearing in every photograph in the albums. He kindly escorts me back to Fort Knox gate. The early evening rain has moved on, leaving a clear night. A frog hops by under the silver moon. As we walk beside the sleeping boats, Ray points out Dave Salt's Loftus Bennett. 'He always wins the annual I.W.A. long-distance boater award on it because he spends at least six months a year afloat.'

'On his own?'

'No. Mostly with his wife. Except on rivers. She always jumps ship then.'

Ray slips the key into the padlock and draws the bolt. I thank him for the invite.

I make an early start, casting off under a shower of spring blossom, and slipping past a field thick with buttercups in which cattle are grazing. Beyond them, hills rise up, patchwork fields broken by an occasional farmhouse or crested by the silhouettes of trees or horses. My dhobi sashays on its washing line as I pass a boat named *This England*. The skipper and I greet each other warmly as if sharing a secret. It's hard to resist breaking into a chorus of Dylan's 'New Morning' (which he wrote after recovering from a near-fatal motorbike accident). It's just a couple of miles from 'Desolation Row' to 'New Morning'.

The Caldon Canal, which opened in 1779, transported the mineral resources of the Churnet Valley – and in particular limestone from Froghall – to the developing industries of Stoke-on-Trent and beyond. In its circumlocutory loquaciousness it is a fitting final obituary to England's first great canal engineer, James Brindley. Today, with its industrial past seemingly as mythical as King Arthur's Avalon, it's hard to imagine there can be a greener, prettier canal in the whole country.

Caroline clings halfway up the hillside, offering clear views out over the countryside, as she barrels through a series of grey bullseye bridges. For a while the railway keeps us company before we drift through an enchanted wooded glade carpeted with bluebells. At the Hazelhurst Locks, the canal splits and dances a dainty figure-of-eight. One arm heads for Leek, as *Caroline* and I dip under it on the Froghall main line.

At Basford Bridge I break for lunch and pop into a handsome stone pub with a painted wooden sign that introduces itself as the Boat Inn. Sitting in the low-beamed interior on a bench far too small for them are two boaters with Olympian girths. Tony, his long white hair cascading over a face run with sun gulleys, is wearing a bleached-out T-shirt of a boat rally held in 1997. Ivan appears younger, maybe in his forties; a bruiser with cropped hair and a mutt serving as a rug across his feet. Standing on the table in front of him are four empty pint pots and two that are half empty. The men are single-handed boaters who live year round on their narrowboats. Tony continues to cruise in winter in the north, while Ivan prefers to stay put at his berth at a marina in Aldermaston. Each summer they meet halfway and set off on that year's six-month odyssey.

Over the next hour they provide invaluable tips on sailing the Leeds

& Liverpool – 'Go through the East Lancs milltowns at sparrow's fart to avoid the kids lobbing stones at you'; and the Aire & Calder – 'Get your mooring pins in tight – those bloody barges go by like battleships – and mind you keep a bit of slack for the wash.' Tony then warns me that Isis Lock, where the Oxford Canal joins the River Thames, can be tricky. 'I was in full reverse and still going like a cork out a champagne bottle.' When it's Ivan's turn, he insists the River Trent is more easily navigated upstream than down (that's how I planned doing it anyway). As for the Caldon, they both concur, the single-handed method I employed the day before on the lift bridges is indeed the only way. 'Unless of course someone's about that you can nab – that's always the best bet.'

After my second and their fifth pint, we peer out the window. It's raining. 'Well, that's that then,' Ivan rubs his hands together gleefully. 'Have to stay put in the pub. No point moving in the pissing rain, is there?'

It's a rhetorical question but I decide to answer anyway. 'I think I'll be moving on actually.' They look at me as if I'm mad. As I stand to make my move, Tony offers two more tips. 'Look out for Tommy Beardmore at Froghall Basin. You'll know him 'cos his boat's like a gypsy caravan. Great craic he is. Ex-professional boxer.' And finally, 'Mind you keep to the towpath at Cheddleton drop and don't move out for no bugger. Ivan ran aground and we had to bow-haul some other bugger off too.'

The Caldon is the most testing canal so far to steer, narrow – at times there's no more than a foot or two either side – with tight bends and low bridges. Sometimes all three come together – a bridge on a narrow bend – and when they do, there's no way of knowing if a boat is approaching as you emerge, or emerging as you approach. I have already had several close shaves.

Like the boaters, the bridges all have their stories. There's Cherry Eye Bridge (Bridge 53) named after the inflamed eyes of the miners emerging from the local ironstone mines. There's Botany Bay Bridge (number 14) commemorating deported convicts' landing place in Australia (Cook's discovery of Australia coincided with the digging of the canal). Bridges do not even have to have names to have histories. Bridge 16 used to access British Aluminium Works which was targeted, but never hit, by the Luftwaffe.

At the point I'd been warned twice to watch out for, where the River Churnet joins the canal, there's a hold-up at the lockgate. Beyond it, a boat has run aground. A Captain Haddock in sailor's cap and snow-white beard is trying to reverse, trying to pole, trying anything to escape but to no avail. He's stuck fast on an atoll of silt. I take his bow rope, someone takes the mid-line, and Captain Haddock tries full-throttle reverse. Eventually we free him and this time he keeps as close to the bank as possible when accelerating forward.

The next boat through the lock veers too wide and gets stuck in exactly the same spot. Again, with the help of ropes, we manage to free him too. When it's my turn, I barrel out of the lock, hugging the bank. *Caroline* doesn't even scrape her belly. I discern a slight swagger to our subsequent progress.

Most narrowboats on the Caldon tend to be 40-footers rather than 70-footers and they're not floating garden centres either: practical, cared for, lived in, but lacking the Olde Worlde brass trinkets of the Grand Union further south.

The countryside becomes increasingly wild and dramatic as the canal snakes along the valley floor through steep wooded hills. I moor at Consall Forge beside a weir where a high rope swing is being played by the breeze. Nestling beneath the forest is the Black Lion pub, accessed by the canal, a river, and a pretty railway station that only receives trains on Sundays.

With no sealed road to the pub, the clientele tends to consist of boaters, bikers and railway anoraks. Inside its entranceway are jellybeans and Smarties dispensers; on the wall there are horse brasses, lace doilies and prints of fat animals. Adrift in its time warp, the Black Lion reminds me of a pub I once visited in Pembrokeshire which was still using the Julian calendar.

Two customers mouth the words to taped songs, none of which was released after 1970 – 'Listen to the rhythm of the falling rain', 'Why do fools fall in love?', 'I don't care what they say I won't stay in a world without love', 'Do not forsake me oh my darling',' 'Dream lover so I don't have to dream alone'.

Each song's truckload of memories unfortunately derail at the first bend when I discover an imposter driving the train. Why do cafés and pubs play anodyne covers when it's just as cheap to buy the real thing? There is nothing in the world so unsettling as Timothy Ragabottom singing Del Shannon's 'Runaway'.

Halfway through my pint, the food arrives – a rubber mat of liver served baconless with hard chips and bullet peas. It looks like the plastic display dishes in the windows of Tokyo restaurants. It probably tastes like them too. The Black Lion has yet to catch wind of the Edible English, let alone the Modern English culinary crusade.

Sticky toffee pud with custard clings to my ribcage as I walk past the buttresses of five 50-foot lime kilns being swallowed by undergrowth like abandoned Inca ruins. An owl hoots as I follow a rising track scattered with acorns and fringed by wild rhubarb, buttercups, and ferns. Each tree has its own choir of birds. Reaching a bluff I stand looking out over fields of clover running down the hill to the River Churnet where a heron lumbers into a heavy landing with a frenetic beating of wings.

Somewhere, hidden by the heavily forested hills rising beyond the Black Lion, is the village of Ipstones where James Brindley, massively overworked and with his diabetes out of control, caught a chill while surveying the Caldon and then slept in a damp bed at the Old Red Lion.

Josiah Wedgwood, in a letter quoted by Cyril T.G. Boucher in his biography of Brindley, prophetically wrote as early as 2 March 1767, 'I am afraid he will do too much and leave us before the vast designs are executed, he is incessantly harassed on every side, that he hath no rest either for his mind or Body, and will not be prevailed upon to take proper care of his health.'

Brindley never recovered. He died on 27 September 1772 aged 56 and was buried at nearby Newchapel which I plan visiting in a few days time when I return to the Trent & Mersey.

Beyond the Black Lion the canal constricts as it burrows under a bridge and then dips beneath the overhanging waiting room of the pretty wooden railway station. A little further on is another stone bridge whose lancet arch would grace any medieval cathedral cloister. As the sun dances in the forest, I watch two greenfinches balancing on a lock beam, tails dipping and rising in time with their trilling.

The canal terminates at Froghall, unfortunately I see the winding hole too late and instead of taking the sensible course and reversing before attempting the 180-degree rotation, I try to squeeze in a belated manoeuvre and make a hash of things. Out of nowhere a scruffy but heroic Daniel Boone figure paddles towards me standing upright in a canoe laden with logs and a black lurcher dog. He pushes *Caroline*'s nose off the bank, assisting the turn. I ask if his name is Tommy Beardmore.

'Might be. Who wants to know?' he asks suspiciously.

I explain that I'd been told to look out for him by Ivan and Tony at the Boat Inn.

'How d'you know it were me?'

'I just had a feeling.' I invite him on board for a beer.

'Don't drink alcohol no more but I'll have a tea,' he replies as he holds the Canadian canoe fast to *Caroline* gripping the stern fender. I fear he, the dog and his boat may end up shredded if he gets any closer to the propeller so I moor up and switch off the engine. Opposite me, on the far bank, is Tommy's battered *African Queen* whose prosaic identity as a concrete-hull Davidson with a three-cylinder Lister engine built in 1966 has been made more exotic by sticking a skeleton in a window and a stuffed buzzard on the roof.

Tommy clambers aboard *Caroline* with Heathcliff, his six-month-old whippet-lurcher cross. Again I try to tempt him with alcohol but he's resolute. 'Haven't had a smoke or drink for eight years – twice as long as I've had me boat.' The kettle is duly filled. 'Jumped in a pint pot for ten years after I finished boxing an' nearly din't come out. In the end though I won. I gave up the bottle and now I can whack out most twenty-year-olds like I used to.' Tommy shadow-boxes and launches a couple of jabs to prove his point. He is like the swallows that sweep in past us, swooping and dipping, rolling and diving. His voice too imitates their flight, rising at the end of sentences instead of dipping, like punch-drunks who don't know when to stay on the canvas.

After each raucous laugh, he playfully punches me on the upper arm and then ducks and weaves as if to underline the fact no one catches Tommy Beardmore. Born just four miles away at Cheadle, I doubt he has ever remained still for more than fifteen seconds.

As we sit in the stern cabin drinking tea while Heathcliff sniffs his way through the boat, I ask Tommy how he first got into boating.

'Been a drifter all me life. Knocked about round the Cut for fifteen years, then I bought my *African Queen*. Free and easy. Been a drifter all me life,' he repeats himself. Again the cabin explodes with his laughter as he stands, sways and only catches me with the second jab this time. I'm learning.

'I had twelve months travelling round New Zealand, six years on oil rigs, and the worst seven years of me life down the pits in Stoke, hauling coal. Best thing they ever did closing 'em. No place for humans, mines. No good for nobody. All me family were pit men. That's why I'm small.'

Small, yes, and stocky, and without an ounce of fat. Tommy looks ideally suited to life as a bantamweight and it's this part of his life that still consumes him. 'Three hundred amateur fights and six professional in all I had. Started at ten and finished at thirty-eight and can still take out blokes half my age.' He nods his head several times as if counting an opponent out. 'I've travelled all over boxing. Used to fight in the sporting clubs and smart hotels for people all dressed up in dinner suits – Coventry, Glasgow, Pontypridd, Brum. Reached the A.B.A. semi-finals, was the Midlands Champion and Coal Board National Champion. But it was boxing that got me into smoking and drinking. Too many temptations. Still I've turned my life around now, and without religion too. When I see that kingfisher flying by, that's me, that's my religion . . . Ark, ark, ark.' Tommy screeches. 'That's my imitation of a crow not a kingfisher,' he explains. Tommy runs his eye along the bank. 'I see foxes, an' there's a green woodpecker comes down each morning and a common buzzard that wheels overhead – I got a stuffed one on the *African Queen*.'

He scratches his head and continues his itinerant history. 'Last year I went down to London on the boat with me three dogs – Ragbag, Teabag and Fleabag. I had to get rid of 'em and now there's just Heathcliff. I'm writing a children's book about them though. Three dogs that go to London called Ragbag, Teabag and Fleabag. They live on a boat and have been brought up to always ask policemen the way if they get lost. Well, in London they do of course and Fleabag asks nine different policemen the way and none of 'em knows it. What d'you reckon?' He asks my opinion but can't wait for it. 'I think it'll be a big success but if it isn't I'll paint the boat and head north, or maybe south. I'll put my thumb up and see which way the wind's blowing.'

Tommy grins and fumbles inside his jacket, retrieving a cardboard box which he opens before setting its contents, a wind-up rat, scampering across the cabin floor. 'That's how I make a living now. Got hundreds of 'em. Sold 'em outside Buckingham Palace till I got moved on. Sold 'em in Oxford. All over. Wherever there's a crowd that the canal reaches.'

I look at the box, 'Crazed Running Rat', made in Taiwan. 'You can keep that one as a present. I got hundreds on the boat. Helps me keep afloat.' Duck, jab, weave. Tommy's nomadic canal existence fits him like a boxing glove.

Reg, a local farmer, goes past and Tommy knocks on the window and waves. 'He's having it hard. Everyone is. Farmers got foot-and-mouth, and all the steel and pottery industries are finished. Too many cheap imports. Like me rats.' Tommy roars. And then he's off again in another direction like my new pet rat.

'Guess how old I am? Go on. Guess.'

'Forty-five?'

'Sixty!'

I can't believe it. Spry, athletic with curly hair and a lightly lined face, he could pass for a good-looking-in-a-rough-way forty-year-old. Tommy rises and slaps his thighs. 'Gotta go. Moving someone's car from a field they crashed into.' And he's up, out on deck and untying the rope to his canoe in fifteen seconds flat. I watch him disappearing back down the creek, paddling upright.

I set off in the opposite direction, the way I came. Over the next couple of days I rewind the forested hills, the water meadows, allotments, pigeon lofts, suburban homes, bottle kilns, and demolition sites back to the statue of Brindley where I fork right to join the Trent & Mersey once more.

My overnight mooring is alongside the last remaining bottle kiln of the Middleport Pottery, just a few miles north of Etruria. Earlier in my journey at the Wedgwood factory I'd been told, 'Burgess, Dorling & Leigh is a fairy-tale success among the ruins of the pot banks.' When I phoned the current owners, Will and Rosemary Dorling, to ask if I could take a look round, they immediately invited me to overnight at their mooring.

I tie my ropes to an old crane on the dockside and pad past the bottle kiln and a hundred-foot brick chimney, following an alley whose cobbles once rang with potters' wooden clogs and the metal shoes of horses. Having sneaked into the factory from the canal, it takes some time to find the street entrance. The 30-foot-high wooden gate is bolted. I knock on a nearby door. No one appears. I shout 'Hello' a couple of times and eventually the door opens and a floppy-haired man in pale cords and a blue crew-neck sweater greets me.

'Paul?' I recognise the voice from my earlier phone call. Will Dorling ushers me inside the high-ceilinged office where he introduces me to his wife and partner Rosemary, whose long blonde hair and ankle-length skirt look more Home Counties than Potteries. William is 43, and Rosemary 47. They make a good-looking couple.

Our introductions boom through the empty Victorian offices. It's as if the entire workforce abandoned the building at the turn of the century and we are the first industrial archaeologists to venture back inside. On top of an ancient safe is an eight-inch-thick bible. Beside it are a couple of frayed-leather desks bought from the Davenport pottery when it went under in 1997. From the walls, portraits of the founders in their high collars and sober expressions stare down. There's also a framed early black-and-white photograph of the city's bottle kilns stretching to the horizon like the pagodas on the plain of Pagan. It suddenly strikes me that while the rest of Britain has switched to colour, Stoke-on-Trent has retained its monochrome character (Arnold Bennett talks of the 'Grim Smile of the Five Towns').

'This is how we found the place when we moved in eighteen months ago. Look.' Rosemary leads me across to a mahogany cupboard. Inside is a Home Guard First World War helmet, a 100-year-old unopened bottle of cough mixture, and a number of pottery-glaze recipes dating back to 1820.

Will returns, balancing a teapot on a tray. We sit at a large wooden ledger desk, sipping tea from pale blue Asiatic Pheasants cups – the shards of which you discover when digging over any English garden. 'It's one of the oldest Staffordshire pottery designs apart from Willow, dating to 1794 and once made by every pottery in Stoke. Now we're the only ones making it,' Will states matter-of-factly.

Despite their rootless accents and the fact they've yet to celebrate their second anniversary of ownership, the couple seem to belong here. When they first walked into the old small packing room as the anxious new owners, there were just three tin stencils attached to the wall visible through the dust – Salisbury, Rhodesia and Johannesburg. They stared long and hard. Rosemary had been brought up in Salisbury, Rhodesia, and Will in Jo'burg. They knew then and there they'd made the right decision buying the place. It was fate.

Just a few years earlier, when they opened a ceramics mail-order shop, The China Box Company, in the front room of their home in Hampshire, they had £400 in the bank. They started making regular trips to Staffordshire to restock the more popular lines. The Middleport Pottery was always their favourite port of call. One day, however, in June 1999, they were told the consignment they were picking up would be the last. The factory was in receivership. They were devastated.

When Rosemary asked an old man she'd befriended in the pottery how his work was going that day, he replied, 'She's not moving so well today. I'm having to heave it out.' To the potter, the clay was as animate as a dog or cat but his heart was heavy. Unselfconsciously he told her, 'You have in pottery God's four elements – water, fire, earth and air. All you need is man to mould it.' He shook his head, unable to accept the most likely future for Middleport Pottery was demolition and reincarnation as a car park.

For some people, the dousing of yet another pot bank in hell's kitchen may have been a cause for celebration, but the potter was desperate to continue the craft his father had passed down to him.

Rosemary breaks off from her story and rifles through a drawer, retrieving a sheet of paper on which she has copied an extract from an address by William Morris to the people of Burslem on 13 October 1881. She declaims sonorously as if reciting the noblest poetry. 'You who in these parts make such hard, smooth, well-compacted, and enduring pottery, understand well that you must give it other qualities besides those which make it fit for ordinary use. You must profess to make it beautiful as well as useful, and if you did not you would certainly lose your market. That has been the view the world has taken of your art, and of all the industrial arts since the beginning of history.'

She nods her head, still visibly moved by the words. 'The potter knew the truth of this even if he'd never heard of William Morris. He faced a dark future without that creativity in his life. As Will and I drove back to Hampshire that night, we hardly spoke. A verse from Omar Khayyam, written a thousand years earlier and that I'd learned at school kept going through my mind.

> And I remember stopping by the way
> To watch the potter thumping his wet clay;

And with its all obliterated tongue
It murmured, Gently Brother, gently pray.

Back home, Rosemary and Will prayed for a last-minute intervention. Will wrote to English Heritage and to Prince Charles but no white charger appeared on the horizon. Finally he woke one morning with the obvious question for his wife. 'If no one else is going to save the works, why don't we?' With an hour left before bids had to be in the receiver's hands, they managed to put together a package by remortgaging their home and begging and borrowing from family and friends.

Their bid was successful. In August 1999 they inherited eighteen staff, including the potter. Just over a year later, with a full order book – mostly filled by new Japanese, Australian and American clients seeking pure English earthenware – they took on another 30 staff. In 1939 there were 512 people working at Burgess & Leigh. But 48 at Burgess, *Dorling* & Leigh was a lot better than one car-park attendant.

A couple of months after my visit, Will and Rosemary planned marking the 150th birthday of the company by sending a consignment of Burleigh ware to London's Camden Town by canal boat. It will be the first time in over fifty years Staffordshire pottery has been transported by boat.

When we finish tea, Will shows me round the two-acre site. It's the weekend and we have the place to ourselves. In the eaves we pass endless shelves heaving with 5,000 pottery moulds dating back to the eighteenth century. We descend the stairs, running our hands along banisters sandpapered by workers with clay on their hands. On the ground floor the original concertina clay press has the weekend off but a 60-foot-tall mangle, stacked with drying glazed pots, is slowly rotating up through the building before the pots are packed and shipped to Buckingham Palace. Slumbering in the basement, we find the great beast of the building, an enormous Lancashire boiler lagged in its winter coat eating its way through another ton of coal.

'What makes us unique,' Will proudly sums up, 'is that we're the only pottery in the world producing traditional Staffordshire earthenware employing exactly the same methods and machinery that were used in the nineteenth century.'

Will returns to his paperwork and I head out for a Saturday-night curry. Outside the factory gate, I walk past terraced homes pottery workers once proudly owned and no one wants today. The houses rarely sell for more than £12,000 and at a recent auction one went for £4,000.

On the corner of the street a girl aged thirteen or fourteen greets me familiarly.

'Alright luv?' She smiles.

'Hi.' I smile back.

'D'you want some then?'

I'm shocked. At first I think she's offering her body and then decide she's too young and it must be drugs. Declining her offer, I head up the hill past tumbling Coronation Street terraces, a fortified Methodist chapel, two working men's clubs, and two bottle kilns looking like futuristic eco-homes in the twilight.

Eventually I find myself back in Burslem following an aroma of curry to the Gulshan in Swan Street. Beneath the generic prints of Polynesian and Swiss landscapes I polish off a saag aloo and a bhuna gosht and then continue my walk, looping back to *Caroline* by a more circuitous route.

Sitting on the kerb of a busy road another thirteen-year-old is screaming at her friend twenty yards away, 'Shirley, cum back 'ere.' Cars hammer along the street no more than eighteen inches from her shriek of desperation. The friend ignores her. The girl's head slumps into her hands and for a moment I fear she might roll into the road. Above her, on a lamppost, is a flier, 'Re-elect Joan Walley Labour'. Three teenagers, aged maybe fifteen, who seem to know the girl, get her to her feet. She staggers off in the opposite direction to her friend, bouncing off one wall and then the other down a long narrow alley. I wonder if Joan Walley deserves re-electing.

I reach the gates of Burgess, Dorling & Leigh around 11 p.m., ring the bell and Roger, the nightwatchman, lets me in. He's been working nights more years than he cares to remember. 'Before security I did nights on the railways.'

'Doesn't your wife mind?' I ask.

'Don't care really. Have to get the work you can. Anyway I like nights.'

Hearing our voices, William appears in a crack of light at the office door.

'Thought it might be you. We're still clearing out paperwork but I've got a bottle of wine if you fancy a nightcap.' We take the wine and three glasses into the display room. I break out a bar of chocolate on the table and we share the food and drink with the ghosts.

I compliment Will on how handsome the building looks from the street. 'When we arrived all three hundred and fifty-three panes of glass – I know, I counted them – were broken. Darren, our young security guard, said he'd sort it out. And he did. When he caught kids breaking a new pane, he promised to take them to free football coaching sessions if they didn't break any more. Then we got the boys in on Saturdays doing odd jobs for us, clearing and burning wood. One midweek evening three of them turned up and asked if I had anything they could do. I told them I was already over budget that month and couldn't offer them anything. They said they were bored and didn't mind not being paid. A couple of weeks later some outsider broke a window and they threatened to break his legs.' Will smiles. Rosemary pours more wine.

On the shelves surrounding us are hallmarked Staffordshire earthenware designs. Recessed lights sparkle on the rich cobalt blue of a Calico soup terrine and a pale blue Burgess Chintz Stilton dish. When conversation lags, silence booms through the building. I tell Rosemary about the girl outside offering me drugs. 'She was so young and pretty. Terrible to be dealing at such an age.'

'I know the one. She's always there. She probably wasn't offering you drugs, it was more likely to be sex. If she had drugs she'd have used them herself. She's always on the corner of Furlong Street between half past five and nine at night selling herself.'

'But she's younger than my daughter,' I protest.

'There's about fifty girls working the streets round here – and a high percentage are thirteen and fourteen. Often they're paying for a drug habit, or a partner's habit. A girl was found strangled behind the churchyard two months ago. She was a junkie and prostitute.'

Rosemary takes a square of chocolate from the bar and pours the last of the wine into my glass. 'I know a retired local schoolteacher who says that forty years ago all the children at the local school had mums and dads. And their parents probably had jobs too. Now there are so many teenage single mums coming from teenage single mums that they have no one to show them how to bring children up. They can't cook, sew or look after themselves, let alone anyone else. It's worse than the Third World round here because the family has completely imploded.'

I walk back to *Caroline* over cobblestones slicked with moonlight and a recent shower. A cat prowls past, and disappears through a gap in the brickwork of the bottle kiln. My two mooring ropes are still fastened round the crane like a ribbon on a present. On board I find it difficult to sleep. Maybe it's the chocolate. Maybe it's the curry. Maybe it's the thought of lost souls wandering the street bashing into walls.

D.H. Lawrence in *Nottingham and the Mining Country* rants:

> the real tragedy of England, as I see it, is the tragedy of ugliness. The country is lovely; the man-made England is so vile...The great crime which the moneyed classes and promoters of industry committed in the palmy Victorian days was the condemning of the workers to ugliness, ugliness, ugliness: meanness and formless and ugly surroundings, ugly ideals, ugly religion, ugly hope, ugly love, ugly clothes, ugly furniture, ugly houses, ugly relationships... (Quoted by Jeremy Paxman in *The English*.)

I lie in bed with a mug of tea worrying about my own children tucked up at home, and wondering whether things are better in the Potteries today with rampant juvenile prostitution than in the eighteenth century when child labour in factories was the accepted norm.

Early the next morning I phone home and wake the children. I feel reassured to hear their voices. Just half an hour from Middleport the water turns an orange-rust colour with the approach of the 2,926-yard Harecastle Tunnel and my mind refocuses on the canals.

Although aqueducts grabbed the eighteenth-century glory as the ultimate engineering marvel, it was the tunnel that was the greatest headache. Brindley was renowned for his 'impossible' aqueduct at Barton (which I shall cross over in a week's time) but he never lived to see the eleven-year completion in 1877 of the tunnel I'm now approaching. As Anthony Burton in *The Canal Builders* points out, 'as a piece of engineering, the Harecastle was crude, unfinished, a rough hole through the hillside. But its 2,900 yards all had to be cut through the most difficult ground with only the most primitive of techniques.'

The method was simple: vertical shafts would be dug to the required depth before a horizontal start was made either side of the hillside.

Primitive horse-powered gins were used to lift waste up the shafts and lower workmen down; the only other mechanical aid were the steam engines that pumped water to the surface, keeping the tunnels dry. Rock would be blasted out by gunpowder but the real work was done by pick and shovel.

The canal companies, naturally, rarely mentioned accidents although there were many: collapsing roofs, scaffolding or buckets carrying workers; flooding, fires, collateral blasting damage...there were 101 things that could, and did, go wrong. Life and labour were cheap and turning a blind eye avoided paying compensation.

Up ahead of me I can see a barrier barring entrance to the underworld. My Nicholson guide informs me that tunnel admission times vary as a one-way system operates. I set off to find someone who can tell me when I'll be able to enter.

Gordon Copeland, who's standing beside the entrance, tells me the first northern passage will be in 35 minutes at 8.40 a.m. With his longish black hair that's combed into a fringe and over his ears, he looks far too young to have been working for B.W. the 26 years he claims to have. 'There's a convoy on the way down at the moment. We allow a maximum of eight boats at a time because of the fumes. You're the first up this morning.'

'How long does it usually take to go through?'

'The average is forty-five minutes and the quickest twenty-five minutes – I know because my dad was tunnel keeper on the north end for twenty years and kept records.'

'Is that Brindley's original tunnel?' I point to a half-sunken tunnel to the left of Telford's. Gordon, short, slightly built and with a strong Staffordshire accent, nods his head.

'Took eleven years to dig, opened in 1777 and was called "The Eighth Wonder of the World". There was a railway tunnel too but you can't see that now. It's hidden by trees.' He doesn't seem in a hurry to be off, and with half an hour to go before I sail through, we sit outside the B.W. hut in the sunshine and chat easily. As a schoolboy, Gordon was forever bunking school, hitching rides with various boatmen still working the route up until 1968. 'I'd often go with Freddie Gibbs on the *Shad*. Most boatmen didn't really need the help as they were all big families, like the Lowes of Middlewich who had eighteen children – but they'd let me help anyway. Hardest bit was always the long walk back. I remember once when nothing was moving for weeks walking on ice two-foot thick too.'

'Didn't your parents tell you off for bunking school?'

'Dad didn't mind. He knew it was hard work and thought it good for me. I don't think many would do the hours today though. Bill Atkins up in Wheelock, where I live now, reckons that in the sixties he'd get £4 for running freight all the way from the start of the Weaver Navigation at Weston Point down to Stoke.' Bill lives close to Gordon's lock-keeper's cottage on the Wheelock Flight. 'You'll be passing through later today. If you fancy it, pull in for tea and I'll show you some of my dad's old records. Our house is the one with the narrowboat at the end of the garden and I'll be there after six o'clock. Lock sixty-six.'

Once inside the tunnel, I know through experience it will be a long haul and that the pinprick of light in the distance is as illusory as a carafe of water dangling on a string in the desert. I try to keep to the centre of the tunnel. The bright pinpoint stretches infinitely, playing havoc with perception as I switch my focus from it to something closer at hand. Feeling giddy, I reset my eyesight on the arc of Caroline's headlight and keep it there. The darkness in the tunnel weighs as heavily as a migraine. Water drips and occasionally showers. It is 30 minutes before I finally emerge once more into daylight. When Brindley's parallel tunnel was legged, it took three to four hours.

I moor up at the first lock on the other side and cycle uphill a few miles to St James's Church at Newchapel. Although I've been operating locks and living the outdoor life for weeks now, my legs appear to have lost their strength and I find the cycling hard going. Eventually I reach the graveyard and track down a copper plaque on which are recorded the bare facts of a man who changed the face of a country and in so doing, the world:

James Brindley Canal Engineer
1716 born at Tunstead near Buxton
1742 In business as wheel and millwright first in Leek and Burslem in
    1756
1758 Patented a fire engine for drawing water out of mines
1758 Surveyed a route for the proposed canal from the Trent to the
    Mersey
1760 Cut the Manchester to Worsley Canal for the Duke of Bridgewater
1766 Married Anne Henshall of Newchapel
1766 to 1777 Cut the Grand Trunk or Mersey Weaver Canal through the

Potteries uniting the rivers Trent and Mersey
1772 Sept 26th Died at Turnhurst Hall

Lying beside Brindley is John Henshall Williamson, a land surveyor from Newchapel who the engineer worked closely with on plans for the Grand Trunk from 1762. During his early visits to Henshall's home he took a shine to his colleague's schoolgirl daughter Anne. He courted her assiduously and when she reached 19, and he 50, persuaded her to marry him. He was roughly my age and Anne a few years older than my daughter.

Anne does not share his final resting place. Nor do his two daughters nor his illegitimate son John Bennett. After his death, Anne Henshall, at the age of 26, remarried into Brindley's in-laws, the Williamsons, and had seven more children.

Brindley's grave, tucked in the bottom left-hand corner and engulfed by trees, is probably the least well positioned in the entire cemetery. Other deceased parishioners buried on the high ground get the breeze and the view across the rolling hills to the Cheshire plain. Still, Brindley came from humble origins and probably expected no better. By ten he was working as odd-job man on a farm in Leek, and by seventeen he'd started a seven-year apprenticeship as carpenter and millwright to Abraham Bennett (no relation as far as I know to Mary Bennett or Arnold Bennett) near Macclesfield.

England's first great civil engineer never got close to mastering the art of reading and writing. When summoned to a select committee in London on 16 February 1762 to give details of his plans for the Grand Trunk, he carved out a model of an aqueduct from a slab of cheese; and mixed sand, clay and water in front of the assembled politicians to show how clay-puddling created a watertight base to the canal. 'Unacquainted with the literature of his own or any other country', Brindley, during this his one and only visit to London, attended a performance of *Richard III* and was so shaken by the experience (it 'disturbed his ideas and rendered him unfit for business'), he took to his bed and vowed never to visit a theatre again.

Taking to his bed was a way with him. When confronted by difficult engineering problems, bed and sleep was also his preferred method of seeking a solution.

Right now I doubt he can be getting much sleep for the hymn being sung inside the church is threatening to lift the roof tiles. I leave the ghosts and join the living inside.

On a clock whose hands declare the time to be eight o'clock (it's actually midday) are the words 'I am the alpha and the omega. The beginning and the end.' The congregation is wedged into pews and many without seats are standing at the rear. There is a drum and an electric guitar parked in one corner. A man in a wetsuit, holding a sailboard, stands in another. In front of the altar the words to the hymn are being projected askew onto a white screen. 'Let's sing of his love forever.' A few hands wave, bodies sway. But my neighbour is not impressed. In whispered tones, he bemoans the fact the congregation is not as charismatic as it once was. 'Most of the youngsters are at university or have left for jobs in the cities.'

Out in the foyer, assembled ranks of teacups wait for the end of the service one and a half hours down the line. I last about fifteen minutes. Perhaps being forced to attend church and confirmation as a child has created an aversion to prayers and kneeling.

I leave the congregation and the nondescript ribbon village behind and return to the ochre-coloured canal stained with struggle and iron ore. Before I make my rendezvous with Gordon, I have the small matter of 28 locks over a 12-mile stretch, known as Heartbreak Hill.

As usual the flight is easier than expected. In fact I quickly get into a groove and revel in it: *Caroline* moving smoothly in and out of gates, me slipping effortlessly round the lock, chatting and getting the occasional helping hand down life's staircase.

At lock 66 I spot Gordon's diminutive lock-keeper's cottage which dates from the building of the canal and inside which he and Vivienne Copeland have somehow managed to bring up four daughters now aged between 16 and 24.

I knock on the front door and Gordon appears at the garden gate. 'We don't use the front door. Come round the back.' I can hear a couple of children moving above my head in the doll's house as we settle into a sofa with Wilf's treasured notebooks. Vivienne appears with a tray of tea and biscuits as Gordon opens the first small black volume as delicately as if it were the Dead Sea Scrolls. Inside is the meticulous sloping handwriting and ruled lines of a man who liked order. Many of the pages are carefully held in place with sellotape or glue. Each passing boat – *Black Chief*, *James Brindley*, *Water Elf*, *Golden Hind*, *Water Fern* – is recorded in ink either in the northbound or southbound column, their time of entry precisely recorded.

Those bearing piscatory names – *Pike*, *Chubb*, *Roach* – Gordon explains, are part of the old B.W. working fleet. Some days in 1958 there are no more than five boats recorded in a day. By the time we reach 1960 it's down to one or two. Pleasure boats meanwhile take the opposite trajectory: in the last ten months of 1959 two pleasure boats pass through, in 1961 144 boats do the same over a six-month period. Six to seven thousand now pass through the tunnel each year.

When we reach the end of the books and the tea, I thank Gordon for so generously inviting me into his home. He waves my thanks away with a flourish of the right hand. 'I like talking about the old days. My father always had people back too. One day when I was still in primary school, a pleasure craft sank in the tunnel. Dad found the skipper sitting in one of the recesses waiting to be rescued. He brought him back home for a bath and when the man reappeared in his fresh clothes he had a dog collar on. We get all sorts.'

One of the most legendary characters, according to Gordon, was Charlie Atkins – 'Chocolate Charlie. He used to run chocolate crumb from Knighton down to Bourneville. Fourteen hours each way every day and he only ever had Christmas Day off in thirteen years. His son Bill followed him. He's the one I told you about who's still living in the village. I can take you to meet him if you like but he doesn't always talk if he doesn't take to someone. Boatpeople are very private.'

I quickly accept Gordon's offer and we walk along the towpath into Wheelock village. Opposite the Commercial pub, in a street that appears to have no name, Gordon knocks at the door of a small terraced home. From inside I can hear a western on TV. A small frail man appears in the crack of light, his face breaking into a smile on seeing Gordon.

'Hello Bill. I've brought someone to meet you who's writing a book on the Cut.'

The smile vanishes.

'Paul's on a four-month narrowboat trip heading up to the Leeds & Liverpool and Aire & Calder.'

Bill's eyes flicker back to life. 'Come on in.' Bill opens the door wide and shuffles back to a table on which are stacked a pile of letters and bills. Gordon tries to get things moving. 'Bill used to work a horsedrawn pair for years, didn't you Bill? There aren't many of the original boatpeople left in Wheelock – used to be lots, weren't there Bill?'

'There were that old boatwoman lived in Soap Row but she died a while back. There's Kenny Wakefield. He's the only one left now but he's not a real boatman 'cos he were on the bank a long time. Me dad died at seventy-seven and I'm seventy-six so I 'spect they'll be calling time on me soon too.'

'Paul's moored up by the cottage on a fifty-foot hire-boat.'

'At the Knoll? That were where the salt were mined and we collected cargo an'. . .' Bill breaks off as something snags and he launches into a coughing fit. 'Knocked off smoking when a packet cost one and six. I reckon the trouble was when I did three year roofing off the Cut working with asbestos.' Bill has emphysema like my father. His inhalers are in a pile on the table next to the correspondence. He looks surprisingly youthful and dapper: although lounging round the house, he's wearing pressed grey trousers and a white shirt. By his feet is a large rusty dog as broad as it is long.

Bill's square face has been chiselled by exposure to fifty years of weather on the Cut. Born on a boat, he worked his dad's horse-drawn butty as soon as he could walk. 'Me grandfather worked boats too. As soon as I could, I left me dad and joined me brother who was using a motorised boat so it were easier. That were in the thirties. With a horse, you're never done even when you're moored up – you still got the horse to feed and stable.' I notice there's a ready smile at the end of each of Bill's sentences and a raising of an eyebrow as if seeking confirmation or empathy.

Motorised or not, it was still a hard life. Sometimes Bill would be asked to run freight nonstop from Birmingham to Manchester to make a designated sailing departure time. 'Fastest I did it were thirty-nine hours and they paid me an extra ninepence.' From the day he started work to the day he retired, Bill only once went on holiday. 'It were when I was a locksman, I went to Blackpool on a day trip with workmates.'

'Have you ever been abroad?'

Bill laughs. 'Don't even see me son but once a year. He's in

Birmingham in the Black Country Railway Museum. Went to grammar school and college but couldn't get a job around here. No work on the Cut and no work anywhere else. Boating may have been a hard life but at least it were work. And we'd have fun too. If we were waiting for a load, or sometimes of an evenin', we'd put on a pair of boxing gloves and knock hell out of each other in the stables. And of course we'd have a drink up when you'd got the order sorted. Some pubs wouldn't let us in. Ones like the Cheshire Cheese here, the Bells at Kidsgrove, the Donkey at Rhode Heath, they were the boaters' pubs.'

Bill still faces the same sense of exclusion and prejudice he's faced all his life. Recently when he went for an eye test he was given two large forms to fill in. 'I told the lady, sorry I can't read nor write. She yelled at the top of her voice to the others in the waiting room, "We got a man here who don't read nor write." Well, I felt about this tall.' Bill leaves a half-inch gap between his forefinger and thumb and grimaces at the memory. 'I nearly walked out I felt so bad. An old lady came up and said "What's the trouble mister?" I said I can't read this form. "That's no problem," she says, and she filled it in for me. She complained too to the chief optician. I should be used to it really. Town folk always called us gypsies, water gypsies. We made friends amongst ourselves and stuck by each other through thick and thin too. We had no education, no school, but we looked after each other. That were the way.'

Over recent years Bill has made a new friend: a physio at the local hospital. 'He got me walking again after I broke me foot. But it took fourteen months 'cos they messed up the operation. I did it in the last lock on Cheshire flight. Me steel toecaps caught under the rail on the ladder and I fell and my foot stayed. Then they set it wrong. I sued B.W. but got nothing. Usual story – you 'av to look after yourself. Same in the war – they wouldn't let boatmen sign up. Made us carry bombs and shells on the canal but unlike soldiers we didn't get paid – just given clothes and food.

'Anyway, me friend Daley, he's the physio, now reads me my correspondence. That's why I've got me letters out.' Bill indicates the pile in front of him. 'So something good came of it.'

Canal life is largely about coming and going. Long days without plans punctuated by unexpected encounters. Meeting Bill is one of the best.

Robert Louis Stevenson in *An Inland Voyage* said, 'I would rather be a bargee than occupy any position under heaven that required attendance at an office.' I'll drink to that.

As I make my way through Cheshire, fishermen sitting in their canvas chairs appear more prosperous and polished looking, less downtrodden. Cows take dips in the water and then can't get back on the bank, herons swoop, swallows dart like Spitfires, *Caroline* drifts on through the uncluttered landscape towards the longest day of the year.

A silver-haired forty-year-old with a chain round his neck walks his 'crew' of two dogs and two cats past *Caroline* while, up on deck, I'm munching on a slice of Dundee cake left from my wife's visit. 'Cats are much more sensible at sea than dogs,' he informs me and is gone. Another man in a blue kagoul, stooped and using an umbrella as a walking stick, passes by singing 'Love Letters in the Sand'. A white horse flicks its tail in the field opposite. A frog hops on lotus leaves across a pond. Another heron joins me, flapping wings frantically above his pipe-cleaner legs. I call him Bruce. I'm into my sixth week afloat now. The canal swells at times to lake-like proportions where land has collapsed through undermining. In one broad reach lies the rotting wooden carcass of an old butty.

I am into the rhythm of the long journey. A journey slow enough to dredge up England's memories in the muddy language of the boat people – snubbers and podging, wharfingers and lengthsmen. My obsession with the Cut is total. But then the journey had taken over my life even before departure. Two days prior to sailing, I remember waking to find myself already prematurely inhabiting a silent, private world. Both ears were blocked. As the nurse at the local surgery syringed the wax log-jams and I swam back to the surface of an audible world, I bumped into her words, 'Both canals are inflamed.' My head span 180 degrees. Had some bastard closed the canals *again*? Closures due to the nationwide foot-and-mouth epidemic had already delayed my trip departure six weeks. There was more than a hint of insanity to my voice as I yelled 'No!' and the nurse took a precautionary step back.

At Marston I chat with an elderly man with overgrown sideboards that protrude from a camouflage commando hat. He has a large belt with a larger buckle pulled tight round his jeans and he's wearing a blue short-sleeved shirt. His name is Alan and he owns a car scrapyard under the next bridge. He's out as usual, walking a Neapolitan mastiff and bull mastiff cross which just happens to be called Bruce. The studded collar round its neck is almost the width of Alan's belt and the beast appears to be a closer relative to a grizzly than a dog.

'It's just a pup really,' Alan assures me. 'I had two pedigree Neapolitan mastiffs three year ago that weighed fifteen stone each. I were fined £7,000 and 'ad to 'av 'em put down though.'

'How come?'

'They had habit o' jumpin' up on people – just playful – but it'd knock folk straight down. I told this woman not to call 'em because one might jump up but she took no notice and it did an' flattened 'er. Then it took a bite out her neck.'

Alan, who looks a crossbreed himself – half Neapolitan mastiff and half sumo wrestler – returns to feeding the ducks from his packet of Mighty White. Bruce continues sniffing hedges. I leave them to it and enter the grounds of Lion Salt Works.

Europe's last open-pan salt works is slowly being brought back to life as a heritage visitor centre but at the moment it's sleeping. For over a hundred years, right up until 1986, wild brine was pumped from 45 metres underground and was then evaporated in a large iron pan. Workers raked the crystals into tubs to form blocks which they dried in brick stove houses while dreaming of West Africa, India and Canada where the salt would soon be heading by canal and sea. In its heyday, salt was the country's third biggest export. Then it expired like everything else.

Lining the walls at the Salt Barge ('Salt of the earth pub') across the street, there are a number of black-and-white photographs charting some of the spectacular collapses in the area due to brine extraction. One is of a section of the Trent & Mersey which sank in 1907 leaving narrowboats stranded in fields (one entire consignment of cheese vanished mysteriously overnight). Another photograph is of St Paul's Church, so badly damaged through subsidence, it had to be demolished in the 1960s. The most memorable photograph, however, is of a man, visible only from the waist up, standing inside the large crack in a collapsed road. It was to sites such as this that preachers hurried in the twentieth century to preach, warning sinners that this was what awaited them if they didn't mend their ways, only their hellholes would be filled with fire and brimstone instead of water.

Just a mile further up the Trent & Mersey, I pass Witton Flashes and Neumann's Flashes – ex-salt mines that collapsed and were then filled by the River Weaver. The lime lagoons are now nature reserves inhabited by wading birds and colonised by exotic marshland plants such as fragrant orchids.

At nearby Northwich, many of the newer homes now wisely use a box-frame construction that allows for earth movement, and can even be jacked up and moved if things look to be getting too perilous. There are many apocryphal tales, however, where there is no warning. One is of a plough and horse disappearing when a farmer took a break for lunch. On another occasion the local militia were training on their drill field and just as they left, so did the field.

A breeze gets up, rippling through an embankment ablaze with yellow azalea. I watch *Caroline* and my reflection, indivisible in the water. I am the man in the boat. Sometimes I dance with the tiller behind me, swapping hands, jiving with my partner. Nothing too extravagant but a dance none the less. A herd of muscular brown bulls don't know quite what to make of us and stampede towards *Caroline*, leaping high in the air as if in a bullring. Beyond them we slip through a forest of blue rhododendrons. Bruce the heron is fishing again.

At Anderton I greet a tramp dressed in a dark shirt, trainers and black trousers who looks as if he's been walking all his life. His unkempt white hair and matted beard are the unwashed giveaway. His large backpack and holdall, the incriminating evidence that he's carrying his home on his back. The tramp ignores my greeting.

'Going far?' I try again. I've heard of long-distance tramps who never leave the Cut, who walk and walk till they run out of canal and then walk back again. The man keeps his steady pace. He keeps on walking as if I don't exist.

A little further on I encounter him again as I'm making my way to the restoration site of the Anderton Boat Lift. This time he's exiting a walk-in rubbish skip with a bunch of last Sunday's newspapers – the *Sunday Times* and *Mail on Sunday*. Both are neatly folded and he's thrusting them into a Waitrose bag. 'It's all election news. I've already read them. Blair seems to be home and dry.' I do my best to engage him. Not a flicker. He continues on his way in his impenetrable bubble, striding off down the towpath.

Behind him is the Anderton Boat Lift, one of the great wonders of the Victorian Age and worldwide prototype for the hydraulic boat lift. Although it closed in 1983, today it's humming with activity – bulldozers, stevedores, labourers all combing the site, disappearing and appearing from beneath the tarpaulins and scaffolding the iron man is Band-aided in. By spring 2002, after extensive winter trials, the wraps will be off. Two narrowboats will be able to slip off the canal and float together across an iron aqueduct into a 72-foot tomb-like metal caisson that will lower them 50 feet to the River Weaver as simultaneously, a second caisson raises another pair of boats up to the canal.

Having negotiated three more tunnels wide enough for just one boat – two of which are crooked, making it impossible to see if anything is coming the other way – I find myself on the Bridgewater Canal. *Caroline* has arrived at the start of the canal system and the industrial revolution. Well, not here exactly – the true pioneering stretch is nearer Manchester – but still the magic of the name is there. The Bridgewater Canal. There will be no more lockgates now until I join the Leeds & Liverpool.

Sailing day after day through rural England you become a connoisseur of field: the incline and texture, the delineation with forest or wall, the intersection with sky and water. And right now *Caroline* is sailing by an absolute peach: a broad, rising field of buttercups, in the centre of which stands a single buckthorn leaning towards the sun as it breaks through bruised clouds.

I sit at the stern, hand vibrating lightly on the tiller, legs crossed, chin resting in left hand. As I observe the shape-shifting of trees and wheat fields, it strikes me that while we spend an inordinate amount of time feeding different needs that simply set in motion new needs, it is at such moments as this, when we inhabit the aesthetic sense and the divide between the self and the world vanishes, that we really sing. A man in a boat is singing his way through the countryside.

I remember once being on a trek where each night the members of the group chose different poems from a compendium to read. Voices reverberated with heaviness and meaning, words were polished, and only poets dressed in finest tuxedos chosen. The poem I selected by William Carlos Williams left everyone singularly unimpressed:

> so much depends
> upon
>
> a red wheel
> barrow
>
> glazed with rain
> water
>
> beside the white
> chickens.

Williams is right: so much really does depend upon being able to see a red wheelbarrow glazed with rainwater beside the white chickens. And on the Cut there are unlimited possibilities for seeing. Apart from the glorious countryside, there's the simplicity and beauty of the lockgates, the eloquence of original warehouses and boatyards, and of course the rainbow of colourful narrowboats. All have an enhanced aesthetic value because of their aquatic setting.

At Stockton Heath I walk up to the nearby Manchester Ship Canal and stand on a vast cantilevered iron bridge staring out over a succession of spans stretching across this broad muddy waterway. There is no barge nor ship to be seen despite the fact 2,500 vessel movements reputedly take place each year and freight is increasing here rather than decreasing as everywhere else on the network.

Stockton Heath itself is a series of easy stepping stones for its denizens through life, carefully spaced along the high street. In childhood there's Pets Pantry; in the preening years there's the Original Barber Shop; for bored middle-agers, Visions Video; Done Bookmakers provides a final marker of hope; and when this is gone, at the end of the road, Help the Aged charity shop, Brindley Court Retirement Flats, and the graveyard of St Thomas await.

A narrowboat called *Houdini* passes. It's the best name I've come across so far. Whoever's on board is following a fine tradition of revolt against the confinement of life, inaugurated by archetypal escapologists

Daedalus and Icarus, and transformed into an art form and business by Harry Houdini himself.

A couple on old bikes cycle past, a three-year-old daughter bringing up the rear bouncing and gurgling in the baby seat. A cat lies coiled on a shed roof like a rope, a swan nests beside garden gnomes, and lawn tables are left out for sunny barbies or late evening beers now that days stretch beyond 10 p.m. As we move into Manchester's preferred Cheshire commuter belt, homes come with wind chimes and BMWs, and even the hedgerows appear more prosperous.

Approaching Lymm on a raised embankment, for the first time I can see the soft rise of the distant Pennines. There is a wonderful whiff of fish and chips in the air as I slide above the chichiest of black and white half-timbered towns. The sundial above the old stocks, a slingshot from the canal, admonishes citizens 'Waste not time'. Wasting time. The cardinal sin for all those who pay wages. But for boaters it is a confirmation not to hurry, to take our time, make the most of time: time is what we swim in, not something we save and spend.

Dead time, wasting time, spending time, saving time, killing time, time is money, God speed – language and socialisation are against us as we battle the endemic hurry sickness. Unfortunately, on a personal note, the Victorian work ethic, encapsulated in another single succinct maxim, has been drilled with army boots into my life. Throughout my childhood my mother dragged her favourite aphorism like a rosary around the house. Several times a day, every day, there were innumerable occasions where 'Upwards and onwards is the motto' proved invaluable – steering her from the seduction of an afternoon nap, encouraging stoicism in my sister when she fell down and cut her knee, or admonishing my brother and me for dawdling over a banana split that we were attempting to stretch into tomorrow. Upwards and onwards is the motto.

As I inhabit my middle years, the force of the message has not weakened. As I lie on a beach totally blissed out, the earwig crawls into my brain and sends me scurrying off on another task-oriented mission. As a travel writer it has served me well, as a holidaymaker and human being it has been my curse.

A boat, *Stolen Moments*, steals by.

A young girl with a transparent folder containing a ruler, protractor and rubber passes. I suddenly badly miss my daughter. A couple are painting their boat in the dry dock of Lymm Cruising Club. The woman waves a paintbrush. Lymm is clearly a big boaters' centre, the banks lined for a mile or more with moored narrowboats and cruisers, hips slung low, prows proud, sitting on their own shimmering reflections. *Cotton Blossom*, *Zulu*, *Free Spirit II*, *Whistler*.

A group of older school kids walk alongside a bank of brilliant yellow laburnum. Their ties are loose, shirts out. One boy, probably aged sixteen, has his arm slung over his girlfriend's shoulder, casually touching a breast that dances beneath her stiff Persil-white shirt.

Nearby an elderly man is trimming his hedge. I ask how come he manages to make it so straight. 'Rack o' th'eye an' screw o' chops.' I stare blankly at him. He slows the delivery for the southerner – 'A Chinese saying, lad – rack the eye and screw your chops.' I'm still nonplussed. 'Look through one eye and screw up yer face like you do when you're trying to look straight ahead.'

At Ye Olde No 3 pub in Little Bollington I'm mooring the boat just as my parents arrive in their red Micra from their home in Poynton. Dad seems to have made good progress since leaving hospital last week but they both have shrunk yet another inch. Mum moves the provisions on board – a round tin crammed with her home-made mince pies, and one of my father's curries in a Tupperware box. Dad meanwhile scours the boat manual and inspects every nook and cranny. Once satisfied with her upwards and onwardsness, Mum gets comfortable on deck and Dad, a virgin narrowboater, requisitions the tiller as we take a short pootle back the way I came. They both adore being on board and Dad, predictably, wants to stay. Our original plan had been for him to join me at this point for several days, sailing on *Caroline* to the starting point of the canal network at Worsley. The damp environment, however, is too big a risk after his recent hospitalisation and so we reschedule a few days together, health permitting, down on the Oxford Canal in a month's time.

By the time they leave me a couple of hours later, there's nothing more to stop for and I swap Altrincham for Stretford – ugliness for ugliness – as quickly as possible. I know I've entered Manchester's entrails by the 'Death to Liverpool' graffiti on the bridges. After Bridge 42, the canal splits. One section of the Bridgewater burrows into the heart of Manchester past Old Trafford and the revitalised city centre at Castle Quay. *Caroline* and I, however, fork left, slipping through hinterland grunge onto the original stretch of the Bridgewater Canal.

I have finally arrived at the start of the canal story. I'm on the route the Duke of Bridgewater created to carry coal from his Worsley mine into Manchester. I haven't time to get too excited, however, as *Caroline* immediately smashes into a concealed floating beam. Several belly-up bream follow in its wake as we disappear into billowing steam beside the Kellogg's factory.

Adrenalin courses again at the Barton Aqueduct as *Caroline* hangs 38 feet above the Manchester Ship Canal. Despite the fact I've flown in planes and plummeted on 230-foot rollercoasters, it's still a buzz to be sailing way *above* ships bound for the Irish Sea. If it's a rush for me, imagine the bewilderment of those who saw boats passing over boats on a 660-foot-long trough for the first time in 1761 when Brindley's original triple-arched aqueduct opened for business. At the time the *Manchester Mercury* reported, 'A large boat carrying upwards of 50 tons was towed along the new part of the Canal over arches across the River Irwell which were so firm, secure and compact that not a drop of water could be perceived to ouze through any part of them.' (Quoted in *James Brindley* by Samuel Smiles.)

Two miles beyond the local Conservative Club – surprisingly without a black flag this post-election day – I arrive at Worsley where canals first crawled out of the mines and made their snail trail across the country.

Two centuries ago this sleepy rural setting with lilac rhododendrons and half-timber homes rang with the clamour of blacksmiths and boats loading and unloading at the wharves and warehouses. When Josiah Wedgwood visited in 1773 he commented that Worsley already had 'the appearance of a considerable seaport town'.

The only telltale sign of the proximity of mines today is in the rust-coloured water. I follow the trail beneath a handsome humpbacked iron footbridge. Beyond it stands the eighteenth-century Packet House, a Grade II listed building given a Tudor makeover in 1845, and now broken into domestic and business units. It was from a jetty here, in an enterprising early nineteenth-century sideline, that the Duke invented tourism with cruises from Manchester to Warrington. From 1808 two packet boats left daily with 80 passengers on board each. In the coffee room wine was sold by the captain's wife; the first cabin cost 2s 6d, the second 1s 6d and the third 1s.

Beyond the Packet House, I find the gateway to the underworld; the start of canal time. A black goose sits guard outside the mouth of the tunnel that burrows into the exhausted coal mine. Eight goslings are snuggling up on a beached plank of wood nearby. Rhododendron drifts down with the roar of cars from the road above. Inside the tunnel entrance water drips and a bottle of Bacardi Breezer bobs up and down. I look for a message in its neck. It's all a rather inglorious and literal end to the canal and the Duke's mines. The boats that slipped in and out of the entrance, pointed at both ends with vertical sides and a flat bottom, known as 'starvationers', had stripped the mine bare. The entrance to the mine is now impassable.

Without Brindley, and the waterway network he was largely responsible for, there wouldn't have been the industrial revolution there was. Before canals existed it was very much a hit-or-miss affair whether a ton of coal or chain, carried over appalling roads by a pair of horses and a cart, would ever arrive. On a boat, however, you could load thirty tons and predict precisely when it would reach London, Liverpool or Bristol. Anyone who's ever given a piggyback in water and then on dry land will have no doubt which navigation is preferable. The increased efficiency resulted in the price of coal dropping from 7d to 4d a hundredweight when Brindley's first canal opened to Stretford in 1761 (the stretch to Castlefield Wharf in Manchester was open by 1765).

In the same decade that the Bridgewater Canal opened, Watt sold his first steam engine and Arkwright patented his first spinning jenny. By the end of the first decade of the next century, the whole world seemed to have been reinvented – steam ships, gas lighting, high-pressure locomotives, and London's first public docks. The industrial revolution is not a grey tale but a romantic epic of achievement against the odds by remarkable characters.

Josiah Wedgwood was the thirteenth child of a humble potter. He only moved on to great things when he was forced to quit the potter's wheel because smallpox left him with a diseased leg which had to be amputated. Brindley, born into poverty (and to a father more often found at the bull ring than at work) at a croft near Buxton, did not allow a lack of formal education to prevent him inventing a fire engine, a steam engine, as well as aqueducts and lockgates. The Duke of Bridgewater, his benefactor, was considered a feeble child who'd amount to nothing. He lost his father when he was five and the runt of the litter only inherited the dukedom because all four elder brothers had died of consumption by the time he was twelve ('His mental capacity was so defective it was even considered passing him over to the next heir'). At the age of seventeen, his guardians, 'finding him alive and likely to live', sent him on a Grand Tour with a tutor. It was on this tour that he first saw the *Canal du Midi* and started planning something similar to bring down the transportation price of his own coal.

These industrial pioneers reduced the cost of many basic goods by 25 per cent and brought them out of the exclusive domain of the privileged classes.

I quit Worsley – passing through a yellow rush of broom, daisies, buttercups and lupins – and slip into Leigh where front doors open onto the canal and the terraced streets off it are a forest of 'For Sale' signs. A man in a Juventus shirt raises a thumb at me familiarly as he takes down one of the signs and puts up a new house name which I can't quite read. The new boy on the block. Two-thirds of the English own the places they live in. They still invest in futures in bricks and mortar but want the Latin fizz, which is why Brazilian and Italian football shirts are worn in the gardens and streets, not those of Germany and Denmark. The English desire to name and personalise their homes rather than be a number reflects their pride and independence, whatever others may think of that name.

More factory chimneys appear, more washing lines decorated with cotton pants, patterned nylon socks and Fred Bloggs jeans. On a wall 'Chelsea FC' has been daubed in paint. It almost brings a tear to my eye. Just before Plank Lane Swing Bridge the Bridgewater Canal becomes the Leigh branch of the Leeds & Liverpool. At the next bridge graffiti welcomes me warmly with the generous offer 'Fancyashag'.

5

Trans-Pennine Crossing

The Leeds & Liverpool Canal

Far below the high embankment *Caroline* is inching along, sailing boats
and sailboards are zipping and unzipping Scotsman's Flash. How it got
its name, God only knows – maybe it has something to do with trains
bound for Scotland that hammered through before the land fell into
underground mines and filled with water. Whatever calamity created it,
it's as beautiful as any lake this sunny June morning.

Beyond the Flash, a church steeple and a factory chimney give way to
the singed, featureless Pennines. I slip past the hulking walls of a
dismantled railway crossing and into the first of the Leeds & Liverpool
locks, well aware that everything up until now has been pre-season
training: the Leeds & Liverpool are to lockgates what the Himalayas are
to mountains.

The Poolstock Locks are wide enough to take two narrowboats,
require a B.W. handcuff key, and have both outer and inner paddles thus
doubling my workload. Unlocking the fiddly chains to the paddling gear
is a rite of passage no less than getting the hang of cufflinks. I fiddle and
I faddle, curse, drop one end and then the other and curse some more.
Once unshackled, I find myself ratcheting the windlass forty times to lift
one paddle. Already I'm missing the single locks of the Midlands where
it's seven turns and hey presto.

On my way up to Wigan I've been deluged with advice concerning
where to moor and not moor in town. If you dwelled on the scare stories,
you'd never leave home. Fortunately a lifetime's globetrotting has
created a healthy scepticism in me, having survived, and sometimes not
even noticed, invasions, typhoons, earthquakes and terrorist activity.

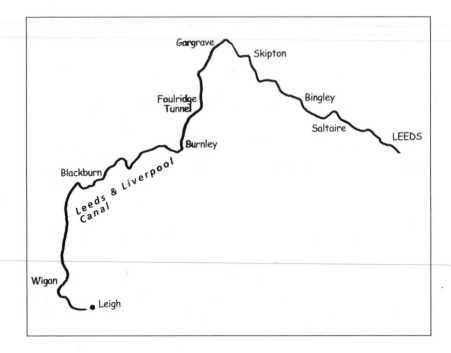

The Trans-Pennine Crossing: the Leeds & Liverpool Canal

And so I ignore all the advice and warnings and sail under Bridge 86 (on which 'I am happy' has been daubed in white paint) to a canal T-junction. One sign points right to Leeds, the other left to Liverpool.

The association of the Cut and Liverpool raises the paddles on a stream of childhood memories. The ancient Greeks believed that the future skulks in shadows behind us, devious, unknowable and therefore not worth concerning ourselves with. The past, on the other hand, spreads out in front of us all the way to the horizon. As I sail across the country, forever bumping into my own history and that of England, I have become a convert to the ancient Greeks' notion of time.

Staring at the sign to Liverpool now, I see Litherland Lift Bridge which separated our Liverpool home on Hawthorn Road from Richmond Sausage Works where my mum worked alongside Ringo Starr's fiancée (he dumped her pretty quickly it has to be said). I was twelve years old and each school morning I crossed the bridge over the Leeds & Liverpool canal to an adjacent sweet shop that was the centre of my universe. Here I'd buy a packet of the newly launched Golden

Wonder cheese and onion crisps (the first flavoured crisps), or if I was flush, a quarter of chocolate chewy nuts. I'd then wait nervously on the bridge until a girl I was dotty about walked by without a glance in my direction. Like clockwork, I'd move into her slipstream and track her along the towpath for a mile until our paths separated for our respective schools. I never did pluck up courage to speak to her. I still don't even know her name.

Today I ignore the left turn to Liverpool and instead turn right in readiness for my climb across the Pennines to the east coast. The Leeds & Liverpool, though the first of the three trans-Pennine canals started, was the last to be completed in 1816. The Rochdale Canal opened in 1804 (due to be fully reopened in 2002), and the Huddersfield in 1811 (fully reopened in 2001).

Mooring at the bottom lock of the legendary Wigan 21 which will lift *Caroline* the first 214 feet 7 inches, I phone the local B.W. office as I've been told that assisted passages are available up through the flight, particularly to those operating boats alone. A man called Pat books me in for 10 a.m. 'We've got another going up single-handed so we'll do you together.'

I feel guilty as soon as I put the phone down. The integrity of the journey somehow feels challenged because I'm cadging a lift. Ridiculous. It's hardly in the Robert Louis Stevenson cop-out league. In *An Inland Voyage* the author freely admits 'partly from the fact that there were no fewer than fifty-five locks between Brussels and Charleroi, we concluded that we should travel by train across the frontier, boats and all.'

Unfortunately, I have a very good friend, another travel writer, who has somehow inveigled his way into tiller position of my superego. Nick Crane, who took the snaps of me sailing under various bridges as I quit the capital at the start of my journey, endured minus-30 temperatures walking across the spine of Europe for eighteen months in *Clear Waters Rising*, and even built a raft to avoid straying more than 1,000 metres either side of the meridian in *Two Degrees West*. Nick is a purist and would be horrified at the thought of an assisted passage even if he'd be too gracious to admit it.

I'm indulging in mental flagellation and pouring milk into coffee when I notice a group of unbelievably scruffy urchins on bikes peering through the cabin window. I hear one telling another to knock and see if anyone's in. I open the fore-end doors and am greeted by the largest boy in the pack. 'Can you do us a big favour, mister?' Before I can answer he follows up with a pre-emptive strike, 'We don't want money, honest.'

'What do you want?'

''Av you got adjustable spanner?' A freckle-faced boy points to a flat front tyre.

I get them the adjustable.

'And a pump, mister.'

I go back through the cabin, through the bathroom, kitchen and rear cabin, to a cupboard beside the stairs where I keep the tools and pump. Eventually I locate it. By the time I return, two of the boys are sitting on deck. 'Can we 'av a look inside?' one asks.

'Aren't you scared livin' on a boat?' Freckles follows up.

'No,' I answer both simultaneously. The boys return to the bank. I meanwhile stand watching the bike repair action going on, sipping from my mug of coffee, and trying to look both nonchalant and hard between gulps.

I guess the six boys all to be aged between twelve and fourteen. They claim to be fifteen to seventeen. 'Do you all go to the same school?' I ask.

'No.' Freckles points to a boy who claims to be sixteen ''E's left.'

'Are you working?' I ask the boy.

'No,' Freckles answers for him. 'He just got out of prison. D'you 'av a TV on board?'

'Yeah, but I don't use it because it never works. It's crap.' I was telling the truth but would have said the same even if it was a 21-inch flat-screen Bang & Olufson. 'What's Wigan like to live in?' I attempt diversionary tactics.

'Alright, but all the cars get broken into. Aren't you worried about your boat? Is that the lock?' Freckles points through the hatch door to a padlock lying on the floor.

'Yes. So where do you live?'

We continue this way, me skating carefully over the thin ice as Freckles chips away at the edges.

'Are you staying here tonight or going on?'

'Haven't decided yet, but I'm probably going to turn the boat round and head for Crooke after I finish my coffee,' I lie.

Eventually saddles are adjusted, tyres pumped and the boys are off with a 'Cheers, mister.' I wash up, clear the table tops and even wash the floors. Finally I snap out of it. 'Are you going to boat-watch for the rest of the trip?' I ask myself rhetorically. 'For Christ's sake let's go.' This conversation is brief compared to some I've been having. Anyway it has the desired effect. I check the baseball bat's handily located for my return and take a little longer than usual securing latches, hatches and the padlock before setting out to find Wigan Pier.

It was local boy George Formby senior who immortalised Wigan's coal staithe in song, drawing visitors to the dreary northwest town seeking out – and being hugely disappointed not to find – the music hall and funfairs of his spiritual home, Blackpool. 'It's very nice is Wigan Pier. Ah've been there many times in my bathing costume and dived off the highboard in 't'water. Nex' time you go on holiday to Wigan, make sure you visit t'pier.' A generation later, his even more famous son managed a Lowryesque sweep of comic realism in the song 'In a Little Garden', ending in a typically suggestive coda.

> In a little Wigan garden,
> With me little Wigan knees,
> Getting stung with bumblebees,
> Between the cabbage and the peas;
> 'Neath the Wigan water lilies,
> Where the drainpipe overflows,
> There's my girl and me,
> She'll sit on my knee,
> And watch how the rhubarb grows.

Like thousands of tourists before me, I ask several people for directions to the pier. Nobody seems to know where it is. A couple suggest I try the Opie Museum of Memories and I eventually find the six-foot black and white coal staithe – Formby's 'pier' – by the water's edge opposite the museum. I would have walked straight by, without giving it a second glance, if I hadn't been looking out for it.

George Orwell, like Formby, fuelled the pier's fame, or rather infamy, in his study of industrial depression, *The Road to Wigan Pier*. But the old Etonian was considerably less popular in town than Formby. When he arrived to write about conditions in northern industrial towns, he immediately alienated Wiganers with his aloofness and then irritated them further with his book's patronising anthropological observations of thumb marks in the butter dish on a kitchen table, and chamber pots under it:

> On the day that there was a full chamber pot under the breakfast table I decided to leave. The place was beginning to depress me. It was not only the dirt, the smells, and the vile food, but the feeling of stagnant, meaningless decay, of having got down into some subterranean place where people go creeping round and round, just like blackbeetles, in an endless muddle of slovened jobs and mean grievances.

At six foot two he struggled to get down a mine and hated the experience. He didn't like Wigan and Wigan didn't like him.

The town does manage a few other claims to fame: Thomas Beecham first manufactured his famous powders and pills in Wigan; and in the sixties and seventies prodigious numbers of other types of pills – dex, bombers, blues – were consumed by mods at Wigan Casino's all-nighters before it burned down in the early eighties. (I once visited it in 1967 when briefly living at my sister's home in nearby Preston – unfortunately I can remember very little of the experience for some reason apart from a frantic itching of the scalp.) Other edibles produced by Wiganers include Heinz Baked Beans, Uncle Joe's Mint Balls and De Roma Ice Cream.

Wigan's greatest asset, however, is the fact that it sits at the intersection of the Bridgewater and the Leeds & Liverpool Canals. While the Bridgewater is without doubt the pioneer, the Leeds & Liverpool, rising 487 feet, is both the most ambitious and, at $127\frac{1}{4}$ miles in length, the longest single canal in the country (the Grand Union doesn't count because it's made up of at least eight separate canals).

I am a little more surprised than usual to wake the next morning having expected a frenzied meat-cleaver attack to have dispatched me for the crappy TV. I vaguely remember wailing sirens during the night but there is no sign of blood or pain, so I warm up the engine in preparation for the start of the next stretch of my trip, the 93-mile journey across the Pennines to Leeds.

At five to ten, Pat, the locksman, knocks on my window. He stands astride his bike, pointing to his watch. 'Time to go.' I raise a thumb and Pat zips ahead to prepare the first lock. As he vanishes, another narrowboat, an immaculate green 53-footer called *Valley Thrush*, passes with a wave of the hand from the skipper. I wave back. We've already passed each other several times over the past few days.

Valley Thrush is already in the lock by the time I catch up. Pat waves me in and *Caroline* and her new bedfellow snuggle up and get acquainted. Painted on the side of the new neighbour, alongside the boat's name, are those of the owners, Alan and Margaret, as well as their intriguingly named home town, 'New Invention'. Today Alan is Margaret-less. 'Me woife's not too keen on boating though she does visit most weekends.'

The strong nasal accent immediately locates New Invention somewhere in the Midlands. 'It's an old Staffordshire moining village which got its noime because it was a centre for lockmaking – Chubb, Yale, deadlatches, that koind of thing. It's not a village any more though – just part of Walsall.' Alan looks behind to check he's not drifting too close to the stern gates. 'There's another town called New Invention soomwhere but I can't remember where.'

As we chat, we simultaneously, and without conscious thought, edge *Valley Thrush* and *Caroline* backwards and forwards like dance partners to compensate for the rushing water. Pat, meanwhile, whistles his way through his work.

At the next lock, to make matters easier for Pat, we only use one gate instead of two; the lead boat squeezing over onto the far side after entry to allow the second boat a straight passage. In the longer Grand Union locks this is fairly easy to manage, but on the Leeds & Liverpool, with just 62 feet in length to manoeuvre in, it's somewhat trickier.

Alan and I compare notes on our progress over the past few days. Yesterday, in Leigh, he'd been threatened by kids who told him that if he didn't give them a ride they'd 'lob stones through yer window an' no messin'.' I briefly mention the visit by the Wigan scalpers. I doubt they were regular churchgoers but as there was no piracy, I give them the benefit of the doubt. 'A few boys popped by, but I didn't have any problems.'

'Surprising that. I were told if I moored where you did I'd be risking more than me property.' Alan is softly spoken, balding, and wearing a snazzy patterned jumper. He smiles often, revealing rather charming buck teeth. He's the typical, laid-back companion I am now familiar with on the canals: an engineer who retired early, built his boat himself – 'Apart from the hull and the lettering; I'm no good at that kind of thing' – and set sail.

In the next lock I clamber up the ladder to chat with Pat who bears a striking resemblance to Charlie Fairhead from BBC1's *Casualty*, only with an East Lancashire accent as deep as Wigan's pits once were. Of indeterminate age, he chain smokes his way round the lock.

We continue our rise up the water escalator, passing a rugby club, some wasteland and a ziggurat of terraced streets. Once the flight was lined with collieries that mined a type of coal known as 'canal'. Weird that. Rose Bridge Colliery, Ince Hall Coal and, biggest of all, Wigan Coal and Iron Co. (which in its glory years was the largest ironworks in the country, employing 10,000 workers) have all closed down.

Somewhere between the sixth and the tenth lock – by now I've lost count – Pat gesticulates to me to join him lockside and again I clamber up the ladder in time to see a swan beating its wings wildly in the pound ahead. After twenty seconds a pen appears from beneath it. While the cob has been satiating itself, the female has been pushed under the water. 'What was it like for you?' Alan drily comments.

As we return to our boats, Pat gets a call on his walkie-talkie and relays a warning. 'Best keep your eyes peeled from here on up. Just heard from me mate Paul who's working another boat down the flight that they've had kids lobbing stones at 'em.'

As we wait in the pound and Pat readies the next lock, four boys aged around sixteen appear out of the scrub. Alan, who's a little ahead and closer to the lock, signals to me but I already have them in my sights and have turned my body square on, staring straight at them, watching their every move. I'm letting them know I know they're bastards. My eyes scan – Adidas, pasty faces, gelled hair. They have adopted the walk of innocents. No swagger. Nature lovers. Out bird-watching no doubt. They walk on chatting. And then one bends down to tie his lace. Here we go. I'm still square on to them. In seconds, stones are raining down, exploding in the water. It's Pearl Harbor on the Leeds & Liverpool Canal. There's nowhere to escape but below deck but then who would steer the boat? Fortunately it's a chilly day and I'm wearing a padded

Rohan jacket which I now raise over my head. I'm lucky, none of the stones hit me. But it's only a matter of feet and fortune. Alan is making a beeline for the now open lockgate. I accelerate in his wake.

Inside the lock Alan is incandescent. 'What's wrong with the little shits? I wasn't like that when I was a boy – wanting to hurt people.' I'm less vocal, but the attack has shocked me too. There is a heightened sense of violation and vulnerability; a reminder that on a boat you are more exposed to the beast as well as beauty.

I try to console myself that violence on the Cut is not new and neither is it confined to our island. On Robert Louis Stevenson's journey, he speaks of 'crop-headed children' who 'spat on us from bridges as we went below'. They lobbed stones too. Unfortunately the knowledge I'm not the first victim of little shits doesn't really help. It's impossible not to take this ultimately impersonal assault personally.

Pat informs us there's to be a meeting between the police and B.W. on Friday to discuss the ongoing problem. 'It's been getting worse again lately.'

Four locks from the top of the flight, Pat leaves us for lunch. He's been working the flight for us continuously for three and a half hours. Alan sets off to prepare the upcoming lock and I sail the two boats in together, breasted up. Alan then closes the gates, drops the lower paddle and heads up to the next lock in the flight while I complete the penning through. We're a good team but it still takes us just under an hour to work our way to the top of the flight.

That evening, low on provisions, I make a magical concoction – a jalfrezi of baked beans, ripped-up ham, tomatoes and sweetcorn. It tastes so good I think of offering Sainsbury's the recipe. I sleep like an angel and wake to rain tap-dancing on the roof. Outside a soggy day awaits.

At 9 a.m. I get a call from an Australian boater on the *Thisbe* I'd met a couple of days back who said he'd heard a section of the canal near Gargrave had been closed due to a fresh outbreak of foot-and-mouth. We promised we'd ring if either of us heard anything more. Good news. The stretch is closing only until Thursday when the cull will be completed. 'There's places you won't be able to get off the boat, mate, but apart from that you should be fine.' The Mobile Towpath Telegraph.

Caroline has the water ridge above Wigan's rooftops to herself. A red bus bumps its way over a bridge followed by a lorry carrying bananas. On a wall someone has scrawled 'Toplock mafia'. I know who they are.

After half an hour I make the first stop of the day and cycle up to Haigh Hall Country Park along a track through steep wooded hillside. On a bluff beside the decidedly drab residence of Haigh Hall itself, two old-timers are chewing the fat, staring over the valley Wigan spreads itself out in. I attempt to inveigle my way into chewing with them. 'Great view,' I comment less than imaginatively.

'*Grand* view,' the most senior of the two upstages me, making the first word sound like a bugle call. 'On 't clear day wi'out this mist you'd see the hills of Wales.'

On the far side of the valley stands what appears to be a large iron suspension bridge beside two mill chimneys. 'What's the bridge?' I ask pointing.

The men look perplexed. 'What bridge?' one finally asks.

'The one there,' I point again.

'That's not a bridge lad, that's the J.J.B. Stadium, that.' He points with his walking stick. 'Rugby, American football, all sorts played there. And that there,' he moves his walking stick to the left, 'that's the J.J.B. factory – see the flat one beside Heinz's, biggest Heinz's in Europe lad.' The man is proud. You'd think he owned the stadium and factories himself. 'Dave Whelan, he's the one what owns J.J.B. He had the Wheatsheaf first, after he broke his leg in the Wembley final playin' fer Blackburn at end of 't war. Then 'e bought a small sports shop in Wigan. Everything that lad touches turns to gold. Chairman of Wigan F.C. now. His brother still sings in pubs. Cracking voice he has, even though he's eighty odd.'

The skyline is broken by a large number of brick chimneys. I wonder if they're masquerading as something else but chance my arm. 'Are any of the mills still operating?' I ask.

'Norrr. Don't be daft. It all comes in from Taiwan or Hindustan or somewhere now. No mines neither. Just light industry like.' The men assume a meditative silence, staring out over the Haigh (pronounced Hay) plantations that were laid out in the 1860s by Lord Crawford's workforce during the cotton famine caused by the American Civil War. Soon my companions are off once more, discussing where so-

and-so did such a thing and who owned what, where and when.

I walk past a sign that declares the nineteenth-century Haigh Hall closed to the public, and enter the front door. The Hall, allegedly, is a hot spot for wedding receptions. All I can say is it must be for the view rather than the building. The reception room is bare apart from an intricately carved teak sideboard depicting the elephant god Ganesh and levitating yogis rising to a pink, blue and white moulded ceiling overhead. A caretaker, pushing a stack of metal chairs on a trolley, tells me there are conferences and meetings going on in all the rooms and it's not open to the public. I leave with my tail between my legs.

Up in the stables block a notice board promises future visits by The Sword and Step Dancers, Pemberton Old Wigan Band, Tyldesley Brass Band, and Standish Brass Band. Inside the Stables Café, I find the route to an upper art gallery blocked by a chain across the stairs. Maybe my B.W. handcuff key would fit. 'Don't think anyone's in yet,' the waitress explains the presence of the chain. She stares hard at me, head to one side weighing me up. 'Yer not the flower girl, are yer?'

I stroll back down to *Caroline* beside rampaging lilac and white rhododendron, and then disappear into a forest of giant oak and beech still dripping from the latest cloudburst. *Caroline* slumbers amid the forest. Nothing stirs.

The first flicker of human life appears at White Bear Marina. Some of the narrowboats moored here appear to have come out of Christmas crackers, others could have been bought with bumper lottery winnings. Each, regardless of its grandness, is strung stern and bow with rope fenders known as Turk's-heads. The roses-and-castles decoration that is the hallmark in the Midlands and further south has been replaced by scallops, scrolls and geometric patterns known as brightwork, which descended from coastal sailing craft. Several of the owners have built decks alongside the boats with benches, sheds, flower boxes and ceramic ducks.

A few fishermen, at least half of whom are wearing flat caps, are trying their luck. I ask one what they mostly catch. 'Coarse fish – roach, tench, bream... You're a long way from home,' he adds, nodding at *Caroline*'s home address. The man's wearing overalls and chomping on a wedge of cheese and is clearly right at home himself.

Friesian dairy cows share the field behind him with a goat and a herd of hardy Lonk sheep with their curlicue horns and black-and-white-minstrels' faces. A sign declaring foot-and-mouth disinfectant procedures to be in operation appears on a hedge so I drop a sterilising tablet into the tray at the stern of the boat.

At Bridge 69 in Adlington I briefly stop for provisions. Ashore, a portly male in a lumberjack's shirt interrupts his waddle with his equally portly dog. 'Come up all 't way from London on that?' I nod and smile. "Mazin. I s'pose they're all connected – the canals. I 'member tugs comin' through wi' forty ton o' coal on board. All went wi't mines in sixties.'

'Did you work in the mines?'

'No but me dad did. He didn't like it and he didn't like retirement no better. Angina, a colostomy, and fifty years of coal dust in his lungs don't make for a happy retirement present. Died last year. Least he's not suffering now.' He puts a finger to his forelock as farewell and waddles off. Behind him, every swing in the local rec is either missing or has been swung round and round the top stanchion until it's strangled.

In the All Days Food Store, beyond the Adlington Independent Co-Op Society Bakery, and Tony Ambrose Bookmaker's, I encounter a policeman with handcuffs and truncheon, waiting to speak with the owner about an incident. I stare long and hard at his handcuffs to see if my B.W. key would fit. It wouldn't. Above him a sign declares, 'All pies, pasties and cream cakes half price after 4 p.m.' I buy a warm Lancashire pasty, smother it with HP, and eat it on deck. The HP doesn't deter the ducks or a swan that are soon queuing for morsels. The uninvited guests are always the deterrent to eating al fresco on the Cut: you settle on deck and start enjoying the view and weather, when the vagrants arrive with their hard-luck stories.

Twenty primary-schoolchildren pass by. They wave, put their thumbs up, brimful of smiles and a liking for the world. Their elder brothers may have been the ones who stoned me.

My next stop, Chorley, is just a very short punt away. When rows of rooftops cross the valley floor, I moor, unlock Rosinante – my bike now has a name to better facilitate conversation between us – and head for the 500-year-old Tuesday Flat Iron Market (so named because formerly they used flat irons to hold down tarpaulins over the stalls). On the way into town a pen drops out my pocket so I stop, leave the bike against the kerb outside a funeral parlour and walk twenty yards to collect it. When I get back to the bike, an undertaker in coat-tails, having a fag, says, 'Thought you'd taken leave o' your senses leaving your bike unlocked. If you'd gone as far as the butcher's [a further twenty yards] it wouldn't be 'ere now tha's fer sure.'

In a pedestrianised lane, someone is playing an Irish jig on a squeeze-box. I readily warm to this town of two-storey redbricks around which crowds of shoppers swirl. The market tradition (there are two a week) has largely protected Chorley from the worst ravages of industrial grimness, the town's textile past consigned to history no less than the memory of greengrocer's assistant Sir Henry Tate (the founder of London's Tate Gallery), declaiming in the streets on the virtues of his potatoes and caulis in the 1830s.

The market patter has probably changed little since Henry was honing his skills. Things have a way of enduring here. When carcoats disappeared from the rest of Britain, they found their way to the Flat Iron Market – line after line of the stiff, cheap jackets stand begging to get behind the wheel. No doubt a spate of copycat specialist coats followed in the wake of their early success, perhaps designated for taking the rubbish out, for cleaning the windows, or climbing a ladder to prune fruit trees. For some reason, however, they never fired the public's imagination quite like the carcoat.

Also on sale are plants, garden shears, cardies, decorator filler, storage hooks, curtains, trainers, ankle pop-socks, 'funky steering wheels' (surprisingly not sold with the carcoat as a set), Man United tops, and reject half-price leather handbags. The stallholders are predominantly white sprinkled with a few Sikhs and Muslims.

I slip between Iceland and Burger King into pedestrianised Market Place, following the plaintive notes of a sax playing Chet Baker's 'Carson City Stage'. My Pied Piper leads me into the infinitely more interesting covered market where Norma's Biscuit's and Cake's, Carter's Cheese, and Barbara's Sweet's Crisp's Pop, await. The town has been hit by a hailstorm of apostrophes. Above another small booth dispensing sweets to school kids is a sign 'Good luck on your wedding day Shelley and Mick'. Blue-rinsers rummage for items for the grandchildren in the Scout and Guide Uniforms stall, a very bored Asian man stands behind his sexy underwear at The Bra Bar, and hawkers bark 'Best bananas 30p a pound...A flat lettuce and a pound of tomatoes 60p.'

Everyone male refers to me as 'maaaate' and everyone female as 'love'. The accent conjures up visions of Tony Culshaw, a nephew who's been living in Sydney for more than a decade now, and Oz hasn't even shone a light into the bottomless mineshaft of his Preston accent.

At 7.30 the next morning I approach the exquisitely sculpted Johnson's Hill flight with something approaching awe. Each pristine gate is freshly painted, each wing of lawn freshly mown, each lock flowing into circular pounds creating an hourglass effect.

I can think of nothing I'd rather be doing this dew-dappled early summer morning than swinging a windlass, working my way through the flight.

Halfway up, I come upon a woman sitting beside a weir, body twisted like the statue in Copenhagen harbour, staring at the rushing water. I try engaging her in conversation but her gaze won't budge from the water. A bag is lying by her side. It's as if on the way to work she suddenly became distracted, and now can't find a way back. Fifteen minutes later, as I leave the lock and move on to the next, she's still there.

At the top of the flight I meet my first Londoner for weeks. Gordon from Erith thanks me for starting out so early; his blue 60-footer *Maisie Jane* now has all seven locks her way. Gordon took early retirement at 55 and bought the boat. The usual wise story. He and wife Jean have been afloat now two years and they've just sailed across from the Yorkshire Dales where I'm heading. Their news isn't good. The foot-and-mouth outbreak is still spreading. There was a cull of thirty cattle yesterday and now a case has been reported in nearby East Morten.

'Keep Out' signs become more prevalent as the route grows increasingly pastoral. According to the news, foot-and-mouth has now cost the farming industry £1.5 billion and tourism £4 billion. With sheep gambolling on the hills there is no visible sign of the grim reaper's cull, no stacked charred bodies, no smell of death. Instead twenty horses graze a high saddle, manes lazily flicking, coats shining with health. Beside a water meadow flecked with pink clover and flickering with white butterflies, cattle are grazing with yellow raffle tickets stuck to their ears. I hope their number doesn't come up.

As *Caroline* arrows through the handsome brick accommodation bridges, I start fantasising about extending my trip indefinitely, transforming it into a business, and taking paying guests aboard under the slogan 'It's not a holiday, it's a life change'. Apart from trainspotting, I can think of few other leisure activities more in need of an image makeover. The predominant perception of the Cut held by the uninitiated may be as anachronistic as peasoupers in London, nevertheless their decidedly unsexy image is the one that persists. To canal virgins, the Cut is what the suet pudding is to the culinary arts. The reality, however, is it's closer to lobster thermidor. As part of the campaign to change its drab profile, my brochure cover will display a fashionably black butty filled with scantily dressed nymphs departing the waterside garden of a divine thatched pub. At the helm will be a youthful strutting gladiator with bulging biceps, toasting the reader with a dainty *mojito* cocktail.

Blackburn may not feature in the brochure. The town announces its proximity with the suburb of Cherry Tree, whose pretty rural name remembers when it was a village before it was gobbled. A modern *Brookside* housing development appears, prim and deadly, followed by infinitely more appealing pebbledash council houses, the back gardens busy with washing, willows, tulips, life buoys, assorted ceramic wildlife, and a toilet urinal serving as flower basket. One garden even has a red post box beside a moored boat named *Grand as Owt*.

Beyond Ewood Park (Blackburn Rovers' football ground), mills rise up and terraced streets run down, and a train clatters over a bridge on which is written, 'Frank rocks the sex bed'.

At the bottom of the Blackburn Flight, an elderly man, maybe five-foot-seven tall with a sliver of hair, bloodshot eyes and Durante's schnozzle, opens the gate for me and I cruise into the lock.

I shimmy up the ladder and start winding the paddles. Bill Melling, aged 77, in razor-creased blue-grey trousers and a blue short-sleeved shirt, watches with an expert's eye. 'I've been on canals since I were one year old – approximately. Grew up in a canal cottage just up the road.' By 'road' he means canal. Many who inhabit the Cut still use the word. With no business appointments ahead of me, no schedule to keep to, and having lost my natural English reserve after a couple of months afloat, I immediately invite Bill for tea and the last of the mince pies my mother left me in Lymm.

I moor up in the pound beyond the lock and together we sit on a

bench beside *Caroline* watching mums pushing prams and patients filing
back from the hospital. 'I were brought up by me aunty in Bank Cottage
at Enam Wharf. Uncle Tom were caretaker, looking after all the horse
boats. And he were on call twenty-four hours a day were that gentleman.
That were in the 1930s when me dad ran a big iron boat for Ranks' Flour
Mills. His mate, Harry Gibbons, were captain. I remember as a young
'un watching 'em go through and all the other boats gathering at
weekend from different ports of call. Then on Monday morning, when
the wharfinger got in and could keep an eye on things, they'd unload
their cotton, cement, you name it. Once school summer holiday arrived
I were off with them too. Me and me dad would sleep head to tail in his
for'ard bunk.'

'Bet you loved it.' I provide encouragement but Bill doesn't need it.

'Ooww,' he rolls the word deliciously, 'I were fascinated. I'd help them
through locks doing paddles. We'd leave Sunday morning at five-thirty
and we never stopped till we got to Lither Land [he separates the words,
making it sound like Disneyland]. You won't have come that way if you
came from Manchester?'

'No, but I spent a couple of years living in Litherland as a kid.'

'Now then. Now we're talking. Now you'll know I'm not spinning it.'
Bill rubs his hands together with enthusiasm. He's in for the long haul.
'It were a wooden bridge then, Litherland Lift Bridge. We were quarter
to ten on Sunday night and we couldn't go no further because it weren't
manned at night. Seventeen hours we'd been going. I can tell you I slept
like a top. Next morning we went across the Mersey to Birkenhead to
load up with corn from Canada at Ranks' Flour Mill. Loaded fifty tons
they did in half hour. They called us yo-yos 'cos we'd be five or six of us
in a line pushed by a tug and if it were rough you'd buck up and down.
It were super to me them trips. Nothing tops that but Blackburn winning
at home.'

Time slips by. I'm back again on the floor in primary school listening
to another myth spun by Mrs Firth. 'Me dad started working on his own
dad's boat at ten. Grandad used to tie a rope round his waist and put him
over the side when the horse were pulling slow. That's how me dad learnt
to swim. Not many boat people could though. Thought it were bad luck.'

Bill is a butterfly flickering with the sunlight through open windows
on his past. 'I were friendly with all the boatmen and they'd give me jobs.

One man would say "Come here Bill. Go ask the man on the *Elizabeth* what time he's setting out in the morning and I'll give thee a halfpenny."' Bill modulates his voice up half an octave to re-enact the conversation between himself and the skipper on the *Elizabeth:* 'Hey up lad, how doin'? Come on board.' Next he skips to the role of Chorus, providing his audience – me – with the necessary asides to keep up with the action. 'Eventually I'd get round to asking the skipper what time he were thinking of leaving in the morning. "Ooh, about six o' clock," he'd say and then I'd excuse meself and nip back to tell the man on the other boat who'd get up half hour early next morning to get all the locks ready and steal a march on *Elizabeth*.' Bill rocks with laughter at the memory and then leans back against the bench to recover.

Surprisingly he never became a boatman himself. In 1938 he started his first job in a weaving shed patterning cloth. 'I fell out with the foreman who was always drinking whisky and kicking me. Then I went in foundry. I were a rivet boy. Big boilers for Russia we built but that didn't pay nowt neither. Used to sneak out half past ten for a cuppa and *Music While You Work* at home. Foreman caught me but I didn't care – nineteen shilling a week wouldn't feed a donkey. Got a job in a timber-yard next, sawing wood for four pound a week. Then I bettered meself again at Mullards where I worked thirty-four year. Took redundancy aged fifty-eight when the place shut down – twenty-second of January it were, 1982, and I used the money to buy me home seven minutes walk from Ewood Park.'

Bill rummages inside the jacket he's carrying and pulls out a season ticket for Blackburn Rovers F.C. With his old-age reduction he'd just paid £170 for it. It seems a snip to me compared to the £700 they ask at Chelsea. Bill, however, complains about rising costs and shows me a £21 stub from the 1987–88 season. It triggers another memory – his first ever visit to Ewood Park. 'I were nine and were scared stiff when me Dad told me to get under his coat between his legs. I were already tired and nervous. It had took best part of an hour to walk down from Enam Wharf and me legs were little – Dad wouldn't take 't tram 'cos it were tuppence for him and a penny for me. Well, I couldn't believe number of people trying to get in the ground and he's telling me to get between his legs, get through the turnstile and when I'm through run like hell up the stairs to the far side of the ground and don't stop for nobody. I did as I were told. Could hear someone calling and shouting but I just kept running as fast as I could. All us kids sat on the duckboards beside the wall. I were hooked.'

Bill screws up his eyes and they water. 'Something in the air,' he explains, looking round suspiciously.

'Do you have any family of your own?' I ask.

'Me wife's dead but I've four grown-up children.' The eldest, a daughter, is a teacher but wants out. A son in Preston does somit clever on computers for the council. Janice works 'at Butlins at Pwelli but it's not Butlins anymore'. The fourth, 'the brightest by a street er lamps', drives a forklift truck in Accrington.

Bill waves me off and I watch him turn homewards. At the fourth lock in the flight, a man sits on a balance beam in pinstriped suit, white shirt and tie, hands on knees, puking, then swigging more from his can of cider before explaining as I pass, 'Just having a break.' It's about 4 p.m. He heads into the office car park, gets into his Vauxhall and drives away.

On a wall before the top lock I read 'NF', and next to it 'Suck my great penis bitch Iqbal'. Parked alongside is a B.W. rubbish collection boat containing an armchair and a caravan retrieved from the canal. I'm now almost 400 feet above sea level, making my way through the stench of a factory, tracked by security cameras. A toddler's three-wheeler bobs past beside a cluster of yellow irises.

Enam Wharf, where Bill grew up, is now home to Granada TV. Beyond it, a handsome teenage Asian couple are hugging and snogging. We exchange big smiles. A little further on, three boys are throwing stones at ducks and I get a bit edgy. Fortunately they've vanished by the time I reach them. I breathe a sigh of relief as pylons stride out across yellow fields to gathering hills dotted with venerable stone farms where a different fear lurks behind every door.

I find myself looking for signs in every animal: a drooling cow, a sheep shaky on its pins. Instead, a carefree boy and girl, aged maybe eight and nine and no doubt called Janet and John, run down the grassy bank from their farmhouse home chased by their pet dog. They stand at the water's edge smiling and waving.

As a high embankment separates the green world below from the blue world above, I slip onto an aqueduct over the M65 and moor a mile further up on the outskirts of Church. I have risen another 120 feet through twelve locks today.

Somewhere – I don't know where – I read that Church is as important as Ironbridge in the evolution of the industrial revolution because in the 1780s an integrated factory on its canal bank was producing both textiles and the chemicals for colouring and patterning them.

There's still a chemical factory on the bank today but much of the towpath is now lined with industrial warehouses and factories decomposing rather beautifully. In the water, reflections of sun–gilded rusting stanchions flicker beside shimmering brick walls. Suddenly, at Foxhill Bank Bridge, I run out of towpath as a frenzy of orange tape bars further progress.

Standing on the bridge two local farmers, John Bailey and Walter Gaskell, are discussing the malignant creep of infection along the Cut. They've just finished replacing the signs and wrapping the footpath in a fresh swathe of tape. There is the fatigue of an already overlong battle about them, a war against an invisible and invincible enemy. It's not just their farm animals that are under threat. John and Walter feel beleaguered by a small but significant section of their local community, by the government, and by Brussels.

At 57 years of age, John is the junior of the pair. He points his shepherd's staff as if at something tangible in the distance. 'It's just there, over the hill near Pendle where the smoke's coming up. The whole of West Bradford's been culled in the last few days.' Black plumes are snake-dancing heavenwards.

Walter's voice trembles with foreboding, 'There's been forty-one outbreaks in Lancashire in the past three weeks since it started again in Settle – before that we'd had seven free weeks.'

'Why do you think it started up again?'

'The knackermen were going farm to farm, vehicles weren't disinfected properly. The wind. Walkers. Dogs. Who knows?' John ruffles his trim silver beard, and then prods, taps and refills his pipe as he continues to ponder the unknowable. Walter fills the silence.

'After the First World War, my grandfather cured the animals of foot-and-mouth by bathing their feet and mouth round the clock and they all recovered.' Walter's words, in such a hurry to get out, start sticking together. 'They'll havetochangetheirpolicysoon on inoculation, because it'llbe everywhere and there'llbenolivestockleftotherwise.'

John returns from his cloud of pipe smoke. 'Many farmers have said they'll replace the animals but the government doesn't want us to. They want to be rid of livestock farming. Costs too much. But we don't want their subsidies. We just want a fair price for our lamb and beef. Eight year ago, the average price for our beef was £1.40, £1.50 a kilo. The same beef's ninety-seven pence now – and it's gone as low as seventy-five pence. That's half what we used to get and all the while costs go up.'

Walter has calmed himself and steadied his words. 'People think farmers are rich. It's a joke. A farmer won big money on the lottery not so long back an' when he were asked what he'd do with it he said he'd keep on farming until it were all gone.'

'Farmers don't retire, they die,' John provides the soundbite.

To keep financially afloat Walter has a second job, working as a contract sheepshearer operating all the way from Preston to Halifax and into Derbyshire. John manages by juggling a second job as a bricklayer.

'When I go to bed at night I'm wondering will I get it tomorrow. You're always on your pins. I've been on Schedule D notice since the twenty-sixth of February. That means I'm not allowed to move animals on or off the farm, yet other folk think it's all right to walk on me farm and let their dogs run riot.'

'Don't the signs keep them out?'

'Dogs don't read signs,' Walter replies. 'On twenty-first of March I had twenty sheep killed, others have aborted and many lambs have died. This year's been the worst lambing I've known in the sixty years I've been breeding sheep.' Walter's lips, dry with emotion, start sticking again, words tripping over each other in their rush to escape. 'I sheared asheeptodaythatwereworriedallrounditsneck. Thewoundsareonlyjust healing.'

John's cattle aren't as highly strung and can better look after themselves when it comes to a fight. At 77, Walter, as you'd expect, is past that stage himself. Recently he was attacked by three men in their twenties when he complained about them coming onto his farm and ignoring his signs. After they finished beating him about his head with their fists and feet, they took his walking stick and beat him with that. The police still haven't taken a statement.

For men involved in such unsentimental work, they share a surprisingly romantic outlook. John believes farmers are 'the guardians of the countryside', and Walter that 'Without us, the country would become wilderness – it would return to wold.' The word is as old as the hills, as old as Walter, almost biblical in its prediction of impending chaos.

On the skyline behind us, Walter points out three trees. 'Silver birch. My grandfather, William Holden, planted those. That were up at Ramsclough Farm in Oswaldtwistle – the southwestern boundary of King Oswald, the Viking King of Northumbria's land.' I stare up into the desolate beauty of the high moorland. 'Tough land to work – twelve-foot-deep peat.'

Walter's finger moves along the skyline to the highest point on the horizon. 'Elm Tree Farm. That's where I started. Could see for miles from there.' Walter, who studied geography fifty years ago at Liverpool University, has a feel for the land and its flocks that city slickers and governments cannot weigh.

In the farm he now owns, bordering the canal, his herd of Lonk sheep, the hardy natives of East Lancashire, graze contentedly. As secretary to the breed for 27 years, he compiles the flock book, a copy of which he promises to pop down to *Caroline* later. Walter claims he can trace his flock back 250 years and 'I've been rearing them meself sixty years. So far, touch wood,' he taps his head with his right forefinger, 'they've been scrapie free.' (Scrapie is the sheep equivalent of B.S.E.) The Lonk, reared on hills, can withstand the wet, harsh winters but not the expediency of government.

John's mind has by now been distracted by two ducklings circling each other beneath us. 'These ducks here, they've no mother and you know why?'

I shake my head. Surely this can't be down to the government too. The paranoia, understandable maybe, but paranoia none the less, is as rampant as the foot-and-mouth. 'Because kids shot their mother. There were five ducklings a month back and now they're down to two.'

'They shot a cob, too,' Walter follows the new clarion call, 'so the pen has to look after her two cygnets herself. You know what? When my daughter was growing up, she canoed all the way from London to Church for her Duke of Edinburgh award. These yobs, they shoot swans and garrotte ducks with fishing gut instead. And yes, I do blame governments. They've created a generation without respect and without respect you're nothing. You reprimand a six-year-old today and they tell you to eff off.'

John abruptly brings the curtain down on the bleak landscape. 'Best be going. Me tea'll be on the table an' I'll be for it.' Walter promises to find me later with the Lonk Sheepbreeder's Flock Book. I am left stilled by the canal after the hurricane has departed.

As I slowly make my way back along the towpath, I think long and hard about these revelations in Church, and wonder how many farmers, and others, up and down the country feel as adrift and disenfranchised as Walter and John. Will they restock if their animals are culled? I have no doubt, but it will break their hearts no less than successive governments have broken their spirit.

Walter and John stand in the same killing field as my Wigan assailants, the yobs who beat up old men, and those that shoot or garrotte fowl. They may be at opposite ends, but it's the same field. It is their sense of disconnectedness and estrangement from the body politic that binds them. I have a strong conviction that if someone was to arrange for the juvenile assassins to canoe from London to Church for a badge they'd jump at the chance. If someone actually shared a narrowboat with them, teaching them to maintain it, they'd stop throwing stones at boaters like me and maybe even learn to respect other property. Alternatively someone should give them a good slap.

The windows at Will and Rosemary Dorling's Middleport Pottery don't get broken any more and its old assailants are now its minders. The hoolies were taken indoors. Meanwhile the rudderless boys of Church are adrift in the sea of meaninglessness like the parentless child prostitutes outside the Middleport Pottery walls. Church needs to take in its flock.

As I think of John and Walter walking the high moorland, two trees bending almost to breaking in the storm, my mobile rings from inside my jacket pocket. I rush, fumble and it squishes like a bar of soap out my hand into the canal. Shit. And double shit.

Three guys in their twenties on the far bank try to suppress giggles as I shed my jacket and top and lower myself into the murky water up to my waist. Bending low, my face hovers above the surface as I stretch a hand into the slime. Miraculously I locate it. Less miraculously it's as dead as a dodo. Back on *Caroline* I administer emergency resuscitation, taking the sim card and battery out, dabbing dry what I can, before popping it into bed in the water-heater cupboard.

It has been an eventful day and I'm knackered. Apart from the unholy baptism, and the emotionally draining encounter with the farmers, I have risen around 120 feet through twelve large double locks, each with outer and inner paddles, each taking about forty ratchets up and down, and each against me.

I turn for comfort to Radio Four and learn that President Bush is trying to convince European leaders in Sweden to go in for a defensive shield like Captain America's to protect the U.S. from rogue nations hurling weapons of mass destruction at them. Maybe all we need to do on the Cut is put umbrellas up on all the narrowboats so the stones don't get us. Maybe. I think not.

That night I dream that the entire workforce in Cherry Wood is to be culled. When I wake, I'm pleased to find the world still in place, and even better, my mobile has risen from the dead. Unfortunately its resurrection is brief and it expires again with a finality that short-circuits any thoughts of returning it to the rebirthing cupboard.

Over breakfast I worry some of the issues that have disturbed my idyll around the bowl with the Nut Clusters. A depressingly large number of the people I've spoken to over recent weeks have pointed fingers at immigrants as the source of their problems, and when they say immigrants they invariably mean Asians. It's a story as old as the Bible – those feeling deprived point to others as the source of their penury, problems or pain. When I used to teach at London comprehensives in the seventies, the hard white kids off the estates picked on the Asians, who looked down on their peers from the Caribbean, who in turn scorned those of African origin, who picked on Middle Eastern kids, who hated the white English kids off the estates. There must always be someone to pass the shit on to.

A sense of persecution is also currently gripping the countryside. Those working the land feel there's a conspiracy to put them out of business, that farming is perceived by urban dwellers and governments as being already past its sell-by date and should be following cotton mills and coalfields into the heritage industry knacker's yard. As higher education is constantly extolled and manual skills denigrated, we devalue things that have been made or grown. So why not throw stones at them? Education is spoken of as the same thing as intelligence but Walter, like the old potter at Middleport, like Chocolate Charlie Atkins and his son Bill, can read the land and the Cut, and can make things that the rest of us are entirely incapable of.

What seems to bind the racism, violence, economic short-sighted expediency, and even snobbery, is the malaise of carelessness: an unwillingness to care for others as ourselves, to care for the land and the environment. The insulated self has no time for husbandry and parenting in the frenzy of personal goal-oriented hurrysickness.

As I slip my moorings, a young man in a Man United top stops on his bike and states in a revelatory tone as if interrupted on his way to Damascus, 'There's a big difference between living in a house and living on a boat. In a house you can't go anywhere, and on a boat you can go where you like.' Having delivered his message, he's off. I'm off too, having first removed a four-foot sheet of malleable wire mesh wound round the propeller.

This morning I'm more aware than ever of the human scale to the Cut. A lengthsman looks after his couple-of-miles stretch of canal no less than a farmer does his acres of England. The canal itself has room for two passing boats that can be operated singly, bridges are little taller than a man standing on water, bankside buildings consist of two or three stories mirroring the human frame, and man-made embankments add to the shapeliness of the land. The Cut blends into the landscape, part of it, not distinct from it. It is the only navigation so far invented to do this.

When engineer Robert Whitworth, a Brindley student, brought navvies down from Scotland (where he'd been working on the Forth & Clyde) to construct the Leeds & Liverpool, they too became part of the land. That's why we have a Nova Scotia in Blackburn where many of the Scots settled. Maybe one day we'll have a Dhaka and an Islamabad too. When the navvies arrived, local men resented them going out with their women, and wanted to send them packing for putting up the prices of food and accommodation. It rings a bell.

Initially canal work was part-time, a break from mining or harvesting, but at the height of Canal Mania contractor gangs were well established and parochial England struggling to accept the itinerant invading armies from depressed parts of the North and Ireland, as well as Scotland.

An experienced professional navvy could dig a trench in good soil three foot wide, three foot deep and 36 foot long in a day but it was a gruelling life. On 2 March 1793 at the height of the Mania when around thirty separate acts for new canals were being passed, *The Times* reported, 'One Hon. Member wished his grand-children might be born webb-footed that they might be able to swim in water, for there would not be a bit of dry land in this island to walk upon.' By the end of the century, it's been estimated, 50,000 itinerant labourers were employed on the canal network. Between them they dug almost 3,000 miles of waterway.

Although the first modern navigational network gave wing to the industrial age, few mechanical devices were used. By 1796 some canals such as the Gloucester & Berkeley Canal were employing conveyor belts, and on the Caledonian Canal steam dredgers were employed. The vast majority of work, however, continued to be carried out by pick, shovel and barrow.

Once filled, each individual barrow had to be steadied at the same time as it rushed a mud-smeared plank run up the embankment. Sometimes, as inclines became more precipitous as they deepened, horses would be brought in to help pull up the load. If the man slipped or the horse suddenly altered its pace, the load and the man ended up back in the trench.

Once the channel had been dug, it would be puddled to make it watertight. Clay often had to be carried long distances to the cutting. It would then be mixed with water and loam and applied in layers eighteen inches to three feet thick. The most common method of puddling the clay was by labourers tramping up and down on it, though occasionally herds of cattle or sheep were used.

Predictably these men who dug and puddled the canal were the poorest paid of the canal workers. When labourers were earning 6s a week in 1768 on the Coventry Canal, a carpenter was earning 11s. Even labourers, however, had their own pecking order. In 1824, while an adult labourer earned 3s 4d a day working on the Harecastle Tunnel, younger boys leading the horses at the gins or doing the barrow runs earned just 1s 6d.

Canal engineers may have been exalted, but their roving worker bees were scorned, as they were here on the Leeds & Liverpool. Drunkenness and violence is pretty inevitable when large numbers of uneducated, poor men come together to work dangerous, gruelling hours away from the stabilising effects of home and their local community. Author Peter

Canal Mania 1760 to (*facing*) 1790

Previous page: A pregnant sky over London as Caroline rides the tidal Thames. Photo. Nick Crane

Above: 'Idle Women' trainees – with Olga on the extreme left – kept the network open carrying essential freight during WW2.

Below left: The traditional 'roses and castles' design on Caroline's doors promise escape to a fairy tale world.

Below: The 200-year old Bulbourne Workshops at the Tring Summit.

Right: Napton Locks from Napton Hill on the Oxford Canal. Photo Robin Smithett

Right, inset: Sue and Kim Russell relax on deck near Hayes.

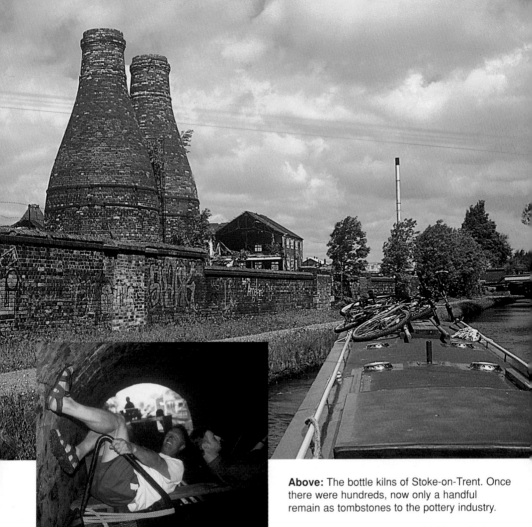

Above: The bottle kilns of Stoke-on-Trent. Once there were hundreds, now only a handful remain as tombstones to the pottery industry.

Below: Tommy Beardmore with his stuffed buzzard on the roof of the African Queen. Froghall Basin, Caldon Canal.

Above, inset: Legging through the Dudley Tunnel. Professional leggers would lie out on 'wing' boards, backs leaning into each other, and walk the boats for up to four hours at a time.

Top: Max becomes a dab hand at fishing.

Above: Alan takes Bruce, a cross Neopolitan mastiff and bull mastiff, for a walk along the towpath outside Lion Salt Works at Marston.

Above, right: One of a pair of motherless blue tits outside the Foulridge Tunnel.

Left: Leonie and Graham Acaster whose family have been freight carriers on the Aire & Calder at Goole for 180 years.

Main picture: White-knuckle riders at Wykewell lift bridge on the River Don Navigation. While the boys bomb, the girls raise the bridge.

Above: The Bingley Five Rise – a working sculpture approaching Saltaire and Bradford on the Leeds & Liverpool Canal.

Below: Nightfall. Another perfect berth on the Thames.

Above: *Caroline* gives canoe hitchhikers a lift from Lechlade.

Below: Medmenham Abbey, once home to debaucheries of the Mad Monks of Medmenham at the Hellfire Club.

Canal Mania 1820 to (*facing*) 1850

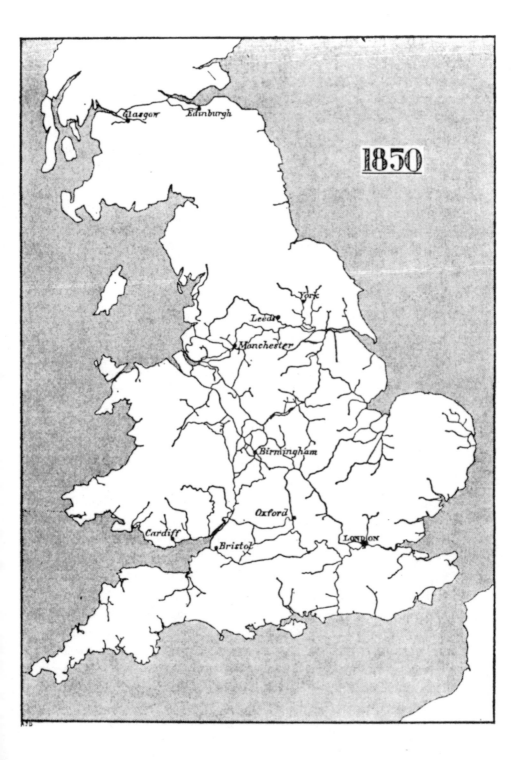

Lecount in 1839 summed up the general consensus:

> These banditto, known in some parts of England by the name 'Navies'
> or 'Navigators', and in others by that of 'Bankers', are generally the
> terror of the surrounding country; they are as completely a class by
> themselves as the Gipsies. Possessed of all the daring recklessness of
> the Smuggler, without any of his redeeming qualities, their ferocious
> behaviour can only be equalled by the brutality of their language.

As I slip out of Church, a couple of men wave to me from a bus stop. On
the next bridge someone has crudely daubed 'NF' in red paint. The
dance of light and dark. After the Epsoms Salts factory, I struggle with
my first swing bridge, which is even worse than the Caldon's lift bridges
because it opens by swinging towards the boat and therefore, with
Caroline tethered to the bridge, there are inevitable collisions. When I get
back on board after the first, I notice the wooden tiller handle has
snapped off and is lying on the deck. For the rest of the trip I'll feel only
snubbed cold metal instead of warm wood in my hand.

At the third swing bridge, Foster's, the farmer has kindly built a swish
wooden jetty. His good deed, however, has not been rewarded with good
fortune. A handwritten sign declares, 'You are entering an infected area.
Boaters please use the jetty side.' The whitewashed farm building is as
silent as a morgue, so are the stables. With the sensitivity one displays at
a funeral service, I creep about undoing the handcuff lock, turning the
windlass as quietly as possible, avoiding smashing the swing bridge
against the bank, and so on. Throughout, I see no sign of life. The stables
and forecourt are spotlessly clean. It's probably only a matter of a day or
two since they were disinfected several times after the cull.

A couple of fields up, a rogue cow among a herd of around sixty makes
a bolt for it by leaping into the canal. It staggers across to the far bank
and then can't get out. A few of its relatives come to take a look, they are
wearing necklaces with yellow tags – is this the sign for culling like the
yellow Star of David once was for Jews? The audience of cows in their
front-row seats flick their tails enjoying the action, watching Patrick
McGoohan cow, number thirty in this case instead of number six, failing
to even make it out of the moat in its bolt for freedom.

I follow the high embankment with hills rolling and lifting beyond blooms of yellow irises and fans of ferns. On the horizon, pylons drop their hands low holding skipping ropes for the Pennine giants. *Caroline* slips between the elephantine legs of the M65 ('Munich '58 Letsalllaugh') and then encounters it again, this time taking her turn on top. No doubt the odd car or two has crashed off the road in shock at seeing boats sailing overhead.

Soon, I'm sliding through an industrial Angkor Wat – abandoned goods yards, orphaned warehouses and mill chimneys. Occasionally a cluster of terraced homes appear, TV aerials all pointing in the same direction like a flock of storks preparing for takeoff. Of all the cities I've passed through so far, Burnley has the greatest potential for regeneration in terms of its industrial architecture. In its present dereliction it possesses a sage beauty that makes Birmingham appear positively frivolous. I have no doubt, however, that Burnleyites would swap a UNESCO World Heritage sticker for jobs and vitality without a moment's hesitation. Pickling the past is the preserve of the comfortably off. Those less flush are more preoccupied with finding a job that pays the bills. Burnley's daunting task is in raising the finance from both the public and private sectors to revitalise their town and get the money going round.

I moor at the Inn on the Wharf with its hanging baskets and nasty snapping wolf. The rain is lashing down as I head off to explore the spinning mills, engine houses, weaving sheds, foundries and depots that sprang up in the mid-nineteenth century when canal water was first utilised to power the mills' steam engines. Fifty years later Burnley was producing more woven cotton cloth than any other town in the world: the cotton textile industry was the driving force behind Britain's economic power in the nineteenth century and East Lancashire was where it began.

In the distance I watch a narrowboat gliding in the mist of rain along the Straight Mile, a raised embankment arrowing above Coronation-Street terraces and beneath the South Pennine moors. To escape the downpour, I pop into the Weaver's Triangle Visitor Centre which straddles two buildings, a passageway linking the old canal toll office with the wharfmaster's house.

It clearly wasn't a bad number, the lot of wharfmaster. With three floors and a plaster-moulding ceiling to impress visitors, it cost £901 17s 6d to build in 1878 for the agent of the Leeds & Liverpool Canal Company, his wife and nine children. Each of the rooms has been restored to how it might have looked at the end of the nineteenth century, down to William Morris wallpaper, peg rugs and dour Victorian furniture.

Over the fireplace in one room is an almanac 'Five o'clock tea' produced in 1879, and lined with notable Victorians such as Florence Nightingale and the philanthropic Earl of Shaftesbury. It was one of the 'gifts' you could exchange for vouchers if you bought Abraham Altham's tea. Altham, a local lad, was possibly the first businessman to give away gift tokens in the late nineteenth century. He was also a pioneer travel agent (the company still exists in town), running day-trip 'Grand Excursions' to Blackpool and Southport for three shillings – a king's ransom at the time.

I pay a very reasonable five times this for tea and fruitcake before moving on to a room containing a working model of Burnley Fair complete with steam yachts and a big wheel. The fair has run annually for 700 years and even lent its name to the annual July holiday which in other mill towns is still known as Wakes Week.

I leave the best room to last. The gallery in the toll house contains Victorian postcards, maps, and early photographs. A large black-and-white print shows a weaving shed the size of a football field in which there are endless lines of belt-driven looms. Another photo, taken in 1910, is of two pretty mill workers in cotton cloth aprons, pinched at the waist by broad weavers' belts from which scissors and other tools protrude. On the far wall is a 150-year-old map of the Leeds & Liverpool Canal. The longitudinal cross section shows the canal dropping from the nearby summit, 487 feet 6 inches above sea level, westwards through the Burnley Pool to Liverpool's Stanley Dock, and eastwards to Bingley and Leeds.

Within the Weaver's Triangle there have been a few brave attempts to draw people back into the old manufacturing area with the opening of the Chicago Rock Café (part of the Foundry Lodge Hotel which either hasn't found the nerve or the funds yet to start operating as a hotel). Another handsomely converted canalside warehouse that was due to open as an upmarket hotel has run out of money and its future hangs in the balance. Beyond the Triangle, towards the M65, a zone has been designated the new entertainment and leisure area for the 25-plus crowd.

So far, however, they've got no further than designating. At present Burnley's waterfront renaissance looks to be some way off.

Back at the wharf, the beast in the pub alley repeats its attempts to tear down the gate and rip my throat out. It's sufficient enticement for me to ignore the continuing downpour and set sail; I'm keen to get on to the next wonder of the inland waterways: Burnley's Straight Mile.

The Mile is actually just three-quarters of a mile, but it's a corker whatever the length. The aqueduct carries me sixty feet above town, parallel with the town hall clock face and its leaded cupola. I have an aerial view of sandblasted terraces and a bus leaving the sprawling station heading for a place called Bleak House. A bare-legged, pasty-faced teenage girl with a rose tattooed on her ankle walks past on the towpath high above the steaming, streaming cars. She smiles and waves as if she's walking a Mediterranean beach in sunshine.

When *Caroline* finally emerges from the urban sprawl, the countryside is a Chinese watercolour of fine muslin mist with hills rising behind hills. By the time I moor at day's end I've only seen two moving boats all day and none, moving or stationary, in the two hours it's taken to circumnavigate Burnley. I'm starting to be a curiosity and I don't think it's just down to the London address of the hire company.

As the rain continues, I cosy down below decks and make a heaving plate of cholesterol – sausages, bacon, onions, mushrooms, baked beans and mash. I switch on Radio Four and learn that a delegation from Oldham (where 12,000 people voted for the British National Party in the election), without a single Asian presence, today met the Home Secretary to discuss the race riots that erupted two weeks ago. The report claims a kind of apartheid exists in Oldham: a deprived Asian community living alongside a deprived white community with only mistrust and hatred mixing. There's also a report on three farmers who recently committed suicide in Welshpool. I lie in bed listening to the rain tap-dancing, listening to news about England. I have one mince pie and two slices of Dundee cake left. Rations are low.

The following morning I find another boat moored beside me at Barden Mill, a shopping outlet on Burnley's borders. The 60-foot narrowboat is waiting to pick up a party of school kids and at the tiller is a handsome olive-skinned 31-year-old by the name of Nigel Warren. I take my mug of tea and head over for a chat. Twenty seconds after our introductions, Nigel changes into a foot-in-the-door God salesman.

'Our trips teach the children to appreciate God's creation, and of course it's also the Lord's blessing to the school when we take the children out for the day.' As an evangelical employee of the Northern Evangelical Trust, Nigel runs canal jaunts for both primary and secondary schools in the Trust's narrowboat, and in the evenings he mans one of two coffee buses that visit the estates. 'The youth come in their droves. They may be mischievous but they're all after love and if you show them that, it turns them round.'

It all sounds so simple I almost believe it. I try to put the brakes on my cynicism, after all it was me ranting about the lack of care in society yesterday. Or was it the day before? Nigel, when all is said and done, is contributing considerably more than I am. Unfortunately I can't quite manage the hallelujahs for there's a fire in Nigel's eyes that I find unsettling. The evangelical one, however, is still galloping on his white horse. 'One child was constantly being sent out of lessons but God did His work through us and he's fine now.'

Nigel's colleague, Alan Scholes, arrives by car. He mainly gives support in the classroom but today is running the trip boat with Nigel. 'We don't charge schools anything. Some make donations but they don't have to. We also do assemblies but our best resource is definitely the canal. Getting the kids out on a boat transforms them.' Alan shares Nigel's optimism but is a little more grounded, his down-to-earth approach picked up during the 32 years he spent in a classroom as a teacher. 'Although there's an increasing number of children who misbehave, it's still a minority. I'm just pleased the government has announced it's backing the opening of more church schools. The young are after attention and if they get it from us, they're going to be better for it than if they're getting it from video nasties and MTV.'

Alan uses a broad brush to paint what's going wrong in society – over-large classes, a packed curriculum with no time for individuals, the swallowing of urban communities with distinct identities into amorphous conglomerations, the break-up of rural communities because there's no work and housing is too pricey for the young... At the end of the litany, he somehow manages to remain upbeat. 'You watch the kids' faces when they arrive this morning, Paul. They won't believe their luck spending the day on the canal. It's going to be a good day.'

'Praise God,' Nigel adds.

I'm gone before the children arrive, continuing my climb towards the summit. There's time in a lock to admire the slippery soaked grasses hanging down from the wooden gate like Indian scalps. There's the eager anticipation when someone works the lock for me and the gate suddenly opens, shazam, a secret door into another new world.

Today, the canal again appears starved of traffic and I realise I haven't seen a single hire-fleet base since first joining the Leeds & Liverpool. Clearly there isn't demand for boat hire in East Lancashire despite its mix of fine countryside and interesting mill towns. Unfortunately its reputation for vandals goes before it and makes some boaters apprehensive. They shouldn't be. The worst they'd get is raining stones. There are worse things. Like raining maggots.

I approach the Barrowford Locks, home to Roger Bannister, the man who first broke the four-minute mile. A mile is currently taking *Caroline* anything between seventeen minutes and an hour. Waiting at the bottom lock is *Muffin 3*. A bell rings in my head. Someone, somewhere, two weeks or a month ago told me to look out for *Muffin* on the Leeds & Liverpool. The owners, Audrey and David Smith, are renowned boaters. David has absented himself from the boat and is up ahead preparing the lock. I introduce myself to his wife, who's at the tiller waiting to steer *Muffin* in once the lock's ready.

'I'm Audrey Smith,' the woman bats the ball back.

'I know,' I reply, 'ex-chair of the Inland Waterways Association. Your fame goes before you.' Audrey laughs and asks where I'm heading. 'London, eventually.'

'We're heading that way too... well, to Milton Keynes for the I.W.A. annual rally. B.W. are planning opening a new canal linking the town with Bedford which will access the network eastwards around the Wash.'

'The I.W.A. and B.W. in bed together! That's the surest sign yet we're living in enlightened times.' We laugh. Audrey has a relaxed manner and absolutely no pomp.

'Yes, it is a different world now. Most of the battling is over. We're working for the same things but you're right, it hasn't always been that way. This is a perfect example.'

'What is?'

'The Leeds & Liverpool.'

'Why?'

'Well, if it weren't for the I.W.A., I doubt very much we'd be sitting here now.' I hardly expected an unbiased view but this seems a smidgen grandiose. Audrey catches the whiff of cynicism. 'Seriously. The Leeds–Liverpool was down for closure but we had a national rally in Blackburn. Boats spread all along the towpath right through the town. It's the method we always use to campaign. It worked. It stayed open. So did lots of other canals round the country.'

I've already read a fair bit about the early I.W.A. campaigns and what I'm really interested in pumping Audrey about is the legendary falling out of two of the I.W.A.'s pioneers, Tom Rolt (author of the highly influential *Narrow Boat* which appeared in 1944 and has never been out of print since), and Robert Aickman – 'A single-minded man, who didn't have much room for anyone else's vision,' according to Audrey. I'd heard that the fall-out was as much over a woman as it was outsized patriarchal male egos.

Audrey will not be drawn on the smut stakes. Although she never met Rolt, she did work with Aickman in the seventies when they were both involved in a campaign to save the Yorkshire Derwent. Audrey is diplomatic but honest. 'I found Aickman very focused, very introverted in many ways. Not a woman's man. Not an easy man to get to know and therefore not an easy man to like. Tom Rolt had a reputation for not relating well to women either but I have no personal experience of that.' She sums up as succinctly and diplomatically as any chair. 'Tom and Aickman did not part amicably as I understand it.'

Having read several books by, or on, Rolt, I know that he is more responsible than any other individual for the survival of the canals. Nevertheless he's a man that's hard to warm to. A few quotes from *Narrow Boat* should suffice in illustrating why. When watching miners passing on the towpath, he speaks of 'their faces so blackened with coal-dust and sweat that as they turned to grin at us teeth and eyeballs flashed like those of a coon at a seaside concert party'. Later, speaking of a neighbour's wife, he muses, 'It would seem the quiet waterways had opened strange doors within her simple mind, to reveal a dimly compre-hended beauty which she could ill express, but which far outweighed the synthetic attractions of the cinema, which constitute the usual refuge of her class.' At times he displays a disdain for his own time that reaches dizzying Victor Meldrew levels of absurdity, for example when railing against 'The evil genius of the can-opener' – a tool that I happen to be particularly fond of. By the spring of 1951 Rolt had already had his come-uppance: his marriage to the beautiful Angela was over (she'd run off with a circus strongman), *Cressy* his beloved narrowboat was critically ill and he had been expelled from the I.W.A. along with the other pioneer Charles Hadfield.

Unfortunately Audrey is clearly not going to dish any dirt, so I move on and ask what it was like to be the first female chair of the I.W.A. 'I was the Grand Mum of the Inland Waterways,' she says with a self-deprecating laugh. 'Just below Mother Theresa and Sonia Rolt, Tom's widow, who's known on the Cut as the Grand Dame.'

Audrey suggests I travel up with her and husband David through the Barrowford flight to the summit but I'm starving and instead return to *Caroline* for lunch. As I bite into a chicken drumstick, I listen to Margaret Beckett soundbites on the radio as the new Secretary of State for the Environment, Food and Rural Affairs makes a promise to place the environment and sustainability at the centre of future agriculture and rural policy. She is keen to cull a system that sees British farmers consume £3 billion a year in subsidies to produce £6 million worth of food. The farming community meanwhile is apparently uncertain – with a Kafkaesque castle of new quangos, ministers and ministries – whether they should be approaching Beckett, Alun Michael (the Minister of State for Rural Affairs), or Lord Whitty (Junior Minister for Food and Farming) to find out exactly what the soundbites mean. A new outbreak of foot-and-mouth was confirmed yesterday in nearby Clitheroe.

Another boat arrives – it's like Oxford Street in the sales today – and I decide it's time to climb the tidy Barrowford flight of seven to the summit. We sail together as far as the entrance to the Foulridge Tunnel where we must wait fifteen minutes for our turn inside (traffic is one-way). I kick my heels on the towpath. Beside the mouth of the tunnel, a blue tit no more than a week old lands on a video camera that's been set up on a tripod to film boats entering and exiting the tunnel. The chick hops off the camera onto the camera operator's head. I take a photograph. Its twin puts in an appearance and we try to feed them bread soaked in milk but they turn up their beaks at it. We decide their best hope for survival is if Mum comes looking for them, so we leave them there in the hedgerow. It's time to be burrowing.

Inside the subterranean passage, I realise too late *Caroline*'s headlight isn't working. I'm being swallowed by darkness, when the second boat, a Samaritan with a light, enters behind me. There's no reversing now; I've no option but to travel in its beam. In 1912 a cow fell into the Cut, stumbled into the tunnel, and similarly found it couldn't turn around or reverse. It swam the mile to the other end in total darkness before being revived with alcohol by the kindly publican of the Hole in the Wall pub. I mentally file the older parable along with my own for the marketing department of an evangelical canals trust I'm considering founding. Our slogan will run along the lines of 'Though you be blind, yet you will see: miracles and Samaritans await all who enter the Cut.' It needs polishing I know.

As *Caroline* emerges into daylight, a couple of ducks are playing with a pair of fluffy car dice in the water. The summit is a yellow carpet rolling to a patchwork of fields divided by dry-stone walling. I am at the summit of the trans-Pennine and the apex of my journey. The stretch I'm now travelling along is unsurpassed by anything I've seen so far: voluptuous folds of land stretching through the Dales National Park up into high moorland. A sign declares that I'm now in Yorkshire. I celebrate, like the cow, by pulling in at the nearest canalside pub in Salterforth.

At the bar of the Anchor, I ask a waitress by the name of Ginny, with piled brown hair and a slinky green satin dress, if she'll show me the cellar. She is an old hand and realises it's not a grubby come-on but just another punter who's read in his Nicholson guide that stalactites and stalagmites are growing there in the darkness.

'As you're the first today I s'pose I will. The record is ten I took down in one session and another waitress had eleven the same day.'

'Are there stalactites there?' I ask, as Ginny collects the keys.

''Old on luv, don't get the cart before the 'orse. I'll tell you it all but we 'av to leave best bit to last, don't we?'

I comment on her accent, as creamy thick as Lancashire Boddingtons despite the fact I've slid into Yorkshire. 'Don't you be fooled by 't bloody Yorkshire sign on 't bloody Cut. That's just 't bloody farmer. His farm's always been in Yorkshire and he's not having no bureaucrat tellin' 'im he's now in Lancashire. He put a sign on road too. Yorkshire? Bloody Yorkshire my arse.'

We traipse down cold stone stairs and along the rabbitlike warren that is the cellar. One room is under two feet of water. 'This were cellar of original pub when 't packhorse went through. See 't pub dates back to 1655 but canal came through in 1788. That's when they built the pub on top of the first pub so you could get to it from canal.'

I think I get her meaning. Ginny, as promised, leaves the best to last, a barrel-vaulted room with three-foot limestone straws hanging from the ceiling and stalagmite mushrooms stretching up to them from the floor. In its stark beauty, it could be a modern art installation in a contemporary gallery. I'm about to step down to get a closer look when Ginny pulls me up. 'Don't you be going down there or I'll belt you. They're right fragile them and we can't buy new ones with the beer.'

As we leave the room and I stoop again along the passageway, Ginny asks if I saw the photograph on the wall in the pub entrance. 'Yes,' I reply, conjuring up again the sepia-tinged picture of a family in Victorian Sunday best.

'See the young 'un? The young boy in picture?'

I nod.

'That's Joseph. He 'ung himself from that beam when he were fourteen,' she points to a joist in another underground room we're passing. 'He's a right pest too; forever turning barrels off and on, and messin' with the pressure.'

By the time we return to the bar, my steak and kidney pie is waiting for me. I order a pint of a local dark mild called Black Cat. The wall is festooned with canal memorabilia including a cautionary note of a £2 17s fine for a motorboat captain in 1921 who went through the Foulridge Tunnel without a permit; and a 1967 local newspaper article commemorating the retirement, after 55 years, of Tom Draper who, like his father before him, spent virtually his whole life towing horse-drawn boats through the tunnel.

Pubs tend to slow progress and I travel at a snail's pace after lunch. At nearby Lower Park Marina on the outskirts of Barnoldswick, I buy a replacement bulb for the tunnel light and am greeted by a couple of boaters who are drinking mugs of coffee at a table they've set up inside the shop. They tell me the field opposite has been cleared of livestock over the past few days. And the one next to it. 'Entire herds of dairy cows and heifers. Not as bad as Gargrave though – they killed about four hundred there last week. You could hear the gunfire going off all afternoon.'

The fields look well grazed and untroubled, green waves travelling across the ocean of fields to clouds crashing on the high moorland. There is a wild beauty to this countryside with its scattered stone farmhouses standing misanthropically beneath wooded uplands. For a while the Pennine Way dovetails with the towpath. But like much of the 680 square miles of the Yorkshire Dales – designated a national park in 1954 and possessing some of the country's finest walking – it's out of bounds today.

For a short stretch, trees close in to lock fingers overhead in prayer. Their particular fear is not foot-and-mouth, however, but global warming and an inability to migrate northwards quickly enough before Costa weather fries them.

Having slipped onto another high embankment, the television transmitter I've been travelling towards for more than a week is just a hundred yards away on its lonely hill. It's surprisingly small – maybe just 120 feet tall – now that I've risen up through the coronet of hills. Two pheasants are stalking the undergrowth beside it as *Caroline* winds her way round the maypole mast. I'm on top of the world and the views are sensational. On a river, snaking its way along a valley floor, you'd never get such an overview of a landscape.

My next port of call, Gargrave, nestles in Upper Airedale with the River Aire running through its centre and pretty thick stone cottages skirting the village green. In the Mason's Arms four locals are discussing the army of foot-and-mouth soldiers swarming through the area. 'The killing crew were through again today off to clear more fields. Trailers were dripping everywhere like they was haemorrhaging.' One of the group complains he's made £100 a week since February at his equestrian centre because the horses can't be hired and there are few visitors in town to buy anything from the shop. The pub's landlord is down £1,000 a week on the tourist trade.

Looping back through the village, I stop Rosinante outside an estate agent's and peer in the window. A three-storey cottage built in 1647 with mullioned windows, oak beams, two bedrooms and a garden is going for £120,000. Tempting. I pick up some fish and chips and take them up on deck with a glass of Shiraz. Night thickens, I lose track of time, lights fizz across the water from the road above.

As I peer into the constellations arching across the dome of night, I remember an equally infinite African sky in the mid-seventies when, with a French colleague, I was slipping about in a riverbed seeking frogs for our supper. Alain Videau and I were colleagues at a lycée in the Algerian mountain town of Draa El Mizan. As we paused and peered into the star-spangled sky, with a mathematician's precision, my French friend attempted to explain the relationship between time and travel. He claimed that if he left me at that very moment and soared through space fast enough, when he returned to earth, he would have aged less than I had. It might even be possible, he added, to stop time completely.

Twenty-five years later, as I now gaze into the sky we continue to share with Algeria, I know, regardless of its logical or scientific truth, that Alain's view is a delusion, a human conceit. Speed is a false god and the only way to really slow down is to slow your own journey. It has taken me a while but I feel I'm there, sitting up on deck. Lost in the Yorkshire Dales.

Replenishing my glass down in the cabin, I spread out the map to plan tomorrow's slow-time waltz. With horror I notice that instead of heading northeast I'll be heading southeast. I've reached the northernmost point of my journey. Halfway. Tomorrow I start turning homewards.

The second part of the journey I know will be very different – 600-tonners overtaking *Caroline* on the Aire & Calder sending tsunamis my way, the fast-flowing River Trent picking us up like a cork and jettisoning us when it's done playing, the dreaming spires along the Oxford Canal, the sumptuous green sweep of the Thames Valley. And I'm still only half done with the Leeds & Liverpool. Yet there is an inescapable sinking feeling, like halfway through Christmas and it's going too fast.

Reluctant to get going, I set sail a little later than usual the next morning. The sun is shining and the countryside in ebullient mood. A broken line of beech and oaks bisects a field before giving way to a snail trail of dry walling so perfectly laid there's not a glimmer of daylight between its compositional stones. Some forty sheep are slowly eating their way across the rising field, beyond which silhouettes of clouds are passing over brooding moorland. Meanwhile, down on the canal, seven juvenile swans, the colour of fresh snow, glide through the reflected landscape. When you're in an office dreaming of England and escape, this is what you have in mind.

I am now on a seventeen-mile stretch of lock-free canal. The first swing bridge, number 173, is fortunately open but I soon wish it wasn't. There's no animals left to cross and the ghost farm has a strong smell of disinfectant hanging in the air. I slip into Skipton on a high embankment. Eight feet below me to the left, an elderly man in white hadj skull cap is bending, working his allotment garden with his wife. Beneath me to the right are the old outhouses of terraced homes.

Skipton is a handsome town with day-trip boats, hire fleets, delis, tea houses, restaurants and a profusion of pubs. With four dedicated mobile-phone shops alone (in one of which I replace my drowned Motorola), it's the first town I've sailed into since Lymm that appears to be booming.

Despite the seeming prosperity and lively atmosphere, at Craven Cameras a rather breathless assistant tells me she's had a rough day. 'When I popped home for lunch to check on the horses, there was a pile of dead animals at the end of my garden.' The latest cull on the farm next to her home in West Marton destroyed close on four hundred cattle.

The Woolly Sheep Inn, the Fleece Inn and the Royal Shepherd pub, all located within a few hundred yards of each other, bear testimony to the pre-eminence of farming in the area. And it's been that way for more than a millennium: the town's Saxon name, Skip-tun means 'sheep town'.

I walk up through the High Street market stalls and sit on a bench in the raised garden of Holy Trinity Church eating a meat pie from Farmhouse Fare ('Support Rural Britain – Keep Britain Farming') and a latte from Café Java. Directly beneath me on the right is a handsome Georgian building in which David Goldie sells his 'Finest Headwear, Footwear and Walking Sticks'. A couple of doors up is a Lilliputian-sized sweetshop still selling sherbet lemons; a couple of doors down stands a Body Shop and the natural food centre Down to Earth ('Organic Alternatives'). Behind me lie the citizens who've run out of time to do any more shopping. The cemetery is overlooked by the black clock face of a squat Norman church tower that reminds the living that time is still ticking.

In the ticket office of Skipton Castle, a rather dapper young man is pacing up and down in a tweed jacket. I comment on the beauty of the seventeenth-century grotto created out of volcanic stone, coral and oyster shells that we're standing in. 'Some hate it. I'm not sure myself. The only other one of its kind in Britain is at Woburn Abbey,' the jacket knowledgeably answers.

'You must work here.'

'I'm the administrator. I'm just waiting to check another school party in. All the schools visit during the last month of the summer term, so I'm afraid it might be a little noisy.'

At that moment, the woman behind the ticket desk gets a call and passes on a message that the next group is running a little late.

'In that case I can give you a quick whizz round if you like,' the dapper one offers.

'Are you sure?'

'Of course. I don't do it often. I'd only welcome the school group and then scarper anyway. I enjoy showing people round the property – it's one of the most complete and best preserved medieval castles in the country, you know?'

I didn't, but I do now. My guide, who introduces himself as Sebastian Fattorini, and appears to be in his early thirties, is clearly one of that lucky breed who have a passion for what they get paid for. He leads the way beneath vast concentric towers built 700 years earlier by the first Lord Clifford. 'The family came over with the Norman invaders and arrived at Skipton via Wales. Once ensconced, they made their money from farming and the quarrying of limestone – it was used as fertiliser, for lime mortar, lime wash – to give reflected light inside homes, and as a flux for Bradford's iron industry. When the Leeds & Liverpool Canal arrived, an extension was made looping to the rear of the castle so the limestone from the quarry could be dropped thirty metres directly into the water – often it went straight through the boats if they waited too close!' Sebastian laughs. 'Even before its economic success though, Skipton Castle was already of enormous strategic importance.'

By now we've ducked and squeezed through a number of stairways into Conduit Court, an exquisite courtyard with a yew tree at its centre beneath which Lady Anne Clifford no doubt used to listen to Elizabethan madrigals. Over a door, on the coat of arms is a dragon with duck feet known as a wyvern. An aquatic dragon. I decide to borrow it as my own coat of arms if I ever get a narrowboat of my own. Sebastian reels me back in to the strategic importance of the castle. 'Take a line north-south down the country and another east-west and you'll arrive at Skipton, or thereabouts.'

'I thought Birmingham was the centre of the country.'

'Preposterous. Far too southerly. The centre of the country is Skipton.

It was a crucial outpost against the Scots. And just about everyone came through – there's even an old Roman road just outside town.'

As we continue our tour, I notice that whenever Sebastian talks of restoration work that has been carried out – a new roof here, a bricked-up window opened there – he keeps saying 'We'. 'We fortified this, we raised that . . .' I assume it's a case of over-relating to his job, embedding himself in history like bank clerks do in Civil War re-enactments. Reality can get blurred. He does after all live in an apartment in the castle.

When Sebastian confesses to being 35 years of age, a bachelor and still 'on the lookout', I comment that it must be an impressive pulling ploy inviting a girl after a boozy night in the pub back to *his* castle – and with eight-foot-thick walls, the neighbours aren't going to complain if it gets a bit lairy, are they?

Eventually it dawns on me Sebastian has a closer involvement than just work. He's actually a member of the Skipton Castle lineage, son of the present owner Tom Fattorini. Sebastian is diffident, like an interloper, and constantly plays down the family involvement. 'It's the castle that matters, not who lives here. There's been a fortification on the site since 1090. If I live to a decent age and spend the rest of my life living here, it will roughly equate to a quarter of an hour of Skipton Castle's day.'

How the Fattorini family came to inhabit the castle is as romantic a tale as the Cliffords' reign is dramatic. Baldisaro Porri was one of six children born to Innocento and Antonia Porri living in Appiano, a small town in northern Italy. They owned a substantial property and are recorded as having mulberry trees on which silkworms feed. Following the Napoleonic War there was a severe economic recession and it was decided to send the sons abroad. Baldisaro Porri was apprenticed to a merchant in Halifax named Alfieri. At the age of 24, he moved his belongings by canal to Skipton, married a local girl, Mary Ovington, and started his own business specialising in china, pottery and glass. Their daughter, Mary Jane, married Innocent Fattorini, a Bradford jeweller. Baldisaro built them a fine property with accommodation above a shop on the corner of High Street and Newmarket Street. It remained a jeweller's shop until the death of Baldisaro's grandson, Thomas, in 1934 when the family switched to mail order and the manufacture of badges, medals and trophies out of Birmingham.

The family, however, continued to maintain close ties with Skipton. Hearing of the demise of the castle and its imminent sale, they could not bear the prospect of seeing their home transformed into a funfair or camp site (two of the suggestions being mooted). It was the 1950s, money was tight, and no one seemed to want to take over its roofless state, its damp, and its high maintenance.

The Fattorini family stepped in and over twenty years slowly did what needed to be done – adding roofs, creating a medieval herb garden, restoring the chapel (it had been transformed into a cattle shed). It was a replay of the restoration work carried out by Lady Anne Clifford after the Civil War, when as punishment for being the last Royalist northern outpost, the castle was slighted and had its top third beheaded like the King.

Sebastian whips us through the banqueting hall, and into the 700-year-old kitchen complete with serving hatch, baking ovens, and waste disposal unit (a hole in the wall and 120-foot drop over the cliff). Adjacent is the garderobe where a similar hole provides a similar drop. Once there were five holes and it was here that the family would gather after breakfast for quality time together, 'with nothing but air between their exposed backsides and the canal below'. Together they'd discuss the state of the nation and decide what time Lucy had to be back that night.

'At the time,' Sebastian warms to his theme, 'there was no toilet paper. A tree branch with one end wrapped in moss did the trick – hence the saying "getting the wrong end of the stick". You definitely didn't want to do that.'

Another saying, 'armed to the teeth', was apparently coined around the same time. 'When the Royalists finally conceded the castle at the end of the Civil War, because of their bravery in holding out longer than anyone else, they were allowed full military honours, which meant leaving the castle unmolested with their guns in their hands and bullets between their teeth so they could charge their rifles quickly if necessary.'

We end our tour at the castle shop where I buy a family-tree chart of the House of Clifford and discover a line of serial killers. Owning all the land from here to Carlisle meant that just about every battle that was going, the Cliffords had some stake in. Each generation on the chart ends with a similar obituary – 'Robert Clifford died at the Battle of Bannockburn in 1314.' Roger Clifford, his son, expired at Boroughbridge in 1327. The next generation lost Robert Clifford at the

Battle of Crecy in 1346; John Clifford died at the Siege of Mieux in 1422; Thomas Clifford at the Battle of St Albans in 1455; John Clifford at the Battle of Towton in 1461 (having first earned the honorary title Butcher Lord for his slaughter of men at Wakefield, his decapitation of both the Duke of York and, for good measure, his son, the Duke of Rutland). Eventually the Cliffords ran out of men and the line passed to the House of Thanet in 1659.

By far the most flamboyant Clifford was the third earl, George, nicknamed the Privateering Earl. Queen Elizabeth's favourite represented her in jousting tournaments and had the reputation of a dandy. Eventually he managed to sink the family fortune before finally sinking himself – having subjugated Puerto Rico, he fell into a harbour in full armour.

Sebastian walks me to the castle gate. As we do so, he enquires what I do for a living. Discovering I'm a travel writer, he asks for advice on where he should go on holiday. 'Nothing over £400 for a week though: remember I'm a Yorkshireman!'

At present a Yorkshireman can buy a three-bedroomed end-of-terrace home with central heating in Skipton for £63,950. A Victorian stone terrace in the most desirable area with six bedrooms and two bathrooms costs £159,950. As I peer into estate agents' windows, it's easy to imagine living here. We English may mostly inhabit cities these days, but our idea of Englishness resides elsewhere: with the crack of a cricket ball on the village green, the welcoming door chime of the craft shop, and the cat's meow in a cobbled alley leading to a thatched pub dripping with brass trinkets. And through each idyll, be it village or rural town, a river or canal passes, carrying our cares away. It may be more mythical than real but then winning the lottery is too, and that doesn't stop us.

Skipton scores high in the quality-of-life stakes: a lively market town that seamlessly marries Victorian and Georgian architecture, and sits in truly spectacular countryside. Its intersection with the itinerant canal is equally beguiling with its humpback brick bridges, and handsome wharves and warehouses sandblasted into stylish flats.

Beside the diminutive Plaza Cinema – which has probably never got closer to gentrification than a lick of paint and a new carpet – an exuberant aerobic class is going on in a first-floor gym. Flanking it, gritstone terraces run up the hill, inhabited by young families, older retired couples and the Express Laundry, Gulshan Takeaway, Bodrum Kebabs, and Westmoreland Fish & Chips. Teenagers look a healthy, happier lot here, lacking the glue-pallor of their neighbours.

My first obstacle when I finally decide to move on is the electric lift bridge at the centre of town. As I raise it, delaying motorised traffic and pedestrians alike, a yellow wagtail makes a jerky fly-past. No one beeps their horn or moans. Out in open country the usual distractions come thick and fast: a chat with a pair of walkers, tracking a buzzard through binoculars, strolling a village, and admiring the towpath garden of pink-headed butterbur (so named because the leaves were used to wrap butter in), purple-tufted vetch, yellow bird's-foot trefoil, and red clover. I may be halfway through my journey but I'm getting slower by the day.

At Silsden Marina I pull in for a pump-out and to top up with diesel. The owners, Richard and Barbara Bradburn, have been there six years. Prior to their present occupation, Richard was an architect and Barbara a nurse. Their thriving operation builds and sells boats; then hires them out, provides them with mechanical first aid and fuel, and with decorative trinkets from the shop. Richard suggests I moor up where I am and join them for their local's weekly quiz night. 'Our team's always short.'

We arrive at the Robin Hood at 8.15. The question master, nicknamed Mr Bean, appears already well oiled and unlikely to make the 9 p.m. starting post. Everybody in the pub takes the piss of him, but then everybody takes the piss of everybody else too. It's hard to tell whether the ridicule is to do with Mr Bean, the ethos of the Robin Hood, or a wider regional malaise. I tend to think it's the latter, that Silsden has adopted the banter of the football changing room for its social interactions.

Somehow Mr Bean manages to stagger to his feet and through all 40 general knowledge questions and 20 picture questions. Our group, The Boaters, scores 51 out of 60, which I'm rather proud of. Among other teams registered is one called The Betty Ford Clinic group, and another called Big Boys consisting of six policemen. The winning team pips us by two and a half points and is given a prize of £10 worth of free booze (it might not be much but it gets a crowd in on a slow Tuesday). We would have got even closer to them if our secretary hadn't put down

Naples for Nepal in answer to the question, 'Which country has the highest number of the world's tallest peaks?' Still, we were the only ones to correctly guess that it was Boy George who briefly sang as Captain Lush in Bow Wow Wow.

Sailing under the bridge out of the marina the following morning, I notice someone has scrawled 'I love you Kurt Cobain xx' on the wall. Graffiti endures longer than rock stars or love. Rock and Roll. 'Hope I die before I get old.' 'Better to burn out than to rust.' My own daughter touts such sixties lyrics today, albeit dressed up differently. I meanwhile have inverted their message, finding the prospect of slowly rusting a far more attractive proposition than burning out as I slip through middle age.

Resplendent gardens of modern bungalows saunter down to show off at the water's edge. A spanking new cruiser, *Back to the Fuschia* – all £35,000 worth – champs at the bit next to an elderly 35-foot narrowboat named *Whay Clanks*. A woman in a dressing gown stares out from a cane armchair through her conservatory window, face unmoving.

Soon another swing bridge is attacking *Caroline*, gouging a quarter-inch scratch through her name on the fore-end. Beyond it, a wild meadow ripples in the wind as if running up the incline. I slip into a rubber-tongued rhododendron forest with its occasional brilliant early burst of purple flowers. Up ahead I can see a swan chasing something, neck craning, feet paddling frantically. It appears to catch whatever it's after. There's a frenzied shaking of the head, a dipping of its long neck into the water and then it's quiet again.

I finally get near enough to make out a duckling on its back, its small chest heaving, its beak open and gasping, its neck loose as if no longer firmly attached to the young body. The swan watches it dispassionately and then prods it away from the boat before embarking on more pummelling. On one bank her cygnets watch; on the other, the mother duck averts her eyes while her soon-to-be only surviving duckling nuzzles into her downy pillow. Soon there will be no competitors on this reach for the cygnets' food store and the pen will relax until next season's race for survival.

The struggle ends. I've now seen a swan copulating – I can't quite bring myself to call the frenzied satiation of want, love-making; and I've seen a swan killing. There doesn't seem an ounce of compassion or gentleness to its beauty. A canopy of oaks conceals the murder from the gods' eyes. As if they give a damn. Capricious and vain, the deities were the role models.

The swan is as Dresden-white as the boys that attacked an Asian taxi driver in Burnley a couple of nights back and left him paralysed before they then attempted to throw a one-legged Asian man onto a burning car. As pasty-faced as the British National Party member that got 11 per cent of the vote in the recent election there. Are they too all simply trying to stop outsiders getting into the pantry?

I wonder whether the boys that shot the swan share a chromosome with the swan that killed the duckling. Is white the colour of evil? Malcolm X thought so. But then he was a bastard too.

I'm white and so far haven't killed anybody. But in my younger days I was as bad as any of the boys that have disturbed my waterborne idyll. There were Nazi experiments at the age of ten when, living in Hong Kong, I was part of a pack following a doctor's son who used a syringe to remove fluids from frogs or snakes before injecting them into chicken, pigeon or mouse. There were gang bangs aged thirteen when living in a village in Hampshire. There was shoplifting and breaking and entering both in Hampshire and Wales. I was off the rails.

I have no doubt my sense of disconnectedness could have been directed against blacks as easily as it was against property, animals or girls, given a different social setting. My own redemption came when I left school (still on probation) and joined my brother in London, becoming part of a group that read, talked, questioned, partied and went out with girls. I became socialised into a different world, and my new peers provided me with something to aspire to that was considerably more satisfying than deviancy.

Swans don't read, discuss ethics, or relate to ducks and geese (though I'd say generally they have a better relationship with the latter than the former). They only ever attempt to make the world a better place for their personal procreation and the survival of their direct family. Our own success as an individual and a species is dependent, however, on relating ever more broadly. In evolutionary terms, delinquent boys, myself included, operate at swan level. The fact that I've been fortunate enough to have subsequently loved women; loved the Indian, Assyrian,

Bangladeshi, Pakistani, Algerian, Ghanaian, Japanese, and Caribbean kids I've taught; sat with Tuaregs in the Sahara and Sherpas in the Himalayas, means that I know all stereotyping, all prejudging (prejudice) is down to ignorance, impoverishment and projection.

That night I dream of the pasty-faced youths at Wigan who rained stones on me a few days back. Their faces swim up from beneath the water, hovering a few inches from the surface. The heads have no body and no depth.

Towards the western end of the Leeds & Liverpool, less than a canal-mile from the towpath I trod as a schoolboy each morning trailing my young beauty, is a railway embankment adjacent to the Cut. It was here in the Liverpool suburb of Bootle that Robert Thompson and Jon Venables, aged ten, stoned James Bulger and then used iron bars to finish him off. Their nearby primary school became the first in Britain to have two murderers on its register. The boys spoke of an unreality, a lack of three-dimensionality to the acts. Like flickering video screens, like reflections on water.

Cruelty is not something new. I think of *Lord of the Flies*, and my own childhood when there was a similar detachment from dark acts, the same disconnectedness from suffering, the same sense of unrelating, of seeing things without depth.

I remember a while back asking my own son, Max, aged twelve, whether he was looking forward to getting older. 'Not really,' he replied, 'I have to get jacked all the time and maybe beaten up.' Being mugged for Max is no less a reality than going to the dentist. Bertrand Russell once asked a boy in a playground why he was picking on a smaller boy. 'The bigs hit me so I hit the smalls,' the boy summed up the history of civilisation. The violence is always there, a darkness seeking the surface. The civilising influences of a loving family and caring community (be it a good school, good friends, or tolerant church, temple, mosque or synagogue – Buddhism believes all faiths can provide a route to enlightenment) are to this darkness what the crucifix is to Dracula.

I switch on the radio to news and comment on the racial disturbances and fighting in Burnley, Oldham and Leeds, that has gone before me, or followed in my wake across the Pennines. There's also an item on Nick Griffin, the chairman of the British National Party, who has suggested that a Berlin Wall is erected to separate Asians and whites in Oldham.

Helpful. Can't he remember the celebrations that greeted the dismantling of the Berlin Wall? Someone once thought that was a good idea too. He has also said he wouldn't countenance his daughter marrying a black man. Maybe he wouldn't countenance her marrying someone with a mole on his ankle or with overly protruding ears either.

On the island of Tana in the Pacific, there exists a little-known religion called John Frum. It is an eminently logical faith that sprang up when Tannese returned home after working for the American forces in the Second World War speaking of laden ships and planes. A cargo cult was born around the pivotal faith that one day freight would come to Tana too, courtesy of John Frum America, and the people would have refrigerators, canned food, Coca-Cola and cigarettes. Each month the villagers would gather on the beach awaiting their cargo, before their chief, with a piece of string to his ear simulating a telephone, informed them the landing was cancelled. It'll come next month.

In England a reverse culture is emerging – Ali Takeaway. Immigrants (and immigrants means blacks) are being held responsible for taking our jobs, our welfare, our identity and our corner shops. It is a religion peopled by demons rather than saints, spurred by fear rather than hope, disintegration rather than integration.

I shake off my dream and the murder of the duckling, and approach the sublimely sculpted Bingley Five Rise, a Grade I staircase as beautiful – despite its 227 years of age – as it is ingeniously simple and effective. As the flight is manned, I moor at the top lock where boaters are sunning themselves at a café, and set off in search of the lock keeper.

Barry Whitelock (what a perfect name), who's maintained and operated the flight for 22 years, is leaning over a balance beam watching white breakers racing through raised paddles. He's wearing a B.W. black sleeveless jacket, half-mast green trousers flapping above black boots, a wide leather belt with a windlass tucked in it, and a peaked check hat beneath which bushy sideburns are bursting out like ivy. Barry does a twelve-hour day, 7 a.m.–7 p.m., though his official hours are 8–6. The Bingley Five Rise is his fiefdom. Strolling through respectful gongooz-lers and obedient boaters, he whizzes his windlass, working two 50-footers up through the last lock, while simultaneously keeping a beady eye on proceedings. There are plenty of opportunities for less than vigilant crews to quickly get into difficulty.

When it's my turn to sail into the top lock, the woman at the tiller of my companion boat tells me of a crew she shared a lock with the week before who almost sank. 'There were five men on board all in their twenties. Four got off to do the lock and the one on board went down to make himself breakfast, leaving no one in charge. The boat banged about in the lock and eventually got caught on the sill. It had tilted thirty degrees and water was rushing in by the time my husband managed to close the paddles and save it from sinking.'

As our own gates open, there's a view straight down the black and white flight to a cluster of honey-coloured mill buildings in the valley. Along with more than 50,000 gallons of water, we step our way down the flight, the bottom oak gate of one lock being the top gate of the next.

The gilded spire, lead cupola, and sandblasted mills of Saltaire are just two miles further on. Beside the Corinthian columns of the United Reformed Church, a guide is regaling her group of visitors, assuming the persona of Sir Titus Salt, the textile mill giant who created the workers' village in the mid-nineteenth century.

'You mucky lot keep yer animals out of my town. I'm warning you an' I won't warn yer a second time. You'll pack yer bags an' go.' Maria Glott switches back to official Saltaire guide. 'See, Titus Salt 'ated animals and wouldn't allow the village to become like Bradford where folk poured in from countryside to the new industrial jobs and 'ad nineteen to the bed with pigs on the floor, chickens on the bedposts and a cow in t'other room for milking next mornin.'

'But he was a great friend of the workers,' someone protests. 'The boook I read said he always 'ad his workers' interests at heart.'

'Course he did. And Stalin was a pussycat!'

'Are you saying he was a bastard?'

'The jury's still out. He cared most for money, that's for sure. Outsiders may have been impressed by the sanitation, hospital and school – Ruskin and Dickens both gave talks here so they obviously were – but don't be kidded he was doing it for love of owt but money. He was fifty and a multimillionaire before he did anything about the terrible conditions his workers put up with.'

'They're all the bloody same. Not a clean penny amongst 'em,' the man has his *Morning Star* suspicions confirmed.

'And don't forget he had children aged six and seven working in the mill all hours. He was the Liberal M.P. for Bradford but didn't know about the 1833 Factory Act which forbade the employment of under-nines in factories? I don't think so.'

'Me neither,' the voluble retired postman keeps up the double act. The lively exchanges continue as I tag into the group's slipstream behind Maria who's dressed in black tights, black calf-length skirt, black raincoat, black shoes, and black sunglasses balanced on black Chrissie Hynde hair.

'Did he make his fortune from cotton or wool?'

'Top of the range wool worsteds and lustre cloths, plus cornering the market in mink, silk, alpaca and mohair. The fact that all the money he paid out to workers came back into his own pocket also helped.'

'How d'you mean?'

'See the old hospital? He owned it. See the dining hall? He owned it. Then there was the wash house with the Turkish baths, sauna and washing machines. Workers paid for that too. And of course there was rent from accommodation, there were the shops, and the undertakers... He owned the lot and he didn't pay his workers with hard cash, oh no, he gave 'em tokens which they exchanged for goods and services.'

'Why di'n't nobody complain?'

'They did. When they eventually discovered they were being paid the equivalent of half what the going rate was in Bradford, they went on strike. Salt gave them half an hour to clear out of their homes and leave town!'

'Bastard.' A couple in the group are getting irritated by the attention Frank the postman is getting and try to squeeze their bodies between him and Maria. He accelerates from the rear and whips back beside Maria who doesn't appear to have noticed anything.

'They backed down of course. This was a man who never gave owt for nowt. At dinner there were three sittings of seven hundred and fifty workers a time. One penny for a bowl of porridge and a glass of milk. Twopence for meat and potato pie. One ha'penny for a cup of tea. He even built underground tunnels from the dining room and school back to the factory so no production time was lost.'

By now we're walking along a neat two-storey Victorian terrace. Someone notices some doors have more panels, and some windows more panes, than others. 'Salt built eight hundred and fifty houses in all, in wide gas-lit streets, no back-to-backs mind, and each with its own toilet and inside water tap thank you very much. How many panes and panels you 'ad depended on whether you were an Ordinary Worker, Improved Worker, Decorated Worker, Improved Overlooker, Executive Overlooker, or Improved Manager.'

'Sounds like *1984* to me.' Frank is off again. 'Maybe Saltaire was where Orwell got the idea? He were in the northern industrial towns a while weren't he?'

'He was but I don't know if he visited. But if he did, he would have noticed there were taller houses at the ends of each street. This was where the senior management lived and kept a beady eye on things. There was no police force, Saltaire was self policed from lookout towers where young men kept a check on things: that you weren't 'anging out your washing or bringing in animals or loitering in groups – no more than five could gather together at any one time. Anyone caught would be arrested and fined.'

The flagstones and cobbles lead us to the Titus Salt Memorial School, now the Saltaire Studies Centre of Shipley College. Each building – the hospital, the office, the Sunday school, the institute and the mill – is built on grand Italianate lines. The Mayor of Bradford, Samuel Smith, declared Saltaire a 'palace of industry almost equal to the palaces of the Caesars'.

'Titus was everywhere. Like Big Brother. Look at these railings.' Maria points to the 'TS' at the top of each rail. 'And look at the gateway.' She points to Salt's head hovering like a Gorgon over us. 'He even assumed the right to go into anyone's home whenever he wanted and if the lady happened to be having a bath in front of the fire and Titus walked in with Prince Albert well, tough, you stood up anyway.'

Salt does not appear to have been prudish despite occupying the Victorian moral high ground. Like his philanthropy, this was probably down to pragmatism and economics more than anything else. 'True he banned pubs from his town but not because he was fighting against the demon drink, but because he feared the planning and plotting that went on in such places.' Instead he sold alcohol and opium in his pharmacy. As to his personal sexual mores, he appears more libertine than prig (in other words a hypocrite). When he died in 1876 he'd fathered eleven legitimate and countless illegitimate children.

The tour winds up, the group return to their coach, and I wander up to Salt's Mill, six stories tall and the length of St Paul's Cathedral. In the 1850s it was considered the longest and largest room in the world. The building, bizarrely, now houses the world's largest collection of David Hockney works. The lower storey of the 1853 Gallery is vast, uncluttered and flooded with light. Stravinsky's *Rite of Spring* is playing to an aircraft hangar full of attentive Hockneys. It's one of the most beautiful spaces for exhibiting art I've come across and yet just two decades back the building was a wreck.

With the decline of the textile industry, the mill, like the rest of Saltaire, was in a parlous state. Lead had been stripped from roofs and many buildings were closed and abandoned. But in 1984 the railway station reopened making Saltaire suddenly a desirable commuter suburb just ten minutes by rail from Bradford and twenty from Leeds. In 1985 the Saltaire Village Society persuaded English Heritage to list the town. And then in 1987 Jonathan Silver, a self-made millionaire who'd made his fortune through a chain of clothing shops, bought the mill. He let part of it to Pace Micro Technology, securing its financial future, and opened the David Hockney Museum on three of the floors. Salt must still be turning in his mausoleum at the overtly homosexual nature of some of the artist's work, and even more so because admission to the museum is free.

Hockney, who was born in nearby Eccles Hill, was already acquainted with Silver prior to the museum project. Jonathan went to the same Bradford grammar school that Hockney had attended a dozen years before him. While editor of the school magazine, Silver got hold of the artist's phone number in California and asked him to design the cover for the next issue. Hockney admired his chutzpah and they met for the first time in a Bradford Wimpy. A long-term friendship was established which only ended when Silver died of cancer in 1997 at the age of 47.

Hockney is still heavily involved with the museum. As I pay for two postcards (one of the artist, and the other of Salt's Mill painted by Hockney in 1999), an assistant informs me that the artist was in just three days earlier. 'He's very deaf now though so you have to shout a bit but he still looks great.'

On the upper floor, things are darker, the ceiling lower and the gallery more intimate. This is where Hockney's various sets from his stage collaborations are on show. The character cut-outs and shop backdrop for *Rake's Progress* are contrasting dense and light interconnecting lines, crosses and squares; the *Rossignol* set is washed in blue and peopled by characters who appear to have walked off Chinese porcelain. There's artwork for *The Magic Flute*, the poster of a clown doing a handstand for *Parade*, as well as a number of the extravagant costumes Hockney designed.

This museum, hidden away in a northern industrial village, is an absolute gem. It would be impressive in New York or London but here it's audacious. Saltaire can in no way be accused of creating a make-believe world, another mothballed village on the industrial heritage trail. This is a genuine Victorian town that's had new life breathed into it. Salt's Mill staged 31 major cultural events between 1987 and 2000; and Pace and Filtonic Comtek have brought new jobs.

The town's grid of streets may have been copied from Chicago and New York but it has more a feel of laid-back New England with its leafy avenues, antiques and fabric shops, tapas bar, bookshop and tea houses. When Maria Glott bought her terraced home here a decade ago she paid £23,000, now it's worth £110,000. Show flats in the recently converted hospital are priced at £140,000.

On my way back to *Caroline* I call in at the Saltaire United Reformed Church (formerly the Congregational Church) where I first ran into Maria's tour. The bald pate of the organist is shining like a halo beneath an overhead light directed onto his musical score. The curtain to the mausoleum where Titus Salt and his wife are interred, rattles backwards and forwards on its metal hooks as visitors step in to take a peek. Suspended above it is a massive gilt chandelier made in Liverpool and transported along the canal.

There is an unmistakable musty smell of churchness to this place, of a century and a half of kneeling and prayer. Backs of pews worn smooth by shuffling hands and backsides. Six grey-haired, bespectacled grannies on a Women's Institute outing rest their bones and listen to the organ. Two or three of them close their eyes transported on the wings of notes that soar and then flutter in panic as they find themselves trapped by the ornate vaulted roof supported on its dark scagiola columns.

Bert Thornton flits like a butterfly, white wisps of Bobby Charlton hair, mustard-brown patterned sweater stretching over the hillock of his belly, the collar of his polo shirt trapped pointing heavenwards. As he approaches, he catches me staring at his limp. 'Only got one good leg now. The other one went at Normandy.' Bert has been a regular church attender and helper for eighty years and claims to have known Hockney's Aunt Phoebe. 'At the Methodist Chapel in Bradford she once confided to me when the children were small, "One of me nephews will make something of himself, but the other is going nowhere; he's forever..."' Bert makes an extravagant drawing in the air, imitating Aunt Phoebe imitating Hockney.

Bert remembers that during his own childhood, the canal was all action. 'In the thirties the bargees carrying coal and wool for Shipley Canal Carriers used to let us ride on their horses occasionally.' Shipley, which Saltaire abuts, was a major transhipment point on the Leeds & Liverpool Canal. 'The boat people mostly lived in New York Street just half a mile from here. The Beehive was their pub and there were often scraps. They got roaring drunk and someone got murdered once.'

In contrast Saltaire was a very safe place. So much so a bookie Bert used to work for would leave any winnings owed to punters under his piano if he had to go out. No money was ever stolen. 'Not like today,' Bert shakes his head. Here we go again, I think, but Bert is no rabid bring-back-the-birch evangelist. 'There is more crime about and disrespect today but what I'm most worried about is the increasing lawlessness and isolation of the Muslim community. They don't want to integrate. There are schools now ninety-five per cent Asian here. We had waves of Poles, Lithuanians, Russians and they all became part of the community. The government don't realise how big a problem they're going to have. We complained about the kids from Shipley College – that must be eighty per cent Asian too – dropping their lunches outside the church and just got a letter back from the head saying we were being racist. My wife used to be head teacher at a village school and she said that in the past there had been a mix across the schools of nationalities, but now the Muslim leaders insist they all go to the same schools. Oldham, Leeds, Wolverhampton – they've all got the same problem. That's why the race riots are happening.'

'I guess if the government is backing more church schools, then it has

to support all-Muslim schools too in our multifaith society,' I raise the
obvious stumbling block to single-faith schools.

'Maybe,' Bert concedes. 'But whatever, it's not working. People aren't
getting on and something isn't working.'

Bert is brandishing a cassette which he says he's just received from his
daughter, a missionary in Peru. His smile wrinkles out from his
spectacles like ripples across a pond as he squints, seemingly attempting
to look inside the cassette for his daughter. 'She's singing on it
apparently.' The organist has stopped playing and Bert shuffles off to
put it on. Soon the notes of the organ are being chased out of the church
by pan pipes. The haunting sounds of the Andes remind me of a
cathedral I visited in Cusco where a painting depicted a black Jesus
carrying a llama instead of a lamb. Maybe there's some Bradfordian
alignment possible with a hybrid Mohammed and Jesus inside the
churches and mosques.

At nearby Rodley I discover the best fish and chip shop in the country.
It may be the size of a shed but the haddock and chips, cooked in beef
dripping, are out of this world. They're served with mushy peas topped
with mint sauce – 'The Parmesan of the north that is,' the barrel-shaped
owner assures me. 'Want Mars-Bars-in-fritters for afters?' I decline,
fearing a hugely enjoyable, but fatal calorie-overload.

'Up to you lad, but if you're worrying about getting fat, don't. There's
hardly any calories in them at all.'

I laugh.

'No, really. Them anorexics might tell you it's fattening but that's
just rubbish.' I look at his frame and decide to side with the anorexics.

Fortified, I slip past a line of waterfront homes, one of which has
four white Buddhas – bearing more than a passing resemblance to
Mars Bar Man – sitting out on its deck. Two Buddhas face the
home and two face the canal. At Newlay Three Locks I'm assisted
by a young muscular B.W. worker called Johnny, a Marlon Brando
on the waterfront. Johnny tells me that five days ago local joyriders
torched two cars and then pushed them down the hill through the
forested woods. 'They stopped no more than twenty yards from
the canal.' When the police arrived, 'The kids torched two more next
to the Cut.'

Johnny knows the gang. They come down to swim in the lock regularly. 'They raise the paddles and swim through them into the lock and call it their private swimming pool. They're usually out of their brains sniffing glue. If a boat arrives, they may or may not let them through. Depends on how they behave – the boaters I mean. One woman tried lecturing them and got pelted with stones, then they hung their arses over the bridge at Kirkstall and crapped on her boat.'

'Have you had any trouble yourself?' I ask, as we raise the paddles on the second lock in the staircase.

'Been threatened by one group once but I swung my windlass and said 'Yes, you'll get me but before you do, one of you will end up in hospital, I guarantee you that.' They didn't try it. Mostly I talk to them. I don't try to stop them swimming. It's pointless.' Carved into the balance beam I'm pushing is, 'Rachel 4 Snowy. Untill death'.

At Forge Locks, another staircase of three, Johnny's friend Danny, assists me through. He points out where the kids had dropped another car through the woods, a flaming comet plunging for the canal. He shows me the pellet marks left in his neck after an airgun attack, as well as those still embedded in the balance beam.

'We had two new huts brought in recently for us workers. Within a couple of days they'd trashed this one and torched the other.' Danny claims things were considerably worse five years earlier when gang wars were raging. 'We had helicopters, riot police, the lot. Trouble is you set a finger on the yobs and you're for it. They walk away from everything. We had police out not so long back and the kids were pushing them out the way to dive in the lock. They know they're untouchable.

'One little sod chucked a rock at the dog – only had three legs and it came in yelping into the hut and Greg, its owner what works with us, went out and thumped one of the lads who was grinning. The boy went to the police and a squad car came round and said, "Now what's been happening here?" Greg says, "I thumped the lad because he chucked a rock at me dog." Policeman says, "No, you didn't thump him, you pushed him." Greg says, "No, I thumped him." Policeman says, "No, you pushed him." Greg says, "I thumped him." Eventually I caught on and said, "No, you pushed him Greg." And policeman says quickly like, "That's a good job otherwise I would have had to take you in".'

'The lad's dad came down next day and asked what the hell was going on. I told him straight. "You need to get the story the right way up, mate. Your lad here chucked a rock at Greg's dog. Look at the bruise on it." The dad asked the son if it were true. The lad went quiet and he got kicked up the arse and told he'd be getting a lot worse when he got home.'

Just before I leave him, Danny points to two estate blocks ahead, to the left of us. 'That's where a lad dropped a lump of concrete from the roof onto a woman's head. Got put away for murder.'

I am relieved to arrive in revitalised Leeds, just a couple of miles from the badlands, without having been torched, crapped upon or stoned again.

I sit in a gorgeous towering Victorian covered arcade with its plasterwork friezes and coloured glass. I know I'm back in a regenerated waterfront city by the £2 price tag on the moccaccino. On my left is the first Harvey Nicks to open outside London and on my right Karen Millen and Hobbs. Sitting at one table is an affluent-looking Asian family, and at another two young beautiful white things speaking into their mobiles while simultaneously peeking at their respective purchases in each other's rustling shopping. We are sitting just a very short joyride from the forgotten estates of the badlands. Where once there were winos, hookers and muggers lining Leeds' waterfront, now there are fashionable restaurants, bars and hotels.

That evening I put on my best black jeans and a creased shirt and make for the downstairs bar of the Malmaison, one of two hip business hotels in the city (the other is 42 The Calls). After one drink I swap its cool cavernous darkness and thirty-something perfect bodies, and cross the street to a brash sports bar where a bouncer eyes my trainers suspiciously but finally gives the nod that tells me my money is welcome even if I'm not. Inside, a hurricane of voices is blowing. Talk, TV, music, all at once equals culture shock. I feel like I've just stepped out of a hayloft. Girls make moves dancing provocatively round boys who only have eyes for two TV screens showing St Helens drawing 12–12 with Wigan.

I give up trying to cope and move on again, making my way to the Prince of Wales, a traditional boozer, ten minutes walk away where a 60-year-old with a curtain hook through an ear and sporting a John Lennon leather cap is patting his knees to 'Love Supreme' by Robbie Williams. The walls are yellow-orange and on an upper shelf is a maharaja's elephant, Chinese bowls, a model of the Titanic, Toby jugs, a

German beer stein and a stag's head.

The Cut has done its trick again, linking the hang-out of the stylishly cool and wealthy to the domain of psychotic youth, and on to the traditional haunt of bohemian old timers. The canal network is not just an extended linear park, it's a cut both through rural and urban England, and across every class and type of society.

At 9 p.m. I treat myself to a pricey meal at Rascasse, one of the city's best restaurants. Over fancy parcels of halibut, I stare out the large window across to *Caroline* moored opposite at Granary Wharf. Behind her, the sun is setting spectacularly above the railway station, a golden comb sweeping through mares' tails. It's my farewell meal to the Leeds & Liverpool Canal. When I get back to *Caroline* around midnight, it's still not fully dark. City lights rake over the water as if searching for someone.

6

Jailbirds, Songbirds and Freight Carriers

Ancient and Pre-industrial Navigations: the Aire & Calder, Trent, Fossdyke, Witham and Soar

Lifting the lid of my sliding bed with one hand, I feel with the other for an envelope stuck to the underside of the half-inch plywood. I maul the sticky tape back, remove the empty envelope and replenish my secret stash with ten £20 notes. As I return it, I notice, lying on the floor inside the empty box, a screwed-up piece of paper upon which the following is scrawled.

> Here lies the body of Michael O'Day
> Who died maintaining his right of way.
> He was right, dead right, as he sailed along
> But he's just as dead as if he'd been wrong.

Is it an omen? Fellow boatmen and a lengthsman have already warned me that on the Aire & Calder, which I join today, there is only one rule, and it isn't keep to the right; it's get out the way. So why did these rhyming couplets choose to slip into my life on the day I join Britain's major freight-carrying channel instead of when I was pootling the Caldon?

On working navigations, freight captains take precedence and their preference is to hug the outside bend – be it the right or the left – to avoid running aground. It's up to me whether I manoeuvre accordingly or follow Michael, 'Who died maintaining his right of way'.

Discovering the note is a little unsettling. Who wrote it? Where was it copied from? And why did I discover it now? I decide meticulous preparations are in order and go up on deck to check the anchor is

182

stowed with easy access in the fore-end. I also replace the short stern rope with a longer one, and make sure I have six mooring pins on board. Unfortunately I haven't found a timberyard where I can replace the long pole that vanished one night and which I'll be in need of in the event of running aground (my remaining pole is just eight feet long). Finally I remove obstructions from the roof (there are a couple of low bridges coming up), and warm up the engine.

I slip through Leeds' reborn waterfront, passing the stylish hotel warehouse conversion of 42 The Calls, the Michelin-starred restaurants, and the glass tube stuffed with assorted weaponry at the Royal Armouries Museum.

Approaching Leeds Lock, I skip the left-hand fork which would lead to a white-knuckle free fall over the weir. I moor instead reasonably close to the lock gate, but not so close that *Caroline* gets sucked down with the 690 tons of water that's required to fill the lock. The first set of traffic lights I've encountered on the Cut is set on amber, indicating a) an altogether different species of lock, and b) that it's too early in the morning for it to be manned. Good. It means I'll be able to get the hang of the monster locks – three times the size of the double locks on the Leeds & Liverpool – without anyone witnessing cock-ups.

I saunter up, slip my B.W. Yale key into the electronic control panel and follow the written instructions. 'Open sluice gates'. I press the appropriate button. Bingo. Whirlpools swirl, and the water surface shudders as if monsters are stirring underneath. Eventually the water calms and I press the 'Open gates' button. Bingo twice. This is child's play. The 64-metre lock may be huge compared to what I'm used to – big enough for a 600-ton tanker barge and a couple of push-tow coal pans – but there's no working the windlass nor straining against balance beams required. (In a couple of days I'll have to negotiate the Pollington Lock which is twice the size of the Leeds Lock and can comfortably accommodate twelve 57-foot narrowboats at one sitting!)

I return to *Caroline* and sail her into the lock, remooring close to the exit gate. She looks lost in the vastness, an audience of one in the Albert Hall. There is a delicious feeling of being part of some secret society that a B.W. Yale key grants access to. I walk back fifty yards to the first gate, close it, drop the sluices, retrieve the key and then repeat the whole performance at the other end. When the exit gates finally open, I bow-haul

Caroline through, retying her just outside the lock. I then shut the gate, drop the sluices, remove the key again and sail off. The whole operation takes around half an hour and is totally gripping. Over the next few days I'm sure I'll refine things but the principles are straightforward enough.

The navigation is already twice the width of what I'm used to but as it's Sunday and most freighters don't work weekends, it's quiet. At Worsley on the Bridgewater, I may have reached the start of the Canal Age but now I'm journeying back further in time to rivers that were widened, deepened, straightened and made accessible considerably earlier.

The Aire and the Calder rivers were first made navigable to Leeds in 1700, accessing the Yorkshire coalfields and establishing the landlocked West Riding woollen industry. They still carry more than two million tons of coal, sand and petrol a year. But today, equally importantly, the Aire & Calder Navigation is writing a new chapter in the unfolding saga of England's inland waterways as the eyesores of exhausted mines that line its banks are landscaped into wildlife sanctuaries.

Having passed a number of ex-collieries and capped spoil heaps at varying stages of repopulation, I sit on deck at Woodlesford Bridge, trying to track bird calls to owners. Prize for the prettiest goes to the song thrush. Common tern are zipping down the green corridor above a legion of fishermen and twitchers as I ask a man, bins bouncing on chest, if he's ever spotted anything rare here.

'Two spoonbills last year. Nobody believed me at first. They hadn't been seen in the area for forty years. But they put it out on the radio anyway and everyone came down and saw for themselves.'

A group of ten fishermen, with enough gear between them for an Antarctic expedition, are setting up for a competition. I continue munching cornflakes, a new lunchtime fad, and watch a charm of chattering goldfinches feeding on thistle heads in the hedgerow. In front of *Caroline* is a line of long-term moorers with barbies smoking between their parked boats and cars. A swan approaches and asks for leftovers but leaves empty-handed. As far as I'm concerned its race still has to expiate a recent murder.

It's the hottest day so far. I strip my Vialli T-shirt off and am down to shorts and sandals as *Caroline* ambles past more grassed-over spoil sites, standing like giant burial chambers beside the bank. At Lemonroyd Lock, the traffic light again is on amber indicating I'm on my own. The

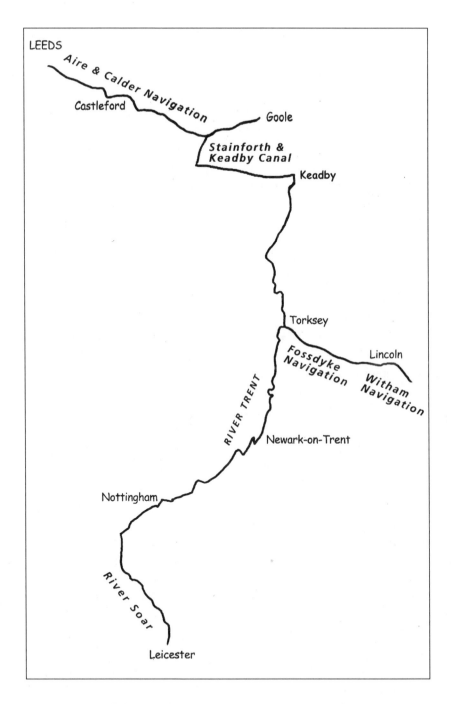

Freight-carrying and Pre-industrial Navigations

fact it's not flashing red also confirms the recent floods are thankfully consigned to history. Locking down is once more a protracted but relaxed affair that climaxes with me bow-hauling *Caroline* out the far gate. Well, she is fifty.

Adjacent to the lock, a bridleway skirts a two-mile lunar landscape known as St Aidens opencast mine. It looks as if a large comet has crash-landed into it. In fact it's the land itself that crashed rather than an alien body. In March 1988 St Aidens collapsed and the River Aire flowed backwards into the depression for half a day filling the mines. Eventually the warrens were pumped out and miners became moles again. They shouldn't have bothered. Now that the mine is virtually exhausted, the plan is to flood it once again over the next few years, transforming another toxic, hostile environment into a shimmer of reed beds, grass-land and open water.

In the hour it takes to reach Castleford, I pass more than a hundred swans. Some are preening themselves, some are hanging their wings out to dry like washing, others snooze with their necks coiled like boat ropes on their backs, some tuck a tired paddling foot under a wing, and still others appear headless as they trawl beneath the water. There are no geese here, no ducks: it is swan heaven.

I try to identify the flapless flight of a bird doing extravagant loop-the-loops above a sloping field. It turns out to be a remote-control model plane. A tern soars, frantically beats its wings to maintain a stationary hover, and then plummets like a gannet straight into the water from twenty feet. It repeats this three times before it climbs away with a struggling fish in its beak.

At Castleford, a dreary, sprawling town, I moor at the junction before the flood lock, and explore a network of paths on my bike. Eventually they lead to Fairburn Ings Nature Reserve where I ask the first man with a name tag on his shirt how the reserve got its name.

'Ings is a Viking word denoting a water meadow subject to seasonal flooding.' The name tag pronounces the man to be Simon Stennett. 'By draining the meadows, the Vikings made it possible to grow hay and graze cattle.'

Simon appears to be in his early to mid-thirties, has black floppy hair and a small silver earring. I comment on his southerly accent. 'I'm from Kent originally, a country boy.' As a country boy Simon used to steal off to the woods at every opportunity, climbing trees, investigating anything and everything. 'I've never really grown out of it,' he confesses. My fiancée says I'm still a big kid. Probably right. But now I get to drive around on tractors and use chainsaws to cut down trees which is even more fun… and I've got a seven-hundred-acre garden to do it in.'

The R.S.P.B. Site of Special Scientific Interest is Beauty and the Beast in one: a refuge of splendour and conservation built on a slag heap. Covering three exhausted mines, the area has become a major nature reserve with 251 different birds spotted by 60,000–80,000 annual visitors.

Armed with a leaflet I leave Simon and make my way along a boardwalk through a swamp. A flock of young starlings wheel above a cluster of yellow irises as I come upon a group of kids belly-down on the boardwalk, pond-dipping for tadpoles. I tiptoe through their excited whispers and a cloud of common blue butterfly before quitting the boardwalk and crossing a meadow ablaze with red campion, meadow vetchling, trefoils and clover.

I climb a steep incline of barren red shale formed by an impoverished mix of slates and clays – the by-products of a century of coal production. Once there were several hundred acres like this here but the reclamation programme added limestone subsoil and other nutrients to the acidic coal spoil to create a new Eden out of the industrial hellhole.

From the top of the hill, I can see thirty miles and more in every direction. The immediate vista sweeps from the soft muddy edges of the old coal-washing lagoons where a lapwing and a little ringed plover are wading, to a grassland fringe ringing with skylarks, to a pond inhabited by a tufted duck and her chicks. Beyond the reserve, looking as far to the northwest as I can, the brooding Pennines slumber. To the south are the towers of a chemical factory rising up like a set from the film *The Fifth Element*. To the east are the steam plumes of power stations – Ferrybridge, Eggborough and Drax (reputedly the largest coal-fired power station in Europe). Linking them is a patchwork of arable farmland and grazing pasture.

Back on *Caroline*, we dawdle together past warehouses, gantries, the Castleford Tigers rugby league ground, and the steaming towers of the chemical works. Beside Hargreaves dry dock I catch a glimpse of my first Tom Pudding, a floating container that when linked into a train and pushed by a tug, was the Aire & Calder's response to railway competition in the nineteenth century. Nearby, kids are leaping extravagantly from a coal staithe twenty feet above the water. I photograph them. They wave. One astoundingly white boy stands on the lip of the chute and shouts 'Wanker' across to me while speedily jerking a cupped hand up and down at groin level before he hits the water.

Once out of Castleford, the river swells and becomes increasingly rural. Every half-mile or so, a barn-owl box perches on a twelve-foot-high post. Over the past fifty years, because of agricultural intensification, the bird's national population has plummeted 60 per cent. The stretch of the Aire & Calder between Castleford and Goole has been selected by B.W. for its first barn-owl regeneration scheme. The banks of the waterway provide a linear foraging area with rough grassland margins where it's hoped the birds will flourish. If it works, the programme will be introduced elsewhere.

As the navigation broadens, a cruiser buffets me: the man at the wheel oblivious, his wife sleeping with her dog on a rug behind him. The scattered rocks, the hundred-yard span, and muddied beaches remind me I'm on a river. In fact it most reminds me of the James River in Virginia which my family, together with an American cousin and an ice-box crammed with beers and food, once rode on six truck innertubes.

Along the banks, debris has been snagged high and dry eight feet above the retreating flood line. On a long, gentle hill three children roll downwards towards the water as if in slow motion. Another lazy Sunday. A handsome blue narrowboat by the name of *Albion* passes. I remember it overtaking me several days earlier on the Leeds & Liverpool. The skipper shouts something from the stern but it's drowned out by our two engines. His wife is holding up a large striped parasol against the summer sun. She waves, too, with her free hand. It already feels considerably further than a couple of days from the Leeds & Liverpool.

Vast brick bridges and the cooling towers of Ferrybridge appear, as striking in their way as Stoke's bottle kilns. Twenty black floating containers (the modern equivalent of the Tom Pudding), shaped like high-sided sardine tins, are waiting in a line for Monday morning beneath a gantry and hoist connecting to a zigzagging coal conveyor. Round the corner, someone is being shown round a new waterfront showroom bungalow. Behind him, bells ring out from a sooty church tower as the river branches to the left, and I fork right for an overnight mooring beyond Ferrybridge Flood Lock.

It feels like I've only just got to sleep, when *Caroline* is shaken as if caught in an earthquake. In the buffeting, I try to remember whether I used mooring pins or a fixed ring. I lift the curtain above the bed and peer out the window into darkness. The moon picks out a seemingly endless train of funereal compartment boats moving in slow motion towards the lock. Eventually, the tug pushing them passes but there appears to be no one at the helm. It's 3.45 a.m.

The following morning, still uncertain whether the water-train driver was corporeal or spectral, I wander down to a nasty parade of shops. The bakery is stocked with ghostly white loaves, and a few 'essentials' – tins of beetroot, dried peas, and processed peas; jars of Branston Pickle and salad cream. I return home empty-handed.

It's a pity I'm not as uncomplaining as *Caroline* on her diet of fuel and oil. As I top up the latter, a brown hawker damselfly, maybe five inches long, lands on the handrail. In the water below, a whirligig beetle ripples the calm while two more dainty damselfly munch on gnats. The canal is a pool of life as busy and sophisticated as London but only those slow enough enter the gate.

I'm just about to sail under Gagg's Bridge when a tug pushing three coal pans in the opposite direction zips round the bend and beats me to it. I bring *Caroline* to heel with a quick burst of reverse, drop into tick-over and hide in the lee of the bridge, timing my re-emergence perfectly to waterski over the coal pans' wake. The tug skipper raises a finger. I'm not sure if it's thanks, congratulations at the fine handling, or the cricket umpire's signal that I'm out.

Yesterday's expansiveness is now more circumspect, this narrower canalised section defined by steel piles driven into the sides of the Cut to resist the softening of the bank through vegetation and subsidence. Beyond Skew Bridge is a line of fine old rust buckets and a vanishing sign 'John Harker Shipyards Marine Engineers'. Next door to the tug graveyard, a man in blue overalls peers out of a matching blue factory door and shouts across to me 'Can I coom too?' Beside Kellingley Colliery ('Tomorrow's energy standards today') red and yellow containers are trundling on rail tracks up to Eggborough Power Station.

A blue and white barge named *Sobriety* speeds past, washing fluttering stiffly in the breeze. Next comes the *Brocodale* from Hull laden with sand. Every passing skipper gives me a wave.

The canal cuts its way across flat arable farmland. Fortunately the water's deep – thirteen foot six at centre – and so *Caroline* positively zips along the rather tedious stretch. After a while, a high-speed train bound for Edinburgh hammers past at around 130 m.p.h. and I'm reminded what zip really is. Soon we're running alongside the Dutch River, an artificial channel cut by drainage engineer Cornelius Vermuyden in the mid-seventeenth century, as part of a web of drains and dykes to protect the area from serious flooding and make the River Don more navigable.

The days are settled and warm. Today is the first day of Wimbledon and apparently a ball girl fainted and fifty spectators had to be assisted because of 95-degree temperatures. It's not that hot up here but my tan definitely has more to do with sun now than weathering.

At lunchtime I moor up at Rawcliffe beside the Black Horse. A group of eight friends in their late twenties or early thirties, with unisex pony-tails, tattoos and black T-shirts, step aboard their barge, returning from a satanic ritual or the pub. One of the group is a bald-headed black guy dressed in cowboy boots. I realise with a shock that he's possibly the first black boater I've seen on the entire trip. Does this put paid to my theory of the Cut being a slice across society?

When the canal officially opened on 20 July 1826, the Black Horse was a farm. Three years later, seeing the rise in passing trade, the farmer converted it into a pub. Not long afterwards his wife died, and he hung himself from one of the beams. That's as much of the history as the current landlord, John Sherrif, knows. John's far more interested in talking about his sea-going cruiser tethered next to *Caroline*. He had the 43-foot, 30-tonner built in the late seventies, and it's been sitting out there pretty much ever since apart from the occasional day jaunt. For thirty years he's been promising himself a proper cruising holiday but for some reason never gets round to it. My guess is he never will.

Alongside John, behind the bar, is a vast belly separating a dishevelled T-shirt from a pair of burgundy Nike shorts. The extravagant girth belongs to John's son-in-law, Tony. For the lion's share of the past eighteen years Tony has lived abroad, occupying a succession of engineering placements through the German company he works for in the world's great coal towns – Pittsburgh, Katowice, Newcastle (South Africa) and Calcutta. Now he's back in his home village of Rawcliffe.

'My eldest daughter speaks with an American accent and my youngest's first word was *danke*. Just a generation back my father never got further than sixty miles from the village.'

I finish my pint but don't leave until Hendrix does. Bizarrely he has followed Tom Jones's 'Delilah' and Sinatra's 'My Way' over the speakers, transforming the Black Horse into the Red House.

Back in daylight, I discover the band of gypsies on the barge have left. I follow suit and just half an hour away, arrive at my easternmost terminus, Goole. In England's largest inland port, moorings are at a premium among the tugs, lighters (unmotorised freight carriers), oil tankers and compartment boats. I ask a bearded man in red overalls working on the *Little Shuva* where there might be space and he points me to the diesel pump outside the marina, 100 yards further up.

As soon as I'm done securing *Caroline*, I walk back along the towpath for a chat. The man with the salt-and-pepper beard and matching shoulder-length hair is now wearing metal-rimmed shades against the sun. Next to him is a woman with an American road-movie hairdo and dressed in identical red overalls to her husband.

Graham and Leonie Acaster appear to be fixing a farm plough to a hoist. The ploughman of the river explains, 'It's for movin' lumps in the river where silt builds up. I run about over 'em wi't plough and disturb 'em then the ebb takes it away.'

'Like a kid playing in the mud,' Leonie provides an alternative perspective.

'I only do it because there's no cargo to move,' Graham attempts to elevate play to work. 'We can't move cargo if we've got nowhere to take it, can we?' It's the same complaint I've heard everywhere concerning B.W.'s earlier haste to sell off moribund assets that has resulted in a waterway system without depots to load or unload at.

Graham, I discover, built the plough himself the day before with his eldest son, Kal. He also converted the old Thames tug we're standing on into a shover (hence *Little Shuva*) to ferry-load pans about. The most surprising thing about this is he does it all with one hand. The other one he lost, aged twenty, when he left it between two barges. 'Got trapped between them like a fender,' he smiles.

'It's only a problem if there's something he doesn't want to do,' Leonie comments, 'like paperwork or housework. Oh, and picking up a brochure to read – we haven't had a holiday in thirty-two years!'

Graham grimaces. It's a domestic wound that's never going to heal. 'I'm on holiday all day, me,' he fends the jibe off as he always has. Leonie smiles indulgently, resigned to the fact that off the boat there are things that Graham refuses to fix.

They finish attaching the plough to the winches. Leonie, who's worked with Graham on the river all her working life, presses a switch and the plough rises with a few shuddering jerks into place. Graham's own family have been carrying cargo from Goole ever since the basin was established in the 1820s.

'As a boy I used to watch everything still coming in by boat – glass, china, sand, wool, pulp. I thought it would go on for ever. I thought then nothing changes. But you do as a kid, don't you? The biggest mistake our country ever made was turning our backs on our Commonwealth partners and joining the European market – once we did that, everything came in from Europe in great containers and that was the beginning of the end.

'I first converted *Little Shuva* when a job I took over from me dad, carrying fifty tons of paper to York weekly, went up to three hundred tons. The company, Westminster Press, were big believers in water transport. Then they got bought by News Quest who weren't, and the carrying was switched to the road. York's supposed to stop lorries entering its centre but it gave way. Money speaks. Holmen, the Scandinavian parent company, built two big roll-on-roll-offs, and they started bringing the pulp in containers from Denmark and Sweden. They'd sail past us down to Tilbury in London and load up the lorries that'd then hammer up the A1 past us again to York to their warehouses fifty feet from the water. Took twice as long and was ten times as polluting.'

Alternative cargo isn't easy to find either: 'Effluent-carrying from Leeds stopped four years back; there's some fuel-carrying to the city but most goes by road. There's still coal to be carried to the power stations but larger contractors handle that, and there's much less of it these days anyway with the demise of the industry since the 1986 strikes and closures. There's some aggregate-carrying, a little sand, gravel, that kind of thing, and that's about it.'

Graham's youngest son threw in the towel after the York paper run ended. His middle son came aboard for just one year and now works in an office at the Drax power station. Now there's just Graham, Leonie and Kal, who came on board after three years at nautical college. There are days when they, too, come close to throwing in the towel, bringing down the curtain on five generations afloat. But they don't.

'We get cheesed off and moan, like we're doing now, but we hang on because we love the life.'

'Graham couldn't leave anyway because he wouldn't know what to do if he didn't have me to boss around – after all I've only been doing it thirty-two years so I still need to be told what's what!'

'Only at work,' Graham laughs.

They were teenage sweethearts and they still appear to be. They must be to spend 24 hours a day together for 32 years. When I ask what needs to be done to create more work for the independent freight carrier, Graham shakes his head vigorously to remind himself first and foremost not to raise false hopes. He's been here too often. 'Needs to be like it were in the 1940s with distribution points for local industries and incentives to businesses to come back to the water. But it won't happen. The government and B.W. are only interested in the leisure side of things.' He points to the long line of rust buckets – tugs, lighters, tankers, barges – on the opposite bank abandoned, taking up space instead of loads. Rust rusting. 'Most of them will never work again. The owners keep on the moorings hoping something will change and they'll be able to come back to the water. But they won't.'

Graham points to two original Tom Puddings moored outside a modern single-storey building. 'You should pop into the Sobriety Centre. The Tom Puddings are theirs and they do all sorts of community projects and educating about life on the water.' I take it as the signal that Graham and Leonie need to be getting back to work and earning a living. I thank them for their time and head for the Centre. As I do, a petrol tanker passes followed by one of the Taylor fleet that's just penned in from the River Ouse taking sand to Leeds.

Inside the Sobriety Centre, which doubles up as an educational establishment and waterway museum, people, mostly in their twenties, roam from room to room. In one, a number of milk churns, low tables and horseshoes have been left with half-completed crude roses and castles designs painted on them. Outside the building, sitting on a bench, a man with goatee and sawn-off army-fatigue trousers is drawing long and hard on a cigarette between sessions. Adie O'Neil lives in nearby Beverley and is a tutor on the 'Environment and Conservation' element of one of the Sobriety courses. His job also involves driving the minibus, and 'fixing things when they go wrong'.

Two young women stop to greet him. 'We've just finished doing steering,' Julie, black, around five-ten and gorgeous, says. Her face has been made even more beautiful by the brightly coloured headscarf that she's tied pirate-style round her head. I guess her to be in her early twenties. It's more difficult to put an age to her companion, Tania, who speaks with an Italian accent diluted by a decade or more living in London.

Adie tutors the two on their *Waterways to Work for Women* course. The Community Boat Association certificate, which will qualify them to take up to twelve passengers out on a boat, requires they put in a total of 300 hours in the classroom and on the boats. The programme started in 2000 and is open to those working sixteen hours or less per week. There are seven students on the current women's course and most of them are doing one or two days a week over six months. I ask Julie where she heard about the programme.

'In London.'

'Whereabouts?'

'I saw it advertised,' she replies evasively. Something doesn't add up. Why has this very well-spoken, glamorous ex-model swapped the catwalk for the tiller? And why are both these women doing a non-residential course in Goole when they profess to miss London so badly?

'I don't know what I'll do later. I may not go on the Cut. I just fancied the course,' she explains, equally vaguely. The friends excuse themselves. Their next session on basic boat maintenance is about to start. I fear for Julie's long nails.

Back inside the building, I appear to be the only one mooching about without a purpose. A couple of women are redecorating a room, others stride about with folders under their arms or flourishing bits of paper. The museum, which opened in 1991, provides an overview of the Aire & Calder Navigation and the port of Goole. It evolved out of a community project Sheffield barge, the *Sobriety*, which I'd passed earlier near

Eggborough Power Station.

Amid the Rinso tin advertisements and sailmaker's tools, there's plenty of hard history. Beside a black-and-white photograph of Goole I read that in the 1930s people were so cold and poor they would slip a few lumps of coal into their pockets and snap tins (lunch boxes) at the docks, risking the sack, to have a little heat in their homes. For virtually the whole of the past two centuries, coal has been gold in Goole.

In Room 58, a floating art gallery converted from an old Rank Hovis grain barge, I ask someone straightening a painting how the Sobriety project is funded. 'It's paid for by European social-fund money, national lottery money, and contributions from Humberside Tech whose only stipulation is that students must be living in the local area.'

'But Julie and Tania are Londoners.'

'Yes, but they're serving their time at Askham.'

'What do you mean?'

'Askham, near York. They're serving their custodial sentences at the open prison there.' I'm dumbstruck. So that's why they were so evasive. 'They're nearing completion of their sentences and volunteered for the course. It gets them out and prepares them a bit for the outside world. We have several inmates here.'

Adie is still smoking on his bench. I sit down next to him, look out over the water, and try to find out a little more about Julie, Tania and others like them who've enrolled on the course.

'I enjoy working with offenders and the disadvantaged best. They're the nicest.'

'In what way?'

'Some are on life sentences and are at the back end of their porridge and this is their decompression chamber, their halfway house. They appreciate it more and are definitely the most honest people I teach. We also take students with mental problems. They're good too.'

'What are Julie and Tania in for?'

'I don't ask so I don't know. It's a personal policy. When they're here, they're all the same.'

But of course they're not. Adie has already pointed to their differences. For inmates, the course has special value: a taste of freedom, an opportunity to restore self-confidence and self-respect. Whether his lack of curiosity is feigned or real, his loyalty and respect for his students' privacy is admirable if frustrating.

Adie, cleverly switching subjects, asks me what I'm doing in Goole. I
tell him I'm on a boat.

'Travelling long?'

'A couple of months.'

'Lucky guy. How can you afford it?'

I tell him about the advance from the publisher of the book I'm writing
on the trip. 'It'll just about see me through the journey.' Adie looks at his
watch and stubs out his last cigarette. He's due to start another session.
He stands, wishes me good luck on my trip and returns indoors.

I wander down to the *Whaledale*, an original B.W. diesel tug beside
which Roy Scott is sorting mooring lines. A Goole boat builder by trade,
Roy spent more than forty years in and about the docks before becoming
workshop supervisor and head boatman at the Sobriety Centre. He has a
creamy Yorkshire accent, bushy eyebrows, generous sideburns, and
a low centre of gravity that would keep him upright in a force-ten
gale.

He will shortly depart on the *Telethon Louise*, a former ship's lifeboat,
for a half-hour tour of the basin. Just as he suggests I join the cruise,
Julie and Tania reappear, agitated.

'Roy, can we just have a word with the journalist in private?' Julie takes
the lead, making 'journalist' sound like a disease. She seems upset.
Tania's face has more fury than distress written across it. They have
discovered, presumably through Adie, that I know their backgrounds.
We move twenty yards from the boat. Roy tells me he'll be leaving in ten
minutes if I want to join the cruise.

'You didn't tell us you were a journalist.'

'Well, you didn't ask, and I'm not anyway.'

'Hmmm. Well, we'd rather you didn't write about us. Tania has been
stitched up by a journalist before.'

'From the *Sun*,' Tania adds.

Who knows whether it's true. It doesn't really matter. The result,
however, is that all three of us feel dissatisfied and frustrated. Julie and
Tania had revelled in the personae of free citizens choosing courses and
lives of their own volition. The fact I now know they are incarcerated
somehow makes them feel diminished. The secret is out and their wings
have been clipped. I am equally disappointed because the little I have

learned about Julie has made her even more exotic, a caged bird. What was she in for? How long? What's it like in prison? I want the whole story. Julie and Tania, unfortunately, are slamming the door shut and retreating back into their cell.

'I'm not a journalist and never have been. I'm a travel writer. There's a big difference,' I attempt to Polyfilla the fissure. They're not buying it. This time it's Tania taking the lead.

'That's what they all say.' It's absurd not because of its fundamental untruth and stereotyping but because it is such a cliché.

'It's a book about a journey by narrowboat through England. It's not about the prison system. The fact that prisoners are doing such a course on the waterfront makes it even more interesting.' I probably shouldn't have said the word. It's too stark. Too divisive. Trust has collapsed and Julie and Tania have fallen into the dungeon.

'Well, don't use our real names,' Julie insists, underscoring what she considers her only inalienable right and possession: that of her own identity.

'I won't,' I agree, and don't.

The girls turn on their heels and return to the building.

'Five minutes,' Roy calls from the bow of the *Telethon Louise*, at the same time raising five fingers on his right hand just in case I don't hear.

I nod. 'OK.' Taking a few deep breaths, I stroll the towpath, composing myself and trying to see through the blanket of guilt enveloping me. It feels like I've ransacked the house of someone who'd trusted me with their keys. It is, of course, absurd. Julie and Tania didn't give me any keys and I haven't even been able to get past the front door. Nevertheless the experience has left a nasty taste of metal in my mouth. I'm also pissed off that I won't learn more about their lives.

I take the tour of the basin. At the start of the nineteenth century Goole consisted of a few cottages dotted across the marshes of the Ouse. By 1828 large vessels were pulling into the port from every corner of the globe and the town was well established 50 nautical miles from open sea, surrounded on three sides by water. It's often claimed Goole was 'born under Victoria and died under her'. In reality the port, operated by a skeleton staff, still makes profits for the larger streamlined companies. Their huge container ships arrive from Germany and Scandinavia via the North Sea and unload here if they don't at Hull or Grimsby (the second largest U.K. port, handling 49 million tons, just three million less than London). They come in on one tide and sail out on the next, attended by no more than a handful of button-pressing operatives.

We pass two old sand carriers. 'They haven't been used for a bit.' Roy Scott tells the four of us taking the tour. There's a Waddington boat awaiting 'a bit of steel to take up to Rotherham, but just a bit mind, not as much as there should be.' At Tim's Grain Yard corn is being unloaded. 'It's imported for Weetabix and the like because our corn isn't considered clean enough. The only cereal crops that are exported – wheat and maize mostly – is for animal feed.'

Roy's commentary is an obituary to our once flourishing agricultural, industrial and shipping industries. A cement tanker is being pumped out by nobody. Beneath a vast all-weather steel canopy four overhead cranes wait to pick the load off a 100-metre long, 3,000-ton ship. It's a ghost port. Once loads have been disgorged, they are transported across the country by lorry.

Those still living amid the redbricks of Goole mostly work at the power stations or in light industry. The dockers have retired, youngsters moved away and Goole is now known as Sleepy Hollows. While Roy's own father worked for the river authority and there was never any question his son would follow him into the docks or onto the Cut, for Roy's own children, times are different: one is a farmer, another is in light industry, and the third in catering.

Later that afternoon, in the local library, I browse the preface to Joseph Priestley's *Navigable Rivers and Canals* (published in 1831), which he wrote while working as head clerk in the Aire & Calder Navigation Office here in Goole (his father had been manager of the Leeds & Liverpool). To the long-forgotten percussion of a typewriter keyboard played by the librarian, I read that before 1820 and the opening of the Knottingley to Goole stretch of the canal, Goole was a hamlet of 450 fishermen and farmers living on a bend in the River Ouse between the villages of Hook and Swinefleet.

I seek out coffee and am about to sit at a pavement table when the café owner apologises, 'Sorry, we're closing.' It's 4.30 p.m. The baker is mopping his floor, his bread vanished. I pass a sign 'Welcome to Goole the U.K.'s largest inland port'. A.K.A. Sleepy Hollows.

Back on *Caroline* I wave to Leonie and Graham on *Little Shuva* as they head out on another job with the tide at 5 p.m. Graham sticks a thumb up on his good hand and gives me a big grin. They deserve a future and so does their son. And his son.

I quit Goole in sunshine, backtracking along the Aire & Calder to the New Junction Canal where I negotiate a series of swing and lift bridges, some moved by arms and bums, others by simply pushing buttons and lifting levers. One at Sykehouse has a lift bridge in the middle of the long lock which has to be raised before preparing the lock itself. Another first. The New Junction joins the River Don Navigation at a 320-degree turn and more lift and swing bridges follow.

As I approach Wykewell Lift Bridge, a towering 60-footer, kids are jumping off the bridge into the water as cars bounce their way across it. I smile and greet two with goosepimples the size of golfballs on the bank. They sort of nod hello. The natives, I think, are friendly. When I reach the grey operating console, three girls aged between thirteen and fifteen run up and ask if they can work it for me. A boy just about to leap off the bridge tells me his aunt has a boat and the girls know how to do it. I give the girls the key (always think the best, I try to remind myself) and return to *Caroline*.

As the bridge goes up, two boys cling to its undercarriage like Spiderman. I grab the camera. They're still hanging there as I steer with my back and take snaps of them. When the bridge is at about 130 degrees and the boys 25 feet above the water, they finally let go and plunge with an exhilarated roar. I'm envious. It's exactly what I'd be doing if I were their age. It's also what Max would be up to if we hadn't inflicted city life upon him.

Once *Caroline*'s through, the girls drop the bridge and dutifully run back with the key. They suggest I give them a lift so they can do the next bridge for me. Danielle Whiting, the most garrulous of the bunch, takes charge of the key. Three of the boys hitch a ride too. They crowd into the deck stairwell for a picture and then split into two camps for the next bridge: the female bridge raisers and the male bridge bombers. B.W. may frown on it, and no doubt accidents have happened, but it beats throwing stones at people, killing swans or sitting in internet chat rooms.

My crew abandon me after Moor's Swing Bridge and I'm left alone to watch my shadow stretching with the early evening as *Caroline* pootles along a seventeenth-century drainage ditch that transformed marsh into the rich arable farmland it is today. For centuries strip farming operated here and a number of homesteads still sit cheek by jowl at the end of their one-acre, 220-yard by 22-yard stretch.

I watch the mohicans of great crested grebe slipping between red-hot pokers and water lilies. An eighty-strong gang of crows pause for a breather, clinging to the telegraph lines like notes on a musical score. A passenger train veers alongside the canal, the driver waving enthusiastically.

Beyond Vazon Swing Bridge, a man appears out of a raised wooden railway hut and waves. It's not a greeting but an imperative. He wants me to keep moving. As I approach the railway crossing, the tracks and the platform they rest on slowly retract, allowing *Caroline* to pass through. Built in 1915, it's one of only three sliding railway bridges in Europe.

Finally, after ten hours cruising, I arrive at Keadby. Tomorrow I enter the seriously tidal Trent with its minefield of sand bars. I've already phoned ahead, as suggested in the Nicholson guide, to book a morning penning out. The locksman has told me to be ready to enter his lock at 9.55 a.m. to catch the best of the high tide.

'Have you sailed the river before?'

'No, and I'm solo, but I do have charts,' I informed him.

'Well, we've got one of our captains looking for a lift up to his tug in Torksey in the morning if you'd like some company.'

I stir to a world of sunshine: sun motes dance through the cabin, water reflections chase across the curtain, birds upstage each other with song, and fishermen chat quietly beside *Caroline*. I slide back the stern hatch and offer the nearest fisherman tea. He thanks me and points to his Thermos. 'Already set up.' He's been here since 5 a.m. and has caught two bream so far. He catapult-sprays bait about thirty feet from the bank – the length of his rod. 'To keep the shoal tight up together,' he explains. 'Get 'em feeding here and they'll stay all day.' I'm dumbfounded to learn his carbon-fibre and graphite roach rod (he calls it a pole) cost £700.

'That's nothing, Dave' – he nods in the direction of his companion perched ten yards away like a shadow – 'has a competition pole that cost £1,300!'

'That's nothing,' the shadow echoes, 'some poles cost £4,000.'

The fisherman I'm conversing with is also named Dave, adding to the double-act confusion. I ask if the rods really make a difference.

Seven-hundred-pound Dave replies, 'Well, another mate reckons they do but I reckon it's more a fashion thing, owning the make, like with a car.'

'What do you use for bait?'

'Maggots, worms, squat, pinkies, casters.'

'Does it matter how deep you fish?'

One-thousand-three-hundred-pound Dave looks at me as if I'm an imbecile. 'Of course it does. We're fishing seven and a half feet, laid on.'

He could just as well be speaking Mongolian. 'What's that mean?'

'Lying on bottom. Best for bream and the skimmers. When it 'eats up, I might lift it a bit.' It takes me a moment to translate ''eats up' to heats up rather than consumes. The fish apparently rise when the water is hotter.

I notice Dave has a rather fine tattoo with the names Cookoo, Cockney and Budgie – 'They were my mates' – over three interlocking hands. 'Means everlasting friendship. Dunno where they are now though.'

Seven-hundred-pound Dave fishes here twice a week and often squeezes a couple more sessions in elsewhere. His father and grandfather started him off at the age of four. 'Loved it straight off.'

'What's your wife think?'

'She reckons it's the fishing what keeps us together!'

I leave £700 and £1,300 Dave to it.

At 9.35 a.m., a wiry man with a face weathered from forty years of sailing the Trent, arrives with a small canvas bag. 'Me sandwiches for lunch – cheese and piccalilli and some scones me wife cooked. More than enough for both of us. Chris Chapman,' he sticks out a hand introducing himself. 'I were told the *Caroline* would give me a ride up to Torksey.'

I invite him on board and we cast off. Five minutes later we're idling outside the lock listening to clanking windlasses. An arm is raised to wave us in. We're the first. Timing is of the essence if we're to have the tide with us all the way upriver. With the charts I managed to get in Leeds, I should be able to steer clear of the sand bars, but with Chris on board they're redundant. No one knows the river more intimately. The only danger today will be if a gravel carrier appears round a bend gunning for me.

As we wait for four more boats to squeeze into the lock before we pen down to the river, Chris shows me photographs in his wallet of his wife, his two daughters and three grandchildren. There's also one of Chris himself thirty years younger looking the spitting image of El Cordobes, the Spanish bullfighter nicknamed 'El Beatle'. His features are considerably more gaunt now. Despite his leanness and the chilly day, he's just wearing thin green cotton B.W. jacket and trousers over a cheesecloth check shirt. Once I've shown interest in the photographs, Chris rummages in his satchel and withdraws a white plastic bag in which there are several more. Only these are not of his family.

Chris has been skipper of the *Arthur A. Bland* B.W. tug 32 years and has worked the river for 40 in all. A lot of that time has been spent dredging and redredging the river, maintaining its minimum six-foot draught. His family have been associated with the river as far as Chris can trace back – his brothers, his father, his father's father and so on. 'I can track relatives on the river a hundred and forty years, but it might be longer.' The photographs in the plastic bag chart Chris's slice of that collective history. He's carrying them because he'd promised to show some to a friend working the lock. The first photo shows what appears to be a snowdrift enveloping a crane boat at Nether Lock near Newark in 1960. 'Used to happen when the detergent pollutants built up foam – Trent's much cleaner now.' In another picture a tidal wave is sweeping over a tug that's run aground. 'It's known as the aegre. On the Severn, they call it the Severn bore. That there's a six-footer crashin' over a five-hundred-ton gravel barge.' I nod my head appreciatively at Chris for so generously sharing this with me just as we're about to enter the river. 'D'you know how it starts?' he interrupts my nodding. It's a good question – the kind *Guardian* readers like. 'It starts as a shadow flashing across the middle of the Atlantic at five hundred m.p.h. We only get aegres on spring tides when high tide hits the ebb and creates a wall. Few year gone I were pullin' hundred-and-fifty-ton hopper and a crane boat and we got caught up by a six-foot aegre. She went under us and lifts us stern up and pushes us with the head down – it's not really a wave, see, it's a wall of water. I were running full astern and we were still surfing foreward like a bat out o' 'ell.'

Chris moves on to other calamitous tales: 300-tonners sinking, yachts blowing up, cruisers ablaze, and narrowboats in their scores running aground. The ones he selects from his back catalogue all involve Keadby Lock and the stretch of water immediately outside it.

'I remember a while back a boat inward bound from Hull for Nottinam coming into the lock overnight and he hit the buttment on a big spring tide and sank and the skipper drowned. Then there were the graveller laid outside lock bound for Hull and the mate drowned 'cos he couldn't get outside the cabin quick enough. When they got the boat up, he were still on ladder.'

It's just the introduction I'm looking for as the gates finally swing open and we're the first boat barrelling out onto the river. As I steer diagonally in the direction of the far mud bank, a tug hammers towards me from Trent Falls. Fortunately it veers off across my wake into the lee of the lock to wait for the last boat to exit before it enters.

We are sailing on twenty feet of water. At low tide it can drop to three feet. The wind has blown up and the river, complete with whitecaps, is as choppy as a sea. We pass Althorpe Church, perched above a fringe of rushes. When Princess Diana died, hordes turned up here looking for the grave. It was the wrong Althorpe but the owner of the local guesthouse didn't mind.

We slip under the M180, with banks rising up to wheat fields and hikers walking ten feet above us. The sky is curdled cream. For two of the other three narrowboats following in our wake (the fifth boat is a cruiser), the Trent is also a new experience. Only one is single-handed. He didn't expect to be but his wife got cold feet just before departure and jumped ship. Sensibly they all have charts with them and one family is togged up in lifejackets. They are all following *Caroline* like faithful dogs because they've heard I have Chris on board. Chris and I in turn follow the Vikings and Romans up Britain's ancient commercial thoroughfare.

Chris has been travelling this river 56 years in total. He made his first journey when just a few months old. 'Dad carried oil. By seven I could splice a rope and were already sculling a cod boat about the docks in Hull an' Grimsby. Sometimes I'd sail on coasters – that were great fun. We were one of the big boating families. It was a fraternity, like the miners, passed down through father to son. There was us – the Chapmans, there were the Johnsons, the Crees, the Ancliffes, the Websters.'

The sun appears as we slip further across to the port side, following the outer bank on a wide arc round the remains of a brick windmill. The river cuts up more as the wind blows directly into the rushing tide. Chris points a couple of hundred yards ahead of us to a white cruiser that appears to have broken down. I hand him my binoculars. 'That's all we need,' Chris confirms we have to play the Lone Ranger. The boat is anchored and facing the oncoming tide, bucking like a trapped horse.

As we come alongside Chris yells across, 'Need help?' A rather large man nods his head. 'Bugger,' Chris exclaims. 'Can't leave him here. Too dangerous.' Chris has done this countless times before, sometimes called out in the middle of the night, and nine times out of ten the breakdown was through negligence of one sort or another. He has, however, never performed a rescue from a narrowboat.

We are now twenty yards beyond the stranded boat. I turn *Caroline* in a wide loop against the tide. We barely move. The will of the river is against us. Eventually, however, we come alongside and Chris expertly lashes the 30-foot Seamster cruiser to our side. As he straddles the boats, securing more ropes and ensuring fenders are strategically positioned between us, waves crash up between the two hulls soaking him. He yells at the man to raise the anchor. At the same time an Alsatian is barking and snarling, mistaking him for a burglar.

Finally we're ready. I turn us back upstream with the tide. Although breasted up and carrying considerably more weight, *Caroline* seems to have slackened her pace very little. The other boats in our flotilla have overtaken us and are long gone. While the man stows assorted nautical bits, his friendly wife plies Chris and me with beers, nuts and biscuits as the Alsatian, which didn't want saving anyway, looks on resentfully.

After a couple of bends, the water calms. Chris is soaked to the skin. I offer him a change of clothes but he prefers to drip dry. With less noise now, he fills me in on the story. 'They came in at Trent Falls early morning with the flood tide. Transmission jammed and he lost power. Says it's the propeller. Sometimes B.W. send a tug out if there's one in the vicinity but I happen to know for a fact the nearest is four or five hours away. He were in serious danger 'cos there's a rise and fall of eight feet and if a seven-hundred-ton graveller came round the corner – and it would sooner or later – it wouldn't have been handy.'

The rescued man's name is Ray Fenshaw and by now he's joined us on board *Caroline*, having left his wife alone on the *Sundowner*. 'We started drifting towards port side bank,' he expands on Chris's shorthand. 'When we were far enough over and away from main channel I put the anchor down and it held. I put a message out on the commercial channel warning boaters to watch out for me. I know the ropes. Been doing it a very long time.'

Chris talks about the need to respect the river. 'Seven year ago I piloted a yacht through Trent Falls all the way up to Newark for an American. In fact he were really English, he were born in Newark and then migrated to Newark, New Jersey. He decided to travel on his yacht between the two. Him and his wife sailed all the way across the Atlantic through thirty and forty-foot waves and didn't bat an eyelid. Sound as a pound out there, but he were frightened to come up the Trent alone!' Chris laughs. 'That's why he wanted a pilot. Maybe he remembered stories from his childhood. Didn't know it and didn't trust it. You have to respect this river.'

After a while, Ray retreats to *Sundowner* and Chris and I are left to enjoy the rest of the journey alone. We turn back to the river and a cormorant perched on a buoy.

'Fifteen years back, you never saw a cormorant above Torksey; now there's loads of them. I've seen a deer swim across the River Ouse at Trent Falls. Last week I saw a porpoise at Torksey Lock. It made the national news. I've seen seals born on the river up at Sutton. I've seen a squirrel swim across the river at Torksey when the wash from me tug swept it off a branch and it swam right across the other side. Shook itself and off it went. You get mink near Nottinam – I saw five at Stoke Lock. And I've seen grass snakes swim across river too. In the 1920s they had a two-and-a-half-ton whale in the river near Keadby. Another smaller one got up as far as Gainsborough. It were put on show and stank the town out.'

As we swing round Jenny Erne Bend, Chris mentions a ghost that walks the flood banks at night. 'Jenny drowned here on a night as black as the inside of yer hat. And now she patrols the watter.' Chris pronounces the final word like batter. It seems that every bend in the river has its tale. At the next, just one wall of a house remains. 'Red House it were called and a hundred year back an old lady what owned it used to wait for boats to ground, then she'd sell them eggs and bake bread for them.'

We pass the remains of a Cistercian priory, old Roman fording points, and the ghost wharves, jetties and gantries of Gainsborough which in the 1960s still bustled with dockers loading and unloading cargo. Unlike everybody else I've talked to concerning freight carrying, Chris is optimistic that traffic will one day return to the Trent. 'Maybe not in my day, but it'll come back eventually, right as ninepence. Stands to reason. Why pay for tax and insurance for thirty vehicles, pay thirty drivers' salaries and thirty lots of petrol when you can use a single three-hundred-tonner to shift your load? It takes twenty hours from Hull to Nottingham. What difference does it make if it's two hours or twenty as long as you can predict with certainty when your stock's going to arrive? When I were younger, bikes – there were a big Raleigh factory in Birmingham – beer, salmon, fruit, they all came by watter. During the war years my dad said there were two hundred boats alone working round the clock into Nottinam.' Chris's accent reminds me of Brian Clough, the city's second best-loved son (Robin Hood still pips him). 'It'll come back sure as eggs is eggs.'

As we continue upstream I ply Chris for more rescue sagas he's been involved in. Eventually he decides enough is enough and sweeps the rest into the miscellaneous box. 'Can't think of a corner or a "showed" where I haven't pulled someone off over the past forty years.'

'What's a showed?'

'A bad place for grounding. Where the bottom's too near the top.'

Finally an exit from the river appears on the port side and Chris instructs me to take it. 'Torksey'. We're here. It's taken the best part of five hours with the carbuncle on our side. Chris phones the locksman and they discuss the problem. It's agreed we'd be too broad for the lock as we are, and that we'll tow *Sundowner* in behind us.

Having risen up into the calm of the Fossdyke Navigation, we untie the tow rope and moor our cargo beside the lock where Ray can attempt to hitch a lift on to Lincoln.

Chris also jumps ship here, the lock keeper having agreed to give him a lift to his home in Newark. Tonight he's off to watch an Irish singer at a pub with his wife. Graciously he thanks me for the ride. I tell him that I'm the one who should be doing the thanking. I moor at the end of a line of well-kept cruisers gracing the permanent moorings, basking in their own reflections.

The *Sundowner* leaves before me the following morning, towed by another narrowboat. They wave as they pass. An hour later I slip my own moorings bound for Lincoln, eleven miles away. In a couple of days I'll return to Torksey to rejoin the Trent but first I'm off to explore the Fossdyke. Built by the Romans about AD 120, it is the oldest artificially constructed waterway in the country that's still navigable.

As I potter, I imagine Charlton Heston and Tony Curtis whipping the boys and yelling to pull harder as they power their galley on and leave me in their wake. How different would the world have looked 1,880 years back when slaves sliced their way through the countryside? Reed huts instead of stucco and redbrick, skirts instead of trousers. No telephone poles.

When the Romans left, Vikings followed in their longships on their rape and pillage breaks. They in turn were followed by Norman barges transporting a cathedral-worth of stone to Lincoln. Since then, the navigation has been straightened and dredged and the surrounding Fens drained. Apart from that, and the steel-pile edging instead of reeds and rushes, it's pretty much as you were.

The Fossdyke owes its navigable existence to the military garrison at Lincoln. First the engineers improved the drainage on the River Witham from Lincoln to the Wash, and then made a cut, as straight as possible, across to the Trent. As you'd expect, the Fossdyke is mostly arrow-straight and banked, meaning you tend to focus on the near at hand – the pretty blue flowers and tall thistles, the water-skiing ducks, the coots cleaning themselves on rocks, a hovering kestrel, and a '50-today' birthday balloon that escaped, drifted, ran out of gas and crashed into the pea-soup water beneath candyfloss clouds.

At Saxilby I top up with water and wander the shopping street that runs parallel to the canal behind a high embankment. The K9 Academy ('All aspects of dog handling') is kennelled between the Ship ('Wednesday half-price cocktails, Music quiz night Thursday') and the Sun Inn ('Karaoke Wednesday, Disco Inferno Saturday, The Sinners Sunday'). There's also a DIY shop, a hairdresser's that thinks it's still the 1960s, Magpie Fabrics, and a charity shop. At Bridge Street Garage an MGB GT is on offer for £695 and a 'Sold' sign has been stuck on the windscreen of a Morris Minor.

Finally the three stone towers of Lincoln Cathedral appear high on the hill beyond a flickering line of silver birch. Brayford Pool, where I moor, is the city's port and the meeting point of the Witham and

Fossdyke Navigations. Equally importantly, Marks & Spencers is just 150 yards away. I take the rucksack and fill up with fruit, salads and other goodies that I've been starved of over the past week or so, restricted to small canalside village shops. I sit on a wall munching a tuna melt baguette, swinging my legs and watching boats out on the water.

In its eighteenth- and nineteenth-century heyday, when it was the fourth busiest port in the country despite being 40 miles from the sea, Lincoln would have been clamouring with grain barges delivering to breweries. Now it's filled with narrowboats, tour boats, cruisers, and a group of kids being instructed in the art of canoeing.

Unlike other cities, Lincoln has not needed rebirth. It is one of our prettiest as well as most historic settlements and appears to be thriving. On the High Street a sales promotion is under way and I'm plied with free samples of a new Kellogg's cereal called Just Right. I eat a box as I walk, passing a crusty in full regimental uniform – army fatigues, boots, mohican and dog-on-a-string – who offers me the *Big Issue*. I notice he has ten boxes of Just Right stacked inside the hood of his sleeping bag.

The aptly named Steep Hill becomes increasingly precipitous but cobblestones provide better traction than the smooth concrete lower down. Hand-painted signs abut the walls and shop fronts come with their own wooden frames as if hung in a gallery. My lungs work double time as businesses become more rarified in the thin air. The only people shopping up here are the tourists drawn to the Georgian, Tudor and Norman buildings requisitioned by Tabletop Ceramics, Wood & Toys, The Cassian Gallery, Chocolaterie, Imperial Teas of Lincoln, and Ancient World (Fossils, Crystals, Antiquities & Coins). Jews Court, now a museum, is the site of a former synagogue and a reminder of the wave of anti-Semitism that led to the persecution and expulsion of the Jewish community in 1290. The history of England is written on this street no less than the history of Christianity is on Jerusalem's Via Dolorosa.

Crowning the city is the Norman castle built on the site of a former Roman fortress by William the Conqueror once he'd evicted the 166 Saxon families still living there. At the time of the Norman Conquest, Lincoln was one of the largest settlements in the country with a population of around 7,000.

Facing the castle is Lincoln Cathedral, rated by Ruskin 'out and out the most precious piece of architecture in the British Isles'. A sign declares it costs £50,000 a week to keep it open. That's about what Roy Keane earns a week playing for Manchester United or three or four times the average *annual* wage.

I try to imagine what it must have been like for a peasant coming from his or her wattle-and-daub single-storey shack, arriving in Lincoln in the thirteenth century and seeing this House of God climbing all the way to heaven. How could you not believe in God?

The cathedral kicks into touch even the achievements of the great civil engineers of the Canal Age. Like the navvies, cathedral labourers and craftsmen did everything by hand; there were no cranes, reinforced concrete nor RSJs. But such gateways to heaven were also trapdoors to hell. Alongside the ribbed arches and soaring columns, the canopied choir stalls, the rose windows, the intricately carved lancet screen, the saints and kings, is a memorial shrine to 'Little St Hugh' and the following written apology: 'Trumped up stories of "ritual murders" of Christian boys by Jewish communities were common throughout the Middle Ages and even much later. These fictions cost many innocent Jews their lives. Lincoln had its own legend, and the alleged victim, Little St Hugh, was buried in the cathedral in the year 1255.' The vague confession concerns a small boy, Little St Hugh, said to have been ritually murdered by Jews. It was, of course, based on nothing more than prejudice. But the Jews died anyway. The apology ends, 'Such stories do not redound to the credit of Christendom and so we pray:

> Lord, forgive what we have been
> Amend what we are
> And direct what we shall be.'

Present parallels are all too obvious in the current stereotyping and superstition that condemns Asians in Oldham and Burnley as free-loading aliens with unsavoury practices. What else can really be behind the barbaric beating of an Asian student by a gang that allegedly included young, preposterously rich footballers?

In the Longland Chantry there are more requests for prayers, this time for those in distress ('I called upon the Lord in distress: the Lord answered me.' Psalm 118). In the Chapel of St Blaise we are informed 'this is an appropriate place to remember in prayer all in the rural economy who are affected by the foot-and-mouth epidemic.' Good prayers.

In the nave, a medieval re-enactment group wander past. They're on the way to the cloister, kitted out in the Lincoln Green that was already being exported from the city about the time they're pretending to inhabit. A Friar Tuck lookalike smiles and greets me, 'Morning. Oh no, afternoon init?'

Leaving the cathedral I stare up again at the Romanesque doorway to a frieze depicting scenes from Genesis that includes the expulsion of Adam and Eve. As the pair of sinners nibble their apple, serpents appear to be devouring their genitals. For some reason it's not the intended Christian moral imperative for chastity and modesty (the serpents play a dual role in covering offensive parts and punishing sinners) that the image evokes for me. Instead I find myself trying to weigh up what would be uppermost in Adam's thoughts: the ecstasy at eating forbidden fruit or the searing pain of having his balls chewed.

Schopenhauer used a different analogy when attempting to quantify the relative weights of the sensate world. He tells us to imagine two dogs, one of which is eating the other. He then asks us to assess whether the dog doing the eating is enjoying more pleasure, or the dog being eaten enduring more pain.

It's a pretty stark notion that could lead directly – as it did with Oblomov – to never leaving your bed. Two things thankfully derail my nihilistic interior monologue. One, unless psychotic, no one believes in a literal heaven and hell, and no one spends 24 hours a day, seven days a week, either in ecstasy or agony. Philosophy would be more gainfully employed addressing the in-between states of automaton semi-consciousness. Two, Schopenhauer was a miserable old Germanic bastard. The *Chambers Biographical Dictionary* sums up, 'throughout his unhappy life, his disposition remained dark, distrustful, misogynistic and truculent...His work is often characterised as a systematic philosophical pessimism.'

The dog analogy does, however, have its shapeliness of thought to commend it. And even if it isn't something you'd exactly plan your life round, it can be instructive on the macro rather than the micro level. Look at Lincoln and look at Burnley, and you can see the two dogs: one well fed, prosperous, confident and sure of its history and power; the other being eaten away.

That night I wake around 2 a.m. to two young men comforting a third at the end of my pontoon. Drunkenly they try to dissuade him from flinging his fiancée's engagement ring into the water. I peer through the curtain. 'Waste 'er money. She's not worth it. You'll regret it. Promise.' It's to no avail. The wretched love-struck one theatrically skims the ring across the water and it sinks to the muddy bottom to join Roman, Viking, Saxon and Norman spurned love tokens. The man has probably not yet left his teens but has already been dumped by a prospective wife for another man. His balls are definitely being eaten at this moment. He just prays his rival's aren't.

At around 10 a.m. I revisit the High Street before pushing off, and supplement my M&S rations with some hooch, a couple of newspapers, a copy of Jim Crace's *Being Dead* (an exquisitely depressing read), and a Chelsea F.C. magazine in which I can anticipate home and the new season. It feels like I have moored up to civilisation.

But civilisation is waking to a hangover. Yates's Wine Lodge is recovering from the night before, its fleet of cleaners sweeping up splinters of glass and broken hearts. The clubby, chummy sofas of Edward's, the Varsity and Cheltenham, three more big breezy bars, are abandoned and ignored. No doubt the three drunks at the end of the pontoon sailed through all four last night trying to drown the spurned one's sorrows. All they managed to drown was the ring. This morning the bars are as quiet as the Roman Empire. The drunken troops have exited and are now awaking under hedges and in foreign beds, no doubt nursing exploding heads.

I buy some local cheeses at the market, an almond croissant at a bakery and a latte at Starbucks and return to *Caroline*. Time to move on.

I exit Brayford Pool on the Witham Navigation burrowing through the aptly named Glory Hole above which a substantial half-timbered building sits (the arch dates from around 1160 and used to be called the Murder Hole). The heart of the city has the feel of Stratford-upon-Avon with its august buildings, milling crowds and swarm of swans. On the outskirts, I pass Doughty's Oil Mill where the grain barges used to dock (now converted into offices) before entering yet another very different lockgate at Stamp End. Instead of paddles, the top gate has a guillotine plate that is ratcheted overhead, allowing the water to rush in, followed by *Caroline*. My involuntary shudder as we pass beneath the slicer is probably a re-enactment of Robespierre's as he knelt before the block.

Beyond Stamp End, I spend the day exploring a short stretch of the 32-mile Witham that runs to Boston (and thus the Wash and the North Sea). It's another topless day. The zephyr bends whispering grasses. I pass hamlets on hills, and banks of pinkish-purple rosebay willowherb. At lunch I pop into the delightfully unmodernised Gypsy Queen at Bardney and share the garden with two Romany caravans. It's one of those balmy days that seem to stretch for ever.

At sunset, I'm moored back at Cherry Willingham, facing Lincoln once more, watching the sky go through its quick-fire final-act costume changes. I have a gibbous moon over my shoulder, and beneath my feet two bowed swan feathers are sailing balanced on a cushion of duckweed. I listen to the long lazy beating of a heron, the scampering run of a swan, and the squawking tight-formation fly-past of four ducks. When darkness descends, my gaze follows the luminous river directly to Lincoln Cathedral, bathed in golden light on its hill. A final peek out the window at 11 p.m., before I switch off my own light, reveals the silhouettes of two cows lazily grazing the raised far bank.

The following evening I'm back at Torksey, watching lock keeper Richard Albery leaping around like a billy goat. A billy goat albeit with a windlass and an abnormally large white beard. I can either piggyback on the tide now up to Cromwell Lock or leave on the early morning tide at 6 a.m. Richard informs me there's a quiz night over in the White Swan and that he's looking for a partner. That settles the matter. I moor *Caroline* on the floating jetty beyond the lockgate, ready for a quick gallop up the Trent tomorrow morning. Because the pontoon is so busy, I have to breast up alongside the narrowboat *My Way*.

The owners, David and Janet, are on their annual fortnight summer jaunt from their base at Pollington, on the Aire & Calder. They both work in Castleford: she in a bank, he as a mechanical fitter. *My Way* was built to their own specifications a year ago in Thorne and cost £40,000. I learn all this in the couple of minutes it takes to moor and secure *Caroline*.

Richard and I arrive independently but simultaneously at 8 p.m. at the White Swan. We swap recent history and Richard, hearing of my jaunt upriver from Keadby, announces himself to be a big fan of Chris Chapman. 'That man and his family is the story book of the Trent. Two hundred and sixty years in total they've worked the river. Maybe more.'

I have a vague recollection Chris himself totted it up at 140 years but who's counting?

'Like lots of boat kids, he wasn't brought up by his own parents but by his aunt and uncle. His dad was always away on the barge. The uncle ran the Newark Nether Lock and that's where Chris grew up. By the age of seven he was skipping school travelling on different barges.'

Richard has been working the Torksey and Holne Locks (near Nottingham) for twelve years. He settles his tall frame familiarly onto a bar stool. He's still dressed in his locksman outfit – blue socks and shorts, brown sandals and a whiter than white shirt whose cuffs have been rolled up to the elbows. He makes a figure 4 by balancing the outside of his left ankle on top of his right knee while the right foot balances the whole assembly on the rail of the stool. It looks precarious but Richard seems practised at the pose. He rams some Clan into the bowl of his pipe and lights it, sucking hard to create a draught. With his enormous white beard he looks like something the Vikings left behind on one of their raids.

A packed house has been drawn from the mobile-home park, the boating community, and surrounding hamlets. Quiz night in the Middle Ages would have been even busier, for Torksey at the time was bigger than Nottingham with at least two monasteries, one convent and three parish churches.

Beneath lace plates and boat club pennants, shoulders are hunched over question sheets as if eyes-down for housey-housey. Richard and I do abysmally. In between rounds I visit the gents where there are two dispensing machines; one offers a Luminous French Tickler for £1, and the other 40 fresh mints for the same price. Call into a pub anywhere in the country and it will confirm there's no regional variation when it comes to the bleary male fantasy of casual sexual conquest. Publicans know their market: well-oiled customers will consider £2 a bargain to dress the todger like a fairground and get a mint refresher for the old snogger.

At 5.30 a.m. my alarm throws a fit. The couple in the next room (*My Way*) are already up. As I push back the hatch doors, their spaniel, Pip, peers in, cocking its head and wagging its rear. Outside it's already light. Another dazzling day lies ahead. Unlike yesterday there's not a breath of wind blowing the smoke signals from the power station cooling towers. Last night, returning from the pub, they looked like eight giant tepees camped round a totem pole.

As I slip my lines, Richard bounds up, beard bouncing on chest, dressed identically to how I left him. He has two last-minute warnings for me. The first seems easier to accommodate than the second, 'Watch out for sunken islands where the old river bank is submerged. And steer clear of the gravellers.'

I'm the first of the five boats lining the jetty to depart but it's only a matter of minutes before *My Way* and *Resolution* overtake. The river is 100 yards wide and its dried mud banks and beaches are lined with greylag geese, oystercatchers and ringed plover. A silky shag sits on the arm of a tree thrusting out of the water like a drowning man. I follow my charts, keeping to the main channel, watching out for the submerged islands. Plastic bags have been left high and dry by recent floods, woven into branches twelve feet above the present river level. Above them a puff of orphan cloud and a trail from a disappearing jet are the only interruptions to an otherwise empty pale-blue sky. This Trent is a very different river from the one I encountered a few days back chopped into white caps at Keadby. The tide is tired, and the countryside unthreatened. Swans and Canada geese share beaches with cattle and sheep.

After four hours I am approaching the final bend before Cromwell Lock (the end of the tidal Trent) when I see someone aboard a tug moored by a mountain of spoil waving at me. I wonder if he's warning of some danger ahead. Then I recognise the slight frame of Chris Chapman aboard the *Arthur A. Bland*. He's back dredging the river. I drop into tick-over and come alongside. Chris is taxiing hoppers between a dredger and a bankside mountain of silt. He invites me to join him on board for an hour and suggests I first moor *Caroline* beyond *Crumwell* Lock (that's how Chris pronounces it, declaring this indeed was once how it used to be spelt).

As I set off to do as instructed, Chris yells, 'Watch out for the weir – that's where one yacht sank, another blew up, and ten soldiers were killed on an exercise.' He smiles, adding the ghoulish postscript, 'I pulled a headless body out the lock one day too.'

Arthur A. Bland, the man rather than the tug, was the younger brother of Little John, and a member of the Robin Hood gang that the Trent B.W. fleet is named after. Built in 1960, the tug weighs seventeen tons, has a 150-horsepower Lister engine and a three-foot propeller, making it the aquatic equivalent of a tank – enormously powerful and able to turn on a sixpence. Chris, into his fourth decade of shoving hoppers about, can think of no better way to spend a day.

Once I'm ensconced on the yellow and green tug, Chris introduces me to his mate, Nick, aged 28. I shake his hand and apologise for delaying them.

'There's nowt spoiling,' Chris replies and Nick provides the translation.

'He means there's no rush. I've only just started understanding him myself after two years aboard.'

We get under way, pushing an empty hopper to the dredger, positioned midstream off Cromwell Lock. Buckets are turning on a conveyor belt, dredging silt from the river bottom and then tipping their contents into an already full hopper. Chris informs me the skipper on board is his brother-in-law, Mick Cassidy, who's been working the dredger 32 years (and started on Chris's tug at the age of sixteen). He seems to be shy of the stranger. 'Don't be fooled by him,' Chris shouts loudly enough for Mick to hear. 'He's got more mouth than Portsmouth.'

Chris is a bundle of such 'Newarkisms', as Nick calls them. Other quirks include the fact he can't swim despite his lifetime afloat; this shortcoming, however, he somehow turns into an advantage – 'You're more careful if you can't swim.' He has no bank account, no credit cards and no cheque book. 'I like to do everything cash. Only have a building society account because B.W. says it has to have somewhere to pay in me wages. Personally I don't see why they can't pay me cash.'

Like many boatmen he's superstitious. While Chris chats with his brother-in-law, Nick whispers to me that I should try to get the skipper to say 'pig'. I don't have time or opportunity to ask why before Chris is back in earshot.

'Do you eat lots of fish, Chris?' I start.

'Lots.'

'And meat?'

'Lots.'

'What's your favourite?'

'Meat? Beef.'

'Do you like bacon?'

'I do.'

I'm beginning to sound like Michael Miles, 'the Quiz Inquisitor', who tried to trap contestants into saying 'yes' or 'no'.

'Pork. Do you like that too?'

'I do.'

'And what animal is that? Goat?' I know it's ridiculous but it's worth a try.

Chris's eyes light up, seeing the trap. 'A grunter!'

Nick explodes with laughter. 'See, you won't get him to say the word. Boaters reckon it's bad luck.'

'Ate that word,' Chris confirms.

'What word?' I try again.'

'The p-i-g word,' he spells it out. 'Grunter's all right but not p-i-g.'

'He won't do much on Friday the thirteenth either,' Nick continues the witchcraft inquisition.

'I go and hide,' Chris sheepishly admits. 'Only time I did something difficult – I had a dredger on in a flood – I had big problems and ended up on the bank. See? Just proves it. Neither do I ever put a boat hook at wrong end – always has to be facing aft. And you don't have playing cards on board neither.'

'Any others?'

'Don't disturb cobwebs or kill spiders. And it's true a sinking ship has no rats aboard.' I laugh. Chris chuckles at himself. 'Daft init? But it's the way I was brought up. Banksiders don't see half the things we do. Like at Stone Bridge at Newark where the ghost of Sally walks. One old lad, Jackie O, he's dead and buried now, and another old lad called Ernie Wheatcroft, were out after midnight. They'd been eeling an' as they were coming across Stone Bridge Ernie says, "Stand aside, Jackie, and let Sally pass," as a joke. But she did pass. An they ran 'ell fer leather.

'On the low side of Hazelford Lock, there are still soldiers walking about from the bloodiest battle of the War of the Roses. It were a terrible slaughter and there were that many killed they called it Red Gutter and they still call it that five hundred year later.'

I'll be passing the area in a couple of days and make a mental note to investigate.

Chris and Nick swap hoppers with Mick. Beyond them is a 240-foot-long, six-foot lip of curling water cascading off the weir that Chris had earlier told me had claimed several boats and lives. As we push the laden hopper back towards the bank, Chris expands on the stories.

'One yacht driven by a petrol engine exploded into flame in the lock. Two people drowned, three were saved.' He makes it sound biblical. 'Another yacht went over the weir with a family of four on board. The two children survived, but the woman was swept away and the man bobbed up to the surface when I freed the boat from the bottom with a dredger.'

'What about the soldiers?'

'Hold yer 'orses. Haven't done with the yacht yet. When it came up I got a six-inch wound on me wrist from the jagged bull rail and had to go to hospital for stitches.'

I wait a respectful time to see if there is more and then repeat the question. 'And the soldiers?'

'Sapper volunteers of the one-thirty-one Independent Parachute Squadron Royal Engineers out on a night exercise on the foulest of nights. It were two a.m. on September the twenty-eighth 1975. A bad, bad storm. And lights were out on the lock because of a power cut.' Chris is enjoying himself. 'It were as black as inside yer hat.' There he goes again. 'They lost their bearings and went over the weir. Mostly aged around eighteen they were. Ten never saw another day dawn.'

Approaching the bank with the hopper laden with 150 tons of spoil, Nick and Chris compete against each other attempting to lasso the bollard from 20 feet with the five-inch-thick rope. Chris wins. His record is 40 feet. With practised ease, the hopper is manoeuvred into position and is instantly pecked at by *Trendis*, a suction dredger, that empties it into a gravel pit.

I say goodbye to Chris for the second time and walk back past the biggest inland tidal lock on the network (24 foot deep and 400 foot long) in which a 60 foot narrowboat is just popping up like a cork. Beside it is a plaque commemorating the loss of the ten sappers during a storm in 1975. Sean the lock keeper shows me the high-water mark of the most recent flood – 'The longest and worst since 1977. It lasted from November to March and our cottage was completely surrounded by water. Chris and Nick off the *Arthur A. Bland* rowed food to us across the fields and we were eventually evacuated with our two parrots.'

Sean has a theory that the floods are a cyclical event rather than a result of freak weather patterns caused by global warming. 'There were the '47 flood, the '63 flood, '77, '83.'

As sensitively as possible I point out that the years seem to be entirely random, that there appears to be no pattern or cycle to them.

'Maybe not,' Sean concedes easily and moves on. 'Anyway one thing's for sure.'

'What's that?'

'It may have been the longest flood I've known, but it won't be the last.'

The twelve months before I set sail had been the wettest since records began in 1766. But even in an average year the most blessed corners of the country endure 60 per cent cloud cover even if it's not raining. Charles II believed the English summer consisted of three fine days and a thunderstorm. Byron's view was equally bleak: 'the English winter – ending in July to recommence in August'. Some years in England it feels like there's no summer: 1816 was one such, and all the encouragement Mary Shelley needed to write *Frankenstein*. No wonder the country shines when the sun does like today.

I make myself a salad, rip off a drumstick from a ready-cooked Lincoln chicken, and take them up onto one of the picnic benches alongside the remote lock. From here on, the river is non-tidal and I'll be able to stop wherever there's mooring.

Some of the cruisers moored at the swish new B.W. Kings Marina in Newark-on-Trent look more suited to the Atlantic than our inland waterways – the equivalent of using a chainsaw for trimming the edges of the lawn. I breathe in and squeeze past them and maybe an hour further upstream, stop at a private jetty where a sign warns anyone not a member of the Trent Water Skiing Club to bugger off. The trouble is there are still very few mooring options between towns and I want to go in search of phantom soldiers at Red Gutter.

And so I moor up anyway, unlock Rosinante and sling her over a gate that leads into the back of a petrol station. The traffic out on the main road is shockingly loud and fast. I zip along, risking life and limb, and at the Pauncefote Arms in East Stoke, turn right down School Lane into Church Lane. This is better. I fill my lungs with unpolluted air, slip under a stone bridge, and abandon the bike to a grassy bank beside the squat tower of St Oswald's Church.

Inside the nave I'm asked to remember the five from the parish who fell in the First World War and the three who did the same in the Second. In a back room, a much longer prayer is required: described on wall panels are the events of the final conflict of the thirty-year War of the Roses between the houses of York and Lancaster that saw 6,000 soldiers lose their lives in a nearby field.

In the cemetery, a life-sized angel proffering an olive branch while concealing a sword behind her back, catches my attention. On the tombstone I read, 'To the dearly loved memory of the Right Honourable Julian Baron Pouncefote of Preston, G.C.B., G.C.M.C. First Ambassador to the United States of America. Born September 13th 1828. Died in Washington, May 24th 1902. "Blessed are the peacemakers." 5. Math. 9.v.' I'll drink to that. So why the concealed sword?

The village itself is as quiet as the battlefield which I reach by following a trail along Humber Lane onto a stony track leading up to the woodland. Eventually I arrive at a tablet known as the Burrand Stone on which is written, 'Here stood the Burrand Bush planted on the spot where Henry VII placed his standard after the Battle of Stoke Field.'

In three hours of continuous fighting on the morning of 16 June 1487, after House of York troops finally managed to cross the Trent, 4,000 of them, along with 2,000 of Henry VII's troops, had sacrificed their lives to the egos of the two men fighting for the Crown. At Red Gutter today an army of wheat marches to the horizon bowing to the wind.

Hazelford Lock is the first I've found operated by a woman since Anna Ward's at Brentford a couple of months back. Both of us have been listening to Tim Henman attempting to beat Todd Martin at Wimbledon and we thumbs-up to each other as she lets me out and simultaneously Tim breaks Martin's serve in the fifth and final set.

Fortunately Tim finishes off his opponent quickly and I'm able to jettison the headphones. It's a perfect English summer day. Behind us is a rare English Wimbledon victory and out in front, the finest stretch of river you're ever likely to find. The Trent spreads itself luxuriously, more than two hundred feet across, on its hundred-mile journey from the Midlands to the Humber ports. On its luminous green canvas the Great

Artist in the Sky has painted a few fluffy clouds, a deep blue sky, and the almost sheer forested cliffs of the Trent Hills. This pristine, transcendently beautiful landscape appears to be only accessible from the water. Butterflies and dragonflies flicker across to check out the intruder. A herd of heifers cool their hooves in the water.

I just wish there was somewhere to moor up and stay the night among the feathery grasses and the Cappadocian mud dwellings of the sand martins. Unfortunately the banks are simply too shallow. I try two or three times but have to reverse each time as *Caroline* starts dragging her belly on the floor.

I work the enormous Gunthorpe Lock myself and then, as it's 7 p.m., call it a day. The pubs on the Trent, following the river's lead, have broad frontages with larger gardens sitting square on to the river. The Unicorn is just a couple of hundred yards away and already mobbed. The sun is balancing on the tree line and the grass bank crowded. Wine glasses clink and empty pint pots stand in clusters on the grass like flowers. I head into the pub to get a drink, passing maybe thirty bikers in full leather regalia on the terrace eyeing up new bikes as they arrive. Boys and girls talk torque – these Nottinghamshire leather-clad maidens are closer kin to Marianne Faithfull than Maid Marian. Helmets and leathers belong to the modern Japanese Manga comic-book canon with yellow lightning flashes and silver explosions. The bikes similarly mostly hail from the Land of the Rising Sun, with a riding position that cleverly slides the pillion passenger tight into the driver.

I am happily now inhabiting the land of the never-setting sun. It will be throwing-out time before darkness falls on the Trent tonight. I rejoin the party on the lawn as all the benches and tables are taken on the terrace. A Mazda sports car has its roof down and Dido is playing on the CD player while the owners lean against the bumper toasting each other with flutes of champagne, the skeleton of a picnic strewn about them. A white woman is voraciously snogging a bald-headed black guy in blue shorts. One of the bikers, head resting in his girlfriend's lap, has a small picture of Munch's *Scream* on the knee of his leather trousers.

The following morning brings more of the same: more beauty, more sun, more Trent, more pubs. Ho hum. Someone's got to do it. The water is dimpled by a westerly wind that billows and blusters the grey-green willows. Swallows dart. *Caroline* continues to take the widest loop with each bend avoiding the silt dumps. We pass another large pub, the stuccoed Ferry Boat, again with lawns sloping down to the river. Big pubs, big locks, big weirs, big river.

And so to Nottingham and the castle Roger Moore persecuted as Robin Hood before he became James Bond. It's the one you'll recognise if you smoke Players, for it's on every packet to remind invalids like my father where their poison was packaged. At the Meadow Lane Lock, a narrowboat is just exiting the gate, allowing me an effortless ride in. Even better, there's a couple waiting their turn to follow their friends into the lock and out onto the Trent. Etiquette demands water is never wasted and so they wave me in first and offer to pen me up as I'm single-handed.

Unfortunately, having help can sometimes backfire. You relax, lose the plot, don't think things through. Or rather I do. For some reason when *Caroline*'s roof is almost level with the lockside, I tie her mid-line from the pulpit rail to a bollard and go off to chat to the woman working the nearside paddles. We exchange tips – she on the Thames where I'm heading, me on the Leeds & Liverpool where they're heading. Suddenly I notice *Caroline* is lurching badly. As she is forced above the lock on the rising water, she's being strangled by the rope. The woman notices impending disaster at the same moment as I do and immediately closes the upper paddle while I dash to the rope only to find it locked tight and impossible to loosen. Any moment the rope will snap, or worse, water flood into the boat. I'm about to borrow the woman's windlass (mine is still on board) and hare down to the bottom paddle when I realise her husband is already letting water out.

Caroline quickly drops to a point where I can unfasten the rope, and we can recommence the locking-up procedure again. As with all such awkward moments, I'm glad to be back on my own and shot of these witnesses to my imbecility. Back on board I apologise profusely to *Caroline* and carry out all necessary subsequent manoeuvres with kid gloves.

The canal is a return to that intimate beauty I left behind a couple of weeks back at Leeds. A return to flowers and bridges you can lean out and touch, and a towpath from which walkers can converse with me as we pootle. The sun is fierce. Burnt bodies fish from the bank and there's not a ripple to be seen ahead.

I moor between the Magistrates' Court and a bevy of alfresco restaurants. In Market Square the Salvation Army are out in force alongside another Christian group kindly willing to waste this gorgeous day saving me from sin. For some reason they do not approach the busker, a curtain rail of earrings, guarded by two mutts on string leashes.

At the heart of the city, in the tourist office, they're flogging Robin Hood capes, hats and egg cups. The city appears prosperous, purposeful and dressed for summer. The women look particularly glamorous to outback Canal Man. Nottingham, along with Caracas, was somewhere I considered moving to in my teenage pimple years, tempted by myths of hugely disproportionate numbers of female inhabitants to males.

On Low Pavement, tables spill onto the pedestrianised street outside the Slug & Lettuce. Mobile phones chatter like birds, as human sparrows pick at low-fat Caesar salads. As all available chairs are taken, I lunch instead with *Caroline*.

A couple of boys dressed in Nottingham Forest F.C. shirts, sitting on steps between me and the Young Offenders Court, ask if I own the boat, if I live on board, what it's equipped with, and how much it costs a week to rent. I presume they're awaiting a hearing at the court and wander across to chat. They ask if I've ever had my ropes slipped at night. I tell them only once so far. 'Bastards,' one explodes rather theatrically, eyeing up the interior for looting potential.

The exit from Nottingham is very green. At the corner of a bend, a menagerie has been assembled in a suburban garden with a skeleton, nautical flotsam and a sign declaring 'Yorkshire Pud, Tatties and Beef £4.50'. In the next-door garden, again facing the Cut, a mannequin has its head trapped in a wooden stock.

At Beeston Lock I rejoin the majestic Trent. Alongside the bank are a number of slightly ramshackle but very desirable wooden summer homes with private jetties. One has a large clock face on a wall with the thought for the day, 'Time Will Tell'. Another of the properties is wearing a 'For Sale' sign. I ring the telephone number and speak to a man called Steve who's handling the sale for his sister while she's out of the country. He tells me it's one of only three original, hundred-year-old properties along the bank. It has two bedrooms, mains electric, phone connection, a septic tank, and 'Only two inches of water got in during the floods.' Unfortunately his sister wants £79,950 for it. If it had been

£950, I'd have it there and then and hang the expense.

It is the hottest day yet and so at 3.45 p.m., with Tim Henman into his second set of the quarter-final, I drift into reeds opposite a weir off Barton Island and the Attenborough Nature Reserve. I tie up to the almost horizontal trunk of a willow tree, strip off and dive into the cool, limpid waters. I swim against the current. The water feels clear and weightless, unlike sea and swimming pools. I duck-dive and do somersaults, rising and falling in the water like the lily pads. A pair of exquisite Beautiful Demoiselle damselflies flit past my head, a rush of blue with a metallic sheen.

The 30 inches *Caroline*'s deck sits proud of the water seems more like 30 feet as I stare up at it, out of my depth and now trying to work out exactly how I get back on board. I try clambering up at the fore-end, and then from midway along the port side, but to no avail. Eventually I get a foot on the propeller, hoist myself painfully to a seated position on the fender (so that my body's wrapped round the stern), and gingerly pull myself up on board with the help of the tiller. I get some ice cream out the fridge, pour a Coke over it and sit on deck, scooping and drinking, watching the river sparkle with life.

At Cranfleet Lock a fatty is leaning over his barby exposing his bum to all who care, or don't care, to take a peek. He has a long ponytail and his boat is called *Zappa*. Whatever did happen to my copy of *Weasels Ripped My Flesh*? One of those mysteries like missing socks. Except in the case of records you always know what really happened. You lent it to someone and the bastard never returned it.

I moor at Trentlock, a major aquatic crossroads that for some reason only attracts a handful of houseboats, a couple of pubs and a tea house. Signs point in four different directions: back the way I came down the Trent, west on to the Trent & Mersey (the start of the canal I joined over a month ago at Great Haywood), due north on the Erewash Canal, or south on to the River Soar where I'll be heading tomorrow.

A middle-aged man, monumentally drunk, is staggering along the towpath followed by a second who appears to be in an even worse state. As the laggard passes he comments warmly, 'Turned out nice.' It obviously has, though tomorrow might not be quite so good.

High above them two peacocks are sitting on the chimney stack of the Tea House. The terrace of the Steam Boat pub is already busy. Opposite, an old man in a white vest is staring out through milk-bottle spectacles from the midship engine hatch of an immaculately kept 43-foot narrowboat called *Trinity*. 'Cracking evening,' he concurs with the drunk. I discover that before Cyril Blood created his little Eden here (complete with shed kitchenette and towpath garden), he served 33 years as locksman at the Cranfleet Lock.

'Don't think they get the nice perks up there now that I used to get. One day I saw some boats coming towards the lock and so I got it ready. There were seven boats in all and it were a baking hot day like today, and they were all Dutchy people. Suddenly a lady walks out the back of the first boat topless. "Vor is dustbin?" she asks. Me mouth were hanging on floor. Then another comes out and two more. All naked, well half like – the top bit. I shouts to the skipper, "What you got there? A nest of them?"' Cyril chortles, his goitre neck rising up and down like a ballcock. 'Seven boats of topless women!' He shakes his head, still not quite able to believe his good fortune. 'Me wife was at work at Shardlow Hospital. When she comes home and I tell her, she says, "You got a bloody good job you have!" He chortles again. 'It were!' He closes the hatch, remembering what he'd gone there for in the first place.

On the horizon, a stream of lava seems to be bubbling up over ploughed snow in a spectacular sunset. At the Navigation, another of those outsized Trent pubs, vintage Vespas and Lambrettas are assembled beside the lawn. Saddles have two-tone or Union Jack designs, and the handlebars are a battery of gleaming chrome lights. The garden has been requisitioned by an army of mods from the Nottingham Scooter Club. Instead of gaudy helmets and the flashy designs of last night's Kawasaki and Suzuki bikers, the scooters and their owners are subtle and stylish – Italian rather than Japanese. The mods, all in their thirties and early forties and dressed in Ben Sherman button-down shirts and zip-up American cotton jackets, are children of the late-seventies Two-Tone revival of the Specials, Beat and Selecter. They probably could drink in the same pub as the bikers these days without a scrap; in the sixties when I was a mod riding a Lambretta (a write-off my dad bought for £15, repaired and resprayed for me), listening to the Skatalites, Geno Washington, and Prince Buster, you wouldn't be caught in the same town let alone pub.

Having biked in both camps, my allegiance is to the bikers' bikes but the mods' style – spiritually I feel closer kin to the boys from the

Nottingham Scooter Club.

I take my pint and sit out in the garden overlooking a couple of heavy petters and the river. On the far side, the Scouts are stacking their canoes in a ramshackle boathouse. Their laughter billows across the water. It's 10 p.m. and still light.

By the time it is finally dark, grey-white smoke swirls from the cooling towers of the power station, rising as if from the boiling pots of cannibals. Above them a full moon sits. Next weekend Max is coming to join me for a full week. Just the two of us. I can't wait.

I breakfast – toast and two cups of coffee (and filter no less) – beneath a parasol on the terrace of the Tea House. Mark and Terry Marsh, who have their narrowboat tethered next to Cyril's, opened the café five years back, finally realising their dream of living and working on the water. They serve cakes, dozens of different teas, lunches, cream teas and of course, coffee. Mark has a ginger moustache and a lean frame. He's dressed in a red polo shirt and blue cargo pants that meet in a broad leather belt on which is written 'Volkswagen'. I tell him about last night's Lambretta and Vespa gathering opposite.

'We get vintage cars and motorbikes too. People just like driving out here.'

David and Janet from the narrowboat *My Way* turn up and stop to chat from the other side of the white picket fence. Janet tells me of a plunge at 8 p.m. last night by the drunk who could barely walk when he passed me.

'They managed to get onto their boat but then one fell in. Once in, he wouldn't get out.' Her blonde bob bounces as she laughs.

'And then when he would, he couldn't,' David adds. He has a closely shaved head and is dressed in a T-shirt, shorts and trainers.

'What happened?' I ask.

'Took us thirty minutes to get him up. We were told later by other local boaters that they're permanently moored here and it's a regular occurrence.'

Soon *Caroline* is following *My Way* up the Soar. Having squeezed through an open flood lock, the river narrows and becomes very shallow. *Caroline* slides across gravel making the percussive sound of the *cabassa*. There is a smell of wild garlic in the air and lining the bank are red poppies, thistle, rushes, and sedge.

Just before the Devil's Elbow, I moor and take another swim. Two cows wander over to see what large animal is cooling itself in the exquisitely chilled water. Common Blue and small Tortoiseshell butterflies hang about too but I think this has more to do with the fragrant water mint than me.

At Bishop Meadow Lock two boys on bicycles, Gary and Thomas, aged around twelve, take a windlass each and lock me through. They then cycle ahead to do the next lock. I thank them. They say it's been fun helping, and grin broadly. The Wigan flight is now completely exorcised.

I pass two narrowboats moored alongside each other. One is called *New Orleans Joys* and the other *Abraxas*. What with *Zappa* yesterday, it would seem musical christenings are popular in these parts. I pull over just before Barrow Deep Lock in front of a handsome brick bridge on a tight bend at Barrow upon Soar.

The smell of curry emanating from the nearby Bengal Tandoori and Balti Restaurant is irresistible. I order the Indian Banquet Menu for One. For £10.95 I get shami kebab, papadum, nan or rice (I get both), bhuna tikka masala, followed by ice cream or coffee (I choose ice cream and they throw in a liqueur). My shoulders have been feeling tight all day from the vibration through the tiller. The curry sorts me out. A good curry, I find, sorts out most things in life.

I share the restaurant with Mohammed Amin who supports his wife and three children on a waiter's salary. Mohammed, like most workers in 'Indian' restaurants, is Bangladeshi. He married at eighteen and claims he still has no regrets as he sails through his thirties. He's dressed in a vermilion shirt and still possesses cherubic good looks. We talk about the racial disturbances in Yorkshire. He finds it inexplicable. He claims to have lived twenty trouble-free years in Leicester. 'Race relations in the city are very good. Though there was an incident a week ago. I was visiting someone in hospital when I saw three fifteen-year-old Bangladeshi boys I know from my children's school. One had part of his leg hanging off, and another had a broken leg and had almost lost an eye.' White joyriders had ploughed into them while they were waiting at a bus stop.

Incredibly Mohammed refuses to believe even this incident was racially motivated. 'There had been a running feud between rival gangs at school. And the police have arrested the boys responsible so it's under control.' Rather than outrage at the whites for this callous and potentially murderous act, Mohammed stormed across to the Bangladeshi boys and gave them an earbashing for being out after 11 p.m. 'I told their father he should have more control. He said he was doing his best.'

Mohammed might be accused of being a toady, like the black teacher I used to teach alongside at Faraday High School nicknamed by the kids Guinness – 'Black, but white on top'. What, however, Mohammed cannot be accused of is succumbing to a siege mentality or avoiding integration. Mohammed claims that he is the norm and that those who have created airtight fortresses are the exception.

By the time the stars are fully out, I'm back on deck listening to electricity buzzing between pylons. A pure white duck looks up at me hopefully and quacks. It has black pinprick pupils, a yellow beak and, I'm shocked to discover, seemingly no ears. 'Where are your ears?' I ask. It doesn't answer. Probably didn't hear. Must write to the *Guardian* and ask where ducks keep their ears.

I feed ducky some bread. Two mallards muscle in. I feed them too and then the male of the pair, not content with filling its own belly, sets off after whitey, to prevent it sharing in the spoils. By the time it gets back, I've finished feeding the rest of the leftover loaf to its partner. It pleads with me. Tough. I'm a strong believer in Skinnerian behavioural conditioning.

Down in the cabin, I browse the Nicholson guide and read the section on Leicester. I discover that close to the canal moorings at Castle Gardens there is a Jain temple. I decide to visit it tomorrow morning to test out Mohammed's claim of ethnic harmony in the city. The most pacific people on earth should be a good barometer of ethnic ease or unease.

I check the navigation notes to make sure there's nothing challenging ahead. As I do, I notice that the River Soar has been swallowed by the Leicester branch of the Grand Union. My run of rivers and ancient navigations is at an end. I'm back on the Grand Union, the canal that first led me out of London.

7

Idle Women, Dreaming Spires and Sky-clad Jains

The Grand Union Canal and the Oxford Canal

The horn of *Muffin 3* provides a dawn reveille, rudely blasting me from the helm of a 60-foot yacht I'm sailing through the calm waters of the British Virgin Islands. I peel back the curtain and find Audrey Smith's face smiling down at me. She and David are still making their way oh-so-slowly south to Milton Keynes. I pull on a pair of shorts and a T-shirt and head up on deck.

'Thought you'd be up by now,' David grins. I feel the upwards-and-onwards earwig moving in the attic. 'We'll moor in front. Come on board for coffee but you'll have to bring the milk – the fridge has died on us.'

On *Muffin* we compare notes on our past couple of weeks and what's in the pipeline. 'We're headed for Stratford-upon-Avon next, to meet up with a couple from the Australian Canal Society.'

'But they don't have canals.'

'Why should that stop them?'

True. A few years back I attended the Henley-on-Todd Regatta, an irreverent version of our own Henley Regatta, held annually in Alice Springs, as far from any meaningful body of water as it's possible to get in Oz. In the Antipodean version boats made from beer cans are raced across a dusty dry river bed, and like all Aussie games, it has been made into a contact sport. The year I attended, however, it rained like never before and the event had to be cancelled because flash floods had transformed the dustbowl into a real river. Aussie regattas and water don't mix. Seen in this context, it seems eminently sane that the canal-less country should have its own canal society.

The Smiths are hooking up with an Australian couple who have swapped their home for an English narrowboat for three months. They plan doing some theatre together and visiting a few of the Shakespeare properties. But first Audrey and David have to pull into the nearest marina and buy a new fridge. I leave them with the milk, and they push on ahead of me.

When I finally wander up to my first lock of the day, I encounter narrowboat virgins grappling with basic water science. An hour earlier the family had never been on a boat in their lives. Now they have to steer into a skinny lock and somehow make the water level drop five feet before they exit. The father, a rockabilly in his early thirties, has an impressive jet-black quiff above an ashen face. He is gripping the tiller like a life preserver. His twelve-year-old son, in a back-to-front N.Y. baseball cap, by contrast is hugely excited but finding one of the balance beams impossible to budge. I tell him to put his back against it and to try rocking it. It works.

I move in after them and slip through Old Junction Boatyard where a dreadlocked woman in her early twenties is emerging from a canalside shop and waving across to a couple with matching silky bum-length hair sitting on their moored narrowboat. As I watch, a scrawny young man with a blond ponytail bangs on his galley window admonishing me for my speed. I'm doing maybe two miles per hour. Tethered to his 40-footer is a rowing boat filled with potted plants collected from the banks. It is clearly a busy alternative community, so busy in fact that there's no space to pull in and investigate further.

At Belgrave Lock, a man stops his bike and sits on a bench watching my single-handed manoeuvre. He doesn't offer help but just observes intently. Beyond the weir, what looks like a 130-foot silver Nike trainer is standing upright on its heel. It is in fact Leicester's much-hyped spanking new Science Museum. I ask my mute observer if he's visited it yet. 'No, it only opened last Saturday. They expected it to be packed but it wasn't. I could've told them that. There aren't many round these parts prepared to pay eight pounds for a museum visit.'

The local community, I learn, have been against the project from the outset. Not so much because of the £56 million it cost to build, but more because a secondary school and college were closed to make way for its car park. 'It was a very popular school – mixed ages, races, sexes, all sorts. Now the kids have to travel miles for their education.'

On the far side of the lock, a sign warns me 'Keep to the main channel and the centre of the bridge.' I do. But the canal bed is still tickling *Caroline*'s belly.

Up ahead, at Abbey Lane Bridge, I can make out a man throwing something across the water to the far bank and then dragging it back. I prepare for invasion. When I get closer I can see that the object being propelled is an anchor and the man throwing it is in his late twenties. Also scouring the water are his girlfriend and father. Yesterday the man popped into a shop for an ice cream and when he came out, his £500 bike was gone. 'A girl told me she saw someone chucking a bike into the canal from this bridge. All we've found so far though is a supermarket trolley.'

I moor at Castle Gardens, a green oasis in the centre of Leicester where the Romans settled 2,000 years ago, the Normans set up home, a medieval town flourished, and Leicester's industrial revolution was kick-started. Now it's a park. I unshackle Rosinante, remind myself to skip the ice-cream shops, double-check the bike lock, and set off.

My first port of call is the Jain temple in Oxford Street. I've had a soft spot for Jainism ever since, twenty years ago, I visited the holy Indian shrine of Sravanabelagola where, alongside a 50-foot naked male statue (I was about the size of his penis), I discovered a nude priest praying on a rock. To say he was naked is actually a slight exaggeration for the holy man was wearing one piece of clothing: a small mouth-mask like those worn by dutiful Japanese office workers on the Tokyo underground to muzzle their cold germs. This man, however, was more intent on protecting passing fragile insects from burning up in the hot air-stream of his breath. Jain priests are not only vigilant breath-watching vegetarians, some also dress in a stylish high-heel sandal in order to reduce the likelihood of splattering ants.

Inside the temple today, thankfully everyone is clothed. I say thankfully because the long-serving white caretaker, Robert Greet, who kindly shows me round (the temple is open to the public between the hours of 2 and 5 p.m.) is already sweating sufficiently for my taste through the exploding buttons of his overstretched white shirt.

Surrounding a forest of intricately carved sandstone pillars shipped from India, the 24 gold *tirthankars*, or prophets of Jainism, sit blissfully in the lotus position. As I stand admiring their calm and wondering how many lifetimes they spent riding the Cochin backwaters or Kashmiri

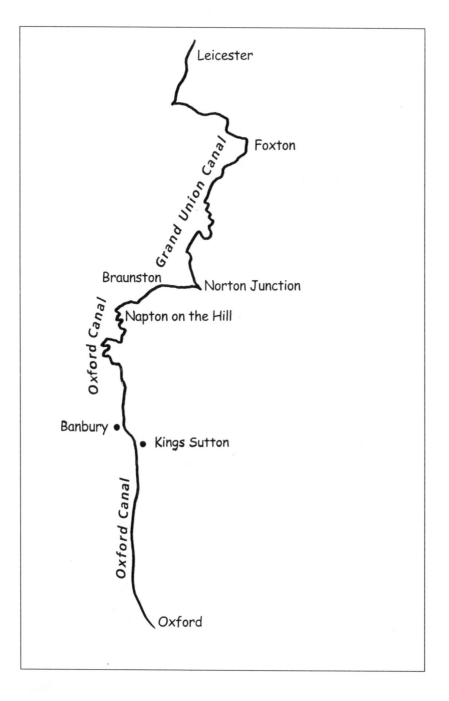

Middle England

lakes to get that way, Dr Rameshbhai Mehta, one of the temple leaders, arrives hotfoot from his surgery. He chats to me about Jainism, explaining that, like Christianity, it has a number of sects.

'In India we would pray at different temples, depending on which branch of Jainism we follow, but in Leicester we all worship at the same temple, albeit in different rooms.'

This, no doubt, is part of the easy coexistence I'd been promised I'd find in Leicester by Mohammed in the Bengal Tandoori. Part of that cohesiveness is due to a Council of Faiths – of which Dr Mehta is a member – which meets regularly to discuss the needs of the different communities. 'We have three presidents – Christian, Muslim and Hindu – and all show mutual tolerance.'

The word that crops up most in its teachings appears to be *shanti* – peace. And nonviolence in the Jain canon relates not just to physical acts but also to thoughts and speech. Dr Mehta chooses his words very carefully therefore when he discusses the current outbreak of racial violence further north.

'Leicester has far more Hindus than Muslims. Home Office records show,' he pauses, mentally plotting his careful statistical course, 'there are fewer Hindu inmates pro rata than any other ethnic group. I don't like to single out any one community, but it's a fact that Pakistani Muslims are unfortunately found at the other end of these statistics. The country has a long history of drug production and dealing and there is therefore a far greater incidence of drugs in that community and therefore a higher potential for crime.' Dr Mehta is only too aware of the damage inflammatory stereotyping can have and immediately qualifies what he has already said. 'Obviously I'm only talking about a minority in any community. The majority of Pakistani Muslims always do their best to crack down on the destabilising elements.'

In Leicester, I learn, ethnic communities tend to be well integrated rather than polarised and so the views of the British National Party and National Front don't take root. Beside the Jain temple, there's a Sikh temple, a Hindu temple, a mosque and several churches. The B.N.P. and N.F. concentrate on more fertile ground where communities are separate and can be more easily fed demonising stereotypes.

'In many of the Yorkshire and Lancashire towns experiencing mistrust and violence, the Muslim-predominant ethnic minority has separated rather than integrated. It is in these towns that the racist message can do most damage.'

A group of devotees, speaking quietly in Gujarati, the language of Jains, asks me if it is my first visit to Leicester. Discovering it is, they suggest I head for the Golden Mile, the heart of Leicester's Asian community. As I leave, I scan the pacific messages of a religion that has flourished for 2,600 years (Jainism is older than Islam, Christianity or Buddhism): 'Relativity in thinking. Nonviolence – live and help to live – reverence for life. Right belief – right knowledge – right conduct – together constitute the path to salvation.' And finally 'Nonattachment to possessions'.

Guiltily I unlock the padlock from Rosinante and cycle up to the Golden Mile where I find Indian jewellers, saree and salwha boutiques, and gulab jamen sticky-sweetshops, providing a splash of colour as brilliant as the illuminations along Blackpool's more famous Golden Mile. Billboards provide a flick-book memory of the subcontinent in spectacular ads for Bollywood films ('The Best in Asian Family Entertainment').

Outside Pipi Printers, a dapper elderly man is looking up and down the street as if expecting someone. Nanu Bhaig Patel opened his shop in 1967 and for 35 years has specialised in Christian, Muslim, Sikh, Hindu, and Jain wedding cards. He originally migrated from Nairobi, a Kenyan Asian like so many in the area. Echoing Dr Mehta, Mr Patel claims Leicester serves as the blueprint for racial tolerance. 'It's a peaceful town where everyone gets on with each other.'

'What's the worst that's ever happened to you here?' I ask.

'Nothing. It's all been good,' he replies and beams. It seems absurd. I have absolutely no doubt his contented disposition is not shared by all members of the multiethnic community, nevertheless the kids wandering the streets – Muslim, Hindu, Sikh and the occasional white – appear relaxed and integrated into their community rather than rampaging through it.

As I cycle back towards the Cut, the city becomes increasingly white. At Eastgates a mixed crowd is watching a Bolivian band busking and attempting to shift some of its stockpile of badly recorded Andean whimsy. Above them, knight Simon de Montfort, in full armour and fresh from crusading in Jerusalem (where he no doubt also managed to line a few chests with booty), stares down, keeping a lookout for infidels.

God only knows what he makes of the Muslims shopping without so much as a ruck.

A middle-aged man who appears the worse for wear, stops a six-foot anorexic beauty to inform her he likes her cropped Union Jack T-shirt. She smiles but seems uncertain how to respond. She scans his face for illumination. Is it her breasts or is it really the T-shirt? Whatever, chances are he's either a lech or a racist. Later I see her exiting a shop with two male rastas. I don't think he'd like that.

I check my watch for the fourth or fifth time in the last half-hour and decide I can finally head for Leicester railway station. It's time, at last, to meet Max's train from London. It's only an hour late, not bad in an era when trains go missing for days. I feel a surge of excitement as Max does his best to run along the platform with his kitbag banging against his leg, and we explode into each other's arms like young lovers. Max is even taller than when I last saw him and his shoe-size has stretched from six to ten in just a couple of months. I've missed a growth spurt.

We have a bite at Burger King before making a leafy exit from Leicester. Max, with an experienced eye, comments on how unkempt this stretch of the Grand Union is. Unchecked vegetation is threatening to consume both *Caroline* and the canal. Every couple of hours we have to lift the weed hatch to clear the propeller of its turban of weeds. As Max tells me how much he's missed *Caroline*, I remember back to his first visit when he confessed to approaching the televisionless couple of days as a survival weekend.

For the next couple of days and nights we manage to moor in isolated spots. Max lies on the bunk downstairs reading magazines, fishing, making snacks, cycling the towpath, sunbathing on deck. It is a Huck Finn existence, as far from his London home as the Mississippi.

Eventually we pull into Foxton, one of the busiest canal villages on the entire network. While Max trawls the canalside shop, I wander up to the Canal Museum and search for exhibits relating to the I.W.A.'s pioneering rally against canal closures in 1950 at nearby Market Harborough. About that time only 1,250 miles of inland waterway remained open – now, largely due to its crusading, we are back up to 3,000.

I can find nothing on the rally and so ask the curator, Mike Beard, if there's anything hidden away. Mike disappears into a back room and returns with a scrapbook that is an anonymous woman's labour of love and diary of the event. The first pen-and-ink entry, written in an exquisite hand, is dated 19 August 1950. In it the author tells us she's travelling on a hotel boat and spells out why she, and others, are heading for Market Harborough. The rally is to be 'a demonstration of the still-neglected potentialities of British waterways for both commercial and pleasure traffic'. There follow postcards, commemorative first-day covers, a map of the network, and newspaper clippings. Together they paint a picture of the gathering that is both a celebration and crusade by more than a hundred boats. Some are the most sumptuous of craft. Other enthusiasts attend on war-surplus floating pontoons. There are pictures of the Ovaltine fleet, Tom Rolt's *Cressy*, and even one of children riding an oil-drum raft. The average length of journey – 215 miles – made by the boats, shows the commitment, particularly as the network was in such a dire condition.

Of the museum exhibits on public display, the most interesting is a model of the Foxton Inclined Plane. The real thing opened in 1900 with two caissons lifting boats sideways on rails up and down the hillside. There is, predictably enough in the New Canal Age, a study under way into the feasibility of reconstructing the Foxton Inclined Plane by 2011, in time to mark the one hundredth anniversary of the abandonment of the white elephant (the planned widening of the Watford flight didn't materialise and so the area remained inaccessible to 50-ton loads).

If the Plane is rebuilt, it will reduce the 75-minute journey time through the flight of locks to just twelve minutes. In so doing, it will entirely miss the point of leisure, as opposed to working, traffic. The Foxton Flight is the Acropolis of the inland waterways, a classically sculpted work of art that only philistines or those suffering the advanced stages of hurrysickness would want to skip. Unlike the Anderton Lift which accesses one navigation from another, a new Foxton Inclined Plane would access nothing new. In my view B.W. could use the money far more usefully in providing first aid to the Cut instead: dredging, replacing leaky gates, cutting back undergrowth, and providing more moorings along the Leicester arm.

By the time Max and I are back on board *Caroline*, we're next in line for our run up the flight, a symphony of ten locks in two staircases with a passing pound in the middle and a total collective rise of 75 feet 2 inches.

A tanned female B.W. lock keeper waves us into the bottom lock. I've been warned that Crystal is a formidable taskmaster and you risk hellfire and damnation if you don't do exactly as you're told. I steer *Caroline* in as clean as a whistle. Crystal smiles. 'Not bad.'

She turns out to be far more kindly and good-humoured than the ogre I expected. With her swept-back black hair, handsome olive skin and shapely figure I guess her to have South American blood in her. I learn later from her aunt who happens to pass while we're doing the fourth lock, that Crystal is actually Italian on her mother's side ('Mafia, if you want to be specific!').

Each lock is just seven feet across and lined with colour-coded paddles. Crystal teaches Max a rhyme to help with the sequence (he's doing one side while Crystal does the other and I lounge about at the tiller watching them) – 'Red before white and you'll be all right.'

At the halfway point of the flight, we're joined by Crystal's husband Mick who shares lock-keeping duties and is just returning from working another boat through. The quieter half of the team, with his muscular legs, low centre of gravity and white beard, looks as if he's just stepped off a fishing smack.

The locks are mobbed and several gongoozlers get in the way but Mick and Crystal are never less than polite. I comment on how busy it is.

'Sunday. Weekends in summer we get around thirty-five boats a day – though we have had fifty.' Crystal was born and bred in nearby Market Harborough and her accent has the soft country drawl that reaches perfection in Norfolk. The couple started working the flight at 7 a.m. and won't finish until 7 p.m. They've been together eleven years, though for part of that time, Mike was extricating himself from an acrimonious first marriage complicated by battles over custody. Mick has been employed by B.W. at Foxton ten years and Crystal six. Despite his seniority, there is no question who's the boss. I ask Mick about Crystal's formidable reputation but before he can reply, Crystal's already in there.

'Men don't like women who tell them what to do. Simple. It's interesting though, they're often more obedient when I tell them than when Mick does. If anyone does get shirty because they're in a hurry, I just tell them to go find a motorway.'

A rogue weed has got a toehold in one of the gates and Crystal comes on board *Caroline* to lean over and rip it out. 'That's been annoying me all day.' A perfectionist. It is a beautifully kept flight. I ask if she'd like a B.W. cottage to live in like other lock keepers so she could grow flowers and weed to her heart's content. 'Nah. If we get fed up with our neighbours, living on a narrowboat, we just move off. Can't do that with a house.'

Each winter, when their B.W. contract expires, the pair take off on *Ragamuffin* for five months exploring sections of the network that aren't closed for repairs. 'Best time of the year,' Mick claims. 'Last year we spent two months on the Birmingham network and didn't see another moving boat. The winter liveaboards are like a secret society. You know everyone but no one bankside knows you're here.'

'It's like the Foxton,' Crystal chips in. 'The best time here is at night when the crowds have gone. In summer it's just magical.'

At the top of the flight, I thank our helpers and they set off to assist another boat down. Max and I head off to play *boules* beside the towpath before taking a football into Foxton Lock Country Park. We are reliving Chelsea's 1998 F.A. Cup victory as someone pulls up into the car park with windows down, and we hear an excited Wimbledon crowd encouraging Tim Henman to hang on in the Wimbledon semi-final. We lie in the grass drinking from a shared bottle of water and listening to the commentary when proceedings are suddenly interrupted by a news bulletin. While swathes of England are buying strawberries and sitting around Centre Court, or their televisions, hanging on to their fantasy that Tim will become the first Brit in 64 years to win Wimbledon, in Bradford cars are being burnt and people beaten up in the worst racial violence since the 1980s.

Twenty years ago a similar parallel universe existed when race riots exploded in Toxteth and Brixton while the rest of the nation sat glued to another televised fairy tale as Princess Diana married her prince (and we all know how that one turned out). News duly filters through that Tim's Wimbledon adventure is over for another year. He has been whipped by a man ranked 150 who he'd beaten on each of their previous four meetings. At this moment, however, his dream of lifting the Lawn Tennis Association's trophy still seems more realisable than England one day creating a harmonious multiethnic society.

Max and I return to the boat to collect our bikes and hare off down switchback country lanes. Ostensibly we are looking for a pub or shop to buy some matches for the gas oven. But it's just an excuse to stretch the legs and the outdoors summer day. Max comments on how quiet the lanes are and how if he lived in the country, he'd do a lot more cycling. He then starts singing Destiny's Child's latest hit, 'I'm a Survivor'.

By 8.30 p.m., still in broad daylight, Max is fishing from *Caroline* as I make supper. He calls down to tell me he's cast the line perfectly twice already. 'I can do it now,' he proudly informs me, 'I haven't tangled it once.' A little later he screams down that he's caught something. A small bream. Within minutes he's caught another and I show him how to remove the hook from its mouth so that he can throw the fish back in the water.

'When I caught my first fish as a boy, I was so scared I threw the rod in the lake,' I confess.

'You wuss.'

'I know. I was. Still am really.'

When Max catches a third small bream, we cannot extricate the hook, however much we try.

'The fish has swallowed it.' I utter the obvious. The hook is now embedded somewhere dark and unreachable. 'I'm afraid I'm going to have to kill it.'

'Why? Why not just break the gut with your teeth and chuck it back in?'

'Because that would be crueller. The fish would just take longer to die and suffer more.' Max stares at the fish and weighs up my words. 'Can you hold on to it for a minute?' I ask and descend into the cabin. When I reappear with the baseball bat, Max laughs. He can't believe I'm going to use something so big on something so small. It's the equivalent of using a pneumatic road drill for a dental filling. I lay the fish out on the bank and bring the bat down. Silver glistens in the grass like a smashed mirror.

Over dinner Max shows me his school report. It's clearly much better than he expected and he peppers his careful recitation of grades with comments like, 'I can't believe she gave me that.' The mood is very mellow. There's more warmth on board now and it's not just because it's summer. I'm sure *Caroline* is enjoying having Max around as much as I am: there's more music playing and Max has even managed to tune in four channels on the television, albeit buried beneath snowstorms.

A news broadcast announces 170 police officers have been injured in Bradford. Max switches to *Survivor*, an American-pioneered show that got the U.S.'s biggest-ever audience for the final showdown when one million dollars was up for grabs for the competitor most endowed with the qualities of connivery, mendacity, betrayal, and ruthlessness. My modest proposal is for Asian drug gangs and any other groups displaying overtly macho or violent behaviour for whatever reason, to be given free passage, along with N.F. and B.N.P. members, to the castaway island where they can hone their chosen values and wipe each other out.

Max sleeps at the fore-end and requests we keep the bathroom doors closed so we can see each other from our bedrooms. He says it makes him feel more secure. In the morning he sees us off solo, unmooring ropes, pushing off the fore-end, and accelerating gently with tiller in hand. I meanwhile catch up with some washing – Max has been on board two days and has already gone through three pairs of shorts, a pair of jeans, a sweatshirt, two pairs of socks and three pairs of pants. Three T-shirts currently last me a week on board. The washing line, a V between the anchor hatch and the two pulpit rails, is more cramped than usual, and several more items are decorating the cabin.

Having swept and washed the floors, I join Max at the stern with tea and cake. Susanna has sent another Dundee cake up and it's disappearing disturbingly quickly. We sit either side of the tiller on the wooden bum-rests welded to the stern rail. The hexagonal pattern imprinted into the compressed wood is the same as that on the decking. Both have grown paler and more mottled over the months since my departure.

Max wants to know more about my book and how it's going. Unusually for a child – and for an adult for that matter – he's always curious about other people's lives. He tells me my working title, which has changed from 'The Motion of Sleep' to 'The Water Road', to 'Watertopia', to 'The Cut', is no good. '"My Journey Along the Cut" would be better,' he suggests. 'People won't know you've actually done it.' He interrupts the analysis to point out a field that 'would be excellent for rolling down'. New eyes. '"The Cut" is better than the others though,' he picks up the thread once more, 'because people might think it's a thriller and buy it by mistake.'

Max spends most of the morning singing, despite the overcast day. He comments on how quiet it all is. The Cut – the navigation, not the book – looks to be in better shape here. From time to time Max returns to the subject of the book: he has clearly decided to give it a fair amount of thought during his sojourn on board. 'Did you include the maggots?'

'Of course.'

'I thought you would. I suppose you could call it "Maggots on the Cut". They'd definitely think that's a thriller – maggots crawling out of a massive wound.'

'It has been for me.'

'What?'

'A thriller,' I pretentiously reply.

'It is fun,' Max concedes.

As we swim along, Max looks for fish. 'Would there be swordfish here?'

'No, that's sea water.'

'That's a big hill,' city boy comments admiringly.

Max has always liked adventures and always been a little anxious about them too. A typical boy then. Until a year ago when his growing bones became heavier and he started sleeping later at weekends, on holiday he'd always be up at the crack of dawn with me, seeking out cliffs, diving off rocks, exploring coves. Health-willing I'll be trekking with him in the Himalayas in twenty years' time. Hopefully, I'll also still be camping out with Larne at Glastonbury.

There are no locks today and no canalside villages. We sail a long pound winding through rolling fields of wheat and grazing pasture. After a couple of hours, we slip under Honey Pot Farm Bridge, moor and walk up to the sleepy village of Husbands Bosworth which, I learn in the general store, got so named because there were two Bosworths: one had a large weekly market and one was mostly involved in husbandry, thus the former became Market Bosworth and the latter Husbands Bosworth. It sounds feasible.

In the graveyard of the church, Max comments intriguingly, 'You can get lime disease standing on dead people.' He follows this up with 'Where's Woolworth's?' We try the church door. It's shut. 'What if someone's just lost someone they love and want to pray?' Good point. Before I can attempt an answer, however, he's distracted by a sign for Butt Lane which cracks him up (his generation have adopted the American slang *butt* in preference to the English *arse*).

It's a very different experience seeing the country through my son's eyes and I love it. Back on board, Max offers to take us through the 1,166-yard Husbands Bosworth Tunnel but quickly finds the pitch darkness and narrowness unnerving. I take the tiller.

'Would there be scorpions here?'

'No.'

'What if the walls close in or the roof collapses?'

'They won't.'

'What if the light goes off?'

'We'd slow and use the torch.'

'What if the bulb in the torch breaks?'

'We'd go slower and bump our way along. Don't worry, we can't get lost, Max.'

'You never know.'

Not long after re-emerging into daylight, we pass a sign: 'Welcome to Northamptonshire. Rose of the Shires.' Seven Canada geese glide by as we pull into Crick Wharf moorings. 'Whoa. Big ducks.' He gets out his rod and catches three more fish. Unfortunately, again we fail to remove the hook from one. Max is mortified to think it will die.

That night, before turning in, Max offers one final piece of advice on the book. 'You must give an introduction saying what the book's about.' I commend him once more on his school report. The shock of generosity from teachers he thinks hate him, clearly still hasn't quite worn off – 'I still can't believe Geography gave me that ... nor French.'

Max sleeps late. I potter. We play *boules* on the towpath and it's eleven o'clock before we're under way, passing a succession of ripe wheat fields that smell as delicious as they look, spreading like soft sand to the horizon. We negotiate the 1,528-yard Crick Tunnel and then the Watford Locks – three singles and a staircase of four. It's a gloomy day. Following last week's blistering summer, we're back to perma-grey.

We lunch at the Stag's Head beside Welton Station Bridge where I ask the Portuguese owners where the nearest shop is.

'Cross the bridge and take the first right – ignore the no-entry sign – one hundred metres and you're there. Can't miss it.'

Following their directions is like slipping down Alice's rabbit hole. From a timeless rustic canal setting, we're thrust 200 yards away into the twenty-first-century bedlam of concrete, tarmac and glass of the M1's Watford Gap service station. In minutes we're standing in line behind truckers and travelling salesmen queuing to pay for fuel. In our basket we have milk, a newspaper, Max's current favourite magazine – *Woman's Own* – and a couple of exorbitantly priced apples.

'Which pump?' the checkout man asks.

'No petrol. No car,' I flummox him before Max, attempting to clear up the confusion but in fact just muddying the water more, adds, 'We're on a boat.' The man looks around the car park nonplussed.

Back aboard *Caroline*, Max sleeps as the rain lashes, and I read the paper. We moor for the night less than an hour away at Norton Junction, where the Leicester arm joins the Grand Union Main Line from the south. Tomorrow, the four miles we travel to Braunston will be a retracing of my earlier water road up the Grand Union. Beyond it, however, instead of heading for the Hatton Flight and Birmingham, this time we'll turn south at Napton Junction onto the Oxford Canal.

I've already warned Max that tomorrow he'll be left more to his own devices as I have an important appointment: a reunion with Margaret Cornish, the 85-year-old canal veteran, along with her friend and former Idle Woman, Olga Kevelos.

Moored just three boats down from us are four old working boats breasted up in pairs. A young man, in tell-tale B.W. green top, is sanding down the hatch cover on one. I leave Max to his fishing and amble across. 'Fine boat. Are all four yours?'

'I wish.' The man pushes his hair off his forehead with the back of his hand before putting down the sandpaper. 'We're minding the *Ara* and *Archimedes* for friends while they're holidaying in Greece. I'm picking them up tomorrow from the airport.'

A second equally petite body, this time female, appears above the hatch and smiles. I introduce myself. Nigel introduces Jo with whom he has been living on the Little Woolwich and river-class butty for the past five years. 'I work as locksman here at Norton Junction.' Jo, aged 31, is a scientist who designs asthma inhalers. Nigel used to design laboratories, 'but in 1997 we sold the house and I retired!' His boss kept the job open for two years, convinced 'the romance with the water couldn't possibly last!' He's now 28 and says he'd never go back bankside.

I ask them what the biggest problem is living year-round afloat. 'The cocky admirals on their seventy-foot floating palaces with wives who never get out of the marigold gloves in case they break their nails doing the locks,' Jo replies.

'Most of the boaters, though, are a friendly, tight-knit community,' Nigel attempts to put the admiral in perspective. 'I suppose it would be njcer, though, if there were more younger people using the canal. There are pockets. But it's basically too expensive for young people. For the price of a week's hire, a young family can fly to the Canaries. And if you want to trade in your house and move full time onto the Cut, it's not so much downscaling these days with the inflated prices of boats.'

'The camping boats used to provide a cheap option,' Jo says, simultaneously picking up the sandpaper and taking over the rubbing. 'But they don't really operate any more and the hotel boats are too pricey.'

'What do the couple do who are holidaying in Greece?'

'They deliver domestic fuel commercially along the Grand Union and into London. In fact the *Ara* used to be a camping boat too before Alan and Trish bought it. Heather Duncan, who lives in Birmingham now, used to run it and sometimes Julia at Stoke Bruerne did. Camping boats were a good way for young skippers to get experience.'

'I've met both Heather and Julia already on my trip,' I proudly declare, expecting them to be shocked it's such a small world. They're not.

'Linear village – that's the Cut. Everybody knows everybody,' Jo sums up. It reminds her, however, of another difficulty facing women on the Cut. 'As you've already found it, it's a time capsule, and so enlightened males are a little thin on the ground. I had one middle-aged weekending boater call out to me a couple of weeks back, "That's a big boat for a small girl. Does your husband allow you to drive the car too?"' I laugh.

'What did you say?'

'"Noooo. You've got it wrong, I'm afraid," I said all sweetly. "I own the boat. He's not my husband, he's my live-in lover and lockwheeler." Then for good measure I added, "That's a nice boat you've got though, but a bit thin for such a fat man."'

'It's the kind of man who invariably leaves a paddle up in the lock too,' Nigel continues when the laughter subsides sufficiently. 'The basic ethic of the canal is community...helping each other. If someone's broken down, you stop. Nowadays unfortunately when I tell someone they've left a paddle up, they'll say, "It wasn't me, son, it was the last boat." So what? Why not just do it? If we're not in it together, the water drains from the pound and it spoils all our boating. It's a microcosm isn't it? It's the same environment question that Bush refuses to face, reneging on the Kyoto agreement, ignoring America's commitment to reduce industrial pollution. We're all in it together.'

A couple of hours after leaving Norton Junction on another cold grey morning, we moor in Braunston at virtually the same spot I interrupted my northbound journey to visit Dannie in hospital in London. Dannie now, thankfully, is fully recovered.

The small town of Braunston is one of the most important on the entire network, not only because it's the starting point for the Oxford Canal and the crossroads of the Midland Canals but because its church on the hill is 'The Cathedral of the Boatpeople'. Its crocketed 120-foot spire dominates the town, soaring from the final Northamptonshire hill before the land dips to the Warwickshire plain.

Although Braundeston is mentioned in the Domesday Survey of 1087, the hamlet only grew to a town with the coming of the canals. By 1780 the wharf was already busy and boatmen enjoyed a reasonable standard of living. By the end of the century a community of leggers settled in Cross Lane, running boats through the nearby 2,042-yard tunnel. Craftsmen and retailers followed, along with boatbuilders, blacksmiths, wheelwrights, harnessmakers, shoemakers, tailors, and rope makers.

There was a cost, however, as the industrial revolution transformed England from a garden to a smelting yard. In the first thirty years of the nineteenth century the populations of both Birmingham and Manchester doubled and the number of households in Braunston rose from 88 to 287. As the town's population soared, living conditions became appallingly insanitary. Boat people washed in the canal alongside long-distance boats carrying night soil from the cities. In 1831 cholera hit our shores for the first time and quickly found its way to Braunston.

A week before the town's biggest party of the year, Wakes Week, a legger from Cross Lane died of 'cold and inflammation', followed by one of his children. As stalls and the fairground were set up on the green, two more children from the same Padmore family died. By the time that night's dancing and revelry was over, Mary Ann Mawby, aged eighteen, had gone the same way. According to David Blagrove, in his booklet, *Braunston...A Canal History*, on the Tuesday the bell in All Saints Church tolled for departing victims, and the local Reverend Richard Miller, the town's Baptist minister, declared the deaths 'a divine warning to the inhabitants not to indulge or tolerate sinful amusements'. By the time the epidemic moved on, seventy people had contracted cholera and nineteen had died.

Max and I take a stroll past a couple of hotel boats (which charge guests £500 each for a week on board), a narrowboat café serving £3.25 breakfasts all day, and on through the marina with its rainbow of brightly coloured narrowboats. After three games of *boules*, Max decides to escort me up to meet Margaret who had designated All Saints Church – where so many boatpeople made their entrances and exits – as the meeting point for our gathering of canal pilgrims.

We're a little early. In the vestry I find Derek Turner, the occasional verger, sorting hymn books. I ask if he knows where any of the old boat-people are buried. He scratches his head with one hand and puts down the two books he's holding in the other. 'Let's take a look.'

Outside, we skirt a stone wall that separates the church from a windmill that's been converted into a home. 'My grandfather knew it when it was a working mill in the late 1800s. One day, when the owner was fitting cloth sails to it, a gust of wind got up so strong, it blew him off the platform and into the churchyard and he died instantly. At least he didn't have far to go for his funeral.'

'How long has your family lived here?'

'As far back as I know about.'

'And do you still live in the village too?'

'Lived here and worked here all my working life. I was postmaster twenty-five years and one month. My mother did it before me and I used to pull her leg that she was a fly-by-night because she only did twenty-four years and eleven months.'

On my left is a memorial stone to three brothers, aged 33, 28 and 19, killed within a year of each other in France in the First World War.

In a second plot across a small treelined road, Derek greets a 60-year-old volunteer with candyfloss-grey hair who's trimming the grass round the graves. Soon Hilda and Derek are playing spot-the-tombstone. It reminds me of a game my children invented when they were younger called 'Blind Man's Grave' which involved feeling a gravestone blind-folded and then attempting to refind it having been spun round several times and led twenty yards away (as Larne attempted to feel her way back by Braille, Max would call out, 'Watch out for the empty grave!').

The first grave Derek and Hilda identify is that of Frank Nurser who

specialised in boat painting and owned Nurser's Yard. We pass three
more Nurser graves before Hilda starts pointing in every direction at
once. 'Mrs Beryl was boatpeople. So were the Kendals – they were a big
boat family. There's Daniel Frederick Kendal over there. He died in 1963
aged sixty-four. There's Fanny and Thomas Hough – they died in their
eighties in the 1960s too...'

'I remember,' Derek interrupts, 'there was one in the Hough family
who, whenever he was heading home, would send a telegram. If it just
said "Home today" it meant he would be around for a few days. If it said
"Home today for away," it meant I want the washing done and I'm off.'
Derek would read out the messages to the family because they were, like
most boat families, illiterate. ('Pension day I'd spend most of the time
certifying crosses.')

Hilda is continuing searching for more graves. 'There's plenty more
Houghs here.' She pronounces it 'Huffs'. 'There's the Littlemores who
ran the Ovaltine boats. There's the Griffens, Wards, Webbs. They're all
here. Most lived in Cross Lane until the carrying finished.'

'This one, Frank Charles Littlemore...' Derek points to a stone on
which the circumference of a life is measured '9 April 1912–2 June 1998'.
'I remember him. Charlie lived with his parents in Cross Lane and came
in and ordered the same thing every Saturday like clockwork. When his
parents died, every Saturday at five p.m. he'd still come in with two
empty pop bottles and get them swapped for full ones – a bottle of
Corona lemonade, and a bottle of Corona lime. He'd order two Lyon's
swiss rolls – one raspberry, one apricot; a four-ounce bar of milk
chocolate; and a quarter of dark stripes – you'd know them as humbugs.
I could have had it all ready, laid out for him, but he wouldn't have it.
Always wanted to order it himself and it never varied once his entire life.'

Hilda herself comes from a long line of boatpeople. Her mother was
born in the Black Country into the Wenlock boating family. And her
Uncle Ted had a boat on the Thames. She shows us to the plot shared
by her grandfather, Edward Wenlock (who died in 1929 aged 69), and
her grandmother Sarah who died in 1942 aged 84. 'They carried
freight,' she provides the headstone for their lives. There are fresh
flowers on the graves.

Hilda notices another woman at a grave, arranging flowers. The beds
of the occupants of the cemetery have been angled on an incline so that

if they should wake and sit up, they can look out to the Oxford North wending its way through the Warwickshire plain. Hilda lowers her voice. 'That's Alice Brown, one of the last proper boatpeople still living in Braunston. She's part of the Kendal family.' Hilda introduces us and returns to her trimming while Derek heads back inside the church to sort out a few more hymn books.

Once we're alone, sitting on a bench, Alice's story tumbles out into an empty grave. Alice was born 'on the *Providence* in a dry dock in 1922 underneath the Cape of Good Hope in West Bromwich'. She wears very thick glasses which her eyes find it daily more difficult to see through. 'Usually when a baby was due, boatpeople moored up close to where a nurse was to hand. But my mother was so used to giving birth – I was the sixth and nine more would follow – it were like shelling peas to her.'

Alice screws up her eyes. 'There's only six of us left now. The last to go was a younger brother just three weeks back. We scattered his ashes at Braunston Lock and flushed them out with the paddles. Most boat people had big families. I knew a father and son who had twenty-two children each. Each time Mum had another baby, the oldest had to go to work another boat. The most children they'd ever have on board was five. I worked with five or six lots of different relatives, walking the horse ahead of the boats. Often you'd work right through the night, taking turns to get your head down. Seven days a week too. And the only time my dad went abroad was in the Boer War aged seventeen – lost his pension through that because he was under age.'

I ask Alice how she was treated by banksiders. She laughs. 'Not good. They called us water gypsies but we were much cleaner than them, I can tell you. You'd have coal on one week and sugar the next. You had to be clean. But people were nasty, especially round the slums at Paddington. Don't know why. They looked down on us. But it's like the black people, they're no different from us are they? Their blood's the same, that's what I always say. I understand that, why can't others?' I feel like hugging Alice. 'What are the parents doing, that's what I'd like to know. I wouldn't allow my children to treat people badly even though they're all grown up and gone now. I still do the stopping. And if my children were doing something on the coloured or Irish I'd stop it there and then. You can't blame the schools. They're not there all the time. They can't rule them. I always told mine if they did wrong, "Just remember you've got

to come back home some time; you can run away all you want but when
you come home there will be a belting waiting." It's not up to the
government to rule your children. You rule them your way. And that's
where the trouble starts, when parents don't rule them.'

'Did you go to school?'

'A few times. They were nice on the Salvation Army barge down at
Bull's Bridge but at the other schools, teachers didn't like us. They'd give
me an Enid Blyton book and sit me in the corner on my own for however
long we had to hang around waiting for our next coal order. I couldn't
read – none of us could. But when I was twelve I bought a tuppence
magazine and learnt myself. If you've got it up here,' Alice taps her head,
'you can work it out. I'm a better speller than my husband. He does the
writing, I do the spelling. He was a farm labourer. Even less money in
that than boats.'

Alice does not sentimentalise her life on the Cut, neither does she
miss it. 'I like the warm of my home in Spinney Hill. It's a lovely
neighbourhood. I'm happy in my house.' As I cross the small road and
return to the church, to see if Margaret and Olga have turned up, I
conclude that a greater affirmation of life than Alice's 'I'm happy in my
house' would be hard to find.

None of the canal pilgrims have yet arrived. It was six months since
I'd visited Margaret at her sheltered accommodation in Felixstowe. As
she doesn't get about as much as she'd like to these days, she decided to
make her outing up to visit me on *Caroline* a bit of a party, inviting
several other scattered friends and family. When I'd asked why she'd
chosen All Saints graveyard as our rendezvous, she typically replied
'Thought it appropriate as I'll be moving in permanently shortly.'

I stand at the church entrance looking into passing car windows, while
Max plays his Game Boy inside. The message board keeps me company:
'Worried? Despairing? Suicidal? Someone to talk to anytime of the day or
night. Samaritans.' There's a small lined card advertising a grave-tidying
service – 'headstones cleaned, grass trimmed, litter removed twice yearly
£30. No V.A.T.' I also learn that a charity cream-tea weekend is approaching. And there's an ad for a salaried post looking after the maintenance of
the village – picking up litter, weed clearing, keeping the bus shelter tidy.

I'm contemplating whether the job would pay for mooring fees down
at the marina when the first of the party appears, a tall man with a shock

of white hair who's dressed in a short-sleeved shirt despite the chill wind. I introduce myself and discover that John White fills one of the middle pieces in the jigsaw of Margaret's life.

After Germany surrendered, Margaret, along with the rest of the army of Idle Women, returned to civvy street. She married a farmer, divorced and lived a nomadic life with her three children, one of whom suffered from Down's syndrome. She taught at a village school, then bought a Romany caravan and moved to Oxford where she became one of the first female clippies on the buses. In the evenings she attended college dances where she pretended to be an undergraduate (it was one of her only regrets that she never attended university). Margaret drifted to Liverpool where she somehow got a post teaching Philosophy on the Master's degree course in Education. She then moved to Ireland, living in the middle of nowhere without electricity or running water. Eventually, she bought a narrowboat, the *Alphons*, and returned full time to the Cut and it was then that John and Margaret's lives crossed.

That was in the late 1960s. At the time there were 802 hire boats operating on the entire inland waterways and seven of them belonged to John's fleet, *The Water Gypsies*. Margaret, discovering a kindred spirit, took up a permanent berth in his basin at Weedon. 'For four years we'd hear her daughter playing the upright piano at all hours – it's the only piano I've ever heard on a canal boat.'

A car slows outside the church and a large woman in the front seat looks up and waves. It's Margaret. She arrives with a basket of provisions and a flask of coffee for Roger, an ex-Gordonstoun, Felixstowe driver, who has become her loyal companion. Roger, a mountain of a man, greets us in a shy, plummy voice. Margaret gives me and Max, who she's never met before, big kisses and together we head inside the church.

Once we've settled into pews, I tell Margaret that I'd met Kevin Jackson on her old boat, the *Hyperion*, a couple of months back and had passed on her blanket for Vicky.

'Did you meet Vicky?'

'No, she was away visiting her mother. I left it with Kevin.'

'I sailed with him last year.'

'Yes, I know. Kevin told me all about your jaunt with Olga up to the I.W.A.–B.W. festival at Crick and how you were the guests of honour.'

'Ridiculous. God knows why they've suddenly decided Idle Women need excavating.'

Margaret's Braunston memories belong both to the 1960s – when living upon the *Alphons* ('I still have my stool and dipper from Nurser's Boatyard') – and her days running essential freight during the war.

'In the forties we were an alien race – women who wore trousers! Scandalous. Even the boat women wore long skirts. Some of our lot were very flamboyant. Olga, who should be arriving soon, took the biscuit. She was the youngest of the recruits and wore colourful trousers topped by an exotic Mexican hat. Initially the boat people treated us with suspicion but once you'd proved yourself and done a little time, they'd give a hand. I expect it helped that we were treated by outsiders as badly as they were – I always had a few lumps of coal ready for when children threw stones or spat at us.'

During her second stint afloat, Margaret switched from coal to a catapult as hooliganism escalated. Although vandals eventually drove her off the Cut, the canals remain her spiritual home. 'On July twenty-first, 1969 I was sitting on my boat when the first man walked on the moon and I remember thinking at the time, that's nothing compared to what Brindley and Telford achieved with the network of canals.'

Max wanders off to resume his battle with the Game Boy. He sits alongside the tombstone of a prone knight with coyly crossed legs. Behind him, above the altar, is a blaze of blue and ruby-red stained glass.

Next to arrive in the doorway are Helen, Margaret's 42-year-old pianist daughter, and her eight-year-old son, John. Margaret kicks away one of the colourful kneelers, so she can get past and plant more kisses.

The final pilgrims are Olga and her brother Raymond, who share a home in nearby King's Sutton. Olga is 77 but with her short brown hair, jeans and blue top, looks to be still in her fifties. Although they worked on different boats during the war, Margaret and Olga occasionally met up. The first time had been at Bull's Bridge, near my own starting point in Southall.

'You were rescuing Daphne at the time from Billy who was threatening to kill her,' Margaret remembers.

'God, yes. She went berserk with an axe, trying to smash her way back into the cabin after Daphne had told her she had to leave the boat. Billy was so strong. I wasn't in the least surprised when she eventually got her sex change.'

'Is that Jo from your book, Margaret?' I ask.

'Yes. I changed her name to protect her but she's dead now.'

'She sounded interesting.'

'Anatomically interesting too,' Olga pipes up in her flat Brummie accent.

'She proposed to me, you know?' Margaret says, matter of factly.

'Was that before or after she proposed to Virginia and Daphne?' Olga asks.

Margaret doesn't laugh. 'She was schizophrenic and paranoid and very strong despite her gamin looks. Excellent boat woman, but volatile. She once picked me straight out the water and I wasn't the smallest of women.'

Among the I.W. recruits there were other characters too, such as Elsie, a cockney who never once changed out of her herringbone jacket: 'a sparrow of a girl who outswore any boatman'. And there was another Margaret, a New Zealander, who crewed her way from Australia to Falmouth on a square-rigged grain ship and was so driven she worked her narrowboat team eighteen hours a day.

'Finding time to eat and sleep were always the preoccupation,' Olga underlines the war-time priorities. 'The overwhelming feeling was hunger and tiredness.'

'Do you remember the national loaf – filled with cement?' Margaret interjects.

'Worst thing I recall was my botched attempts to kill a duck I caught for the pot. Bungled one murder attempt after another. Bashed it on the head, wrung its neck, tried slitting its throat. In the end I just plunged it underwater to shut it up and eventually it stopped fighting. Funny that. Didn't know a duck could drown.'

As we sit in our pews and the rain lashes down outside, the two old friends move on to the rumours that greeted each new delivery order. Sometimes when they were told they were carrying prefab houses, they discovered later they were actually carrying ammunition – 'You'd always know because a bowler-hatted gentleman in a suit would come to check on you en route.'

'Then there were the steel strips. The story was it was really gold from the Treasury heading up to Liverpool and then to the United States because Churchill was concerned if it was left in London, it might receive a direct hit during an air raid.'

Eventually the rain stops and we pile into cars and head down to a canalside pub, the Admiral Nelson, for lunch. With our pints and halves, ploughman's and fish pies, we sit on the grassy bank overlooking Braunston Locks where Alice Brown recently scattered her brother's ashes.

Max and Helen's son feed the ducks leftovers and, as Olga stares into the water rushing through leaky paddles, it triggers another memory.

'Sometimes we'd find babies dumped in the canal, you know? They'd end up banging behind the double locks with the logs.' She catches my sceptical look. 'Well, what would you do? Before the war if you were a servant girl and had an illegitimate child, you'd be put in a mental asylum.' I nod, suitably chastened. 'There was a baby I actually saved once too. That was at the Cowroast near Tring. I was drawing a lock after a long pound and there was a group of children no older than six playing nearby who'd left the baby lying on the bank. The next thing I knew, the baby was face down in the water and the children's laughter had turned to screams. The baby was being drawn to the open paddle but I managed to grab hold of her. She was no more than eight months old. I got her up onto the bank and was removing all her wet clothes to wrap her in a blanket when the mother turned up and started screaming at me. "You bastard gypsies. Stealing our clothes. I've had it before."'

Olga's history is as extraordinary as Margaret's. Her first job was at the Royal Observatory as an astronomer. When war broke out and the Royal Observatory relocated to Bath, Olga answered a Ministry of War Transport advertisement for volunteer female trainees to carry vital supplies on the canals. After the war, she traded the slowest of waterborne transport for the slickest of wheels, becoming a highly successful professional motorbike racer, competing all over Europe for the next 26 years.

Olga and Margaret continue to focus on the part of their lives when their paths crossed – the three years they served as I.W. recruits. The pair seemingly have an unslakeable thirst for mechanical matters. They quote figures, comparing miles per gallon achieved relative to boat load – 'Took two hundred gallons to move a seventy-ton load a hundred miles.' They discuss the most efficient split of that load between the motor and the butty. They then move on to navigational difficulties encountered, particularly the double kink in the Braunston Tunnel. The latter in turn prompts another of Olga's stories – one which captivates

Max. It concerns her most disturbing experience in another tunnel, the Blisworth, down at Stoke Bruerne.

'I was on my own on the motor. The other two crew were on the butty, we were into the middle section of the tunnel and it was getting a little hypnotic with the metronomic putter of the engine and staring into the arc of light. Anyway I saw a light approaching and so I started slowing and moving over to the right. But the approaching boats didn't alter their speed nor direction. When they got closer I could make out a cream and coffee colour of a diamond-patterned bow. I recognised it immediately as one of the Barlow fleet. They weren't slowing and they weren't moving over. It wasn't like normal boatpeople. Just as we were about to collide, I put the tiller right over and we hit the wall. The stern slammed in and the tiller arm got bent. By now the Barlow motor had slipped past and I was going to give an earful to whoever was in charge of the butty. Being freight carriers they should know better. There was a lantern on the cabin top and only boatpeople had that because the old girls were frightened of the bogeys jumping out of the side holes in the tunnel. They'd refuse point-blank to go through in the dark. Anyway there was a hard-looking woman in the butty dressed in the black high-fastening dress of the boat women. She looked horrible. So I held my tongue.

'When we emerged from the tunnel, I pulled over as I had to find the spare tiller to replace the damaged one. The towrope was all over the place and the pair on the butty were yelling at me, asking why I hit the wall. "It was the bloody Barlow boat's fault not mine," I yelled back. They said, "What boat?" They never saw a thing.'

Max's eyes are as wide as an owl's.

'I tell you I was shaken. We went into the pub because we'd knocked some rivets out of the bow that needed replacing, and we sat discussing the incident while these boatpeople at the next table listened in silence to us. They just stared at us. Later a mechanic told us why. They knew the old story.'

'What story?' Max gasps.

'Well. Many years earlier a pair of Barlow boats were moored outside Blisworth Tunnel, breasted up and loaded with sugar. Boatpeople always sealed the boat at night, closed everything, so all the windows and doors were tightly closed. Apparently the stern gland, where the prop enters, leaked water in and filled the bilges. The sugar got wetter and wetter, heavier and heavier. And at some point it reached a critical point and the boat sank.'

'What happened?' Max asks breathlessly.

'The old couple drowned.'

'Was it ghosts you saw then?'

'I don't believe in ghosts,' Olga replies. 'I believe powerful traumas can leave traces, however, and I think that's what we stumbled into.'

'Like the Bermuda Triangle.'

'Maybe.'

The party breaks up. Everyone drifts back to the roads that lead from the water back to their homes. Olga and Raymond are the last to go. As they do so, they invite Max and me for dinner when we pass through their village, King's Sutton, in a few days time.

'I'm afraid Max will have jumped ship by then and I'll have my father on board.'

'Bring him instead. It will be a pleasure to meet him.'

Max and I walk the towpath back to *Caroline*. It's a pleasant evening and we decide to travel on a couple of hours into open country. Eventually we find a totally deserted spot with rolling hills fringed with trees running to the horizon. I cook pasta and we sit out on deck. Max tells me how much he's enjoying the week. I in turn tell him how glad I am he came. I also commend him on his patience through what must have been a long and boring day.

'No, I quite enjoyed it,' he replies. 'I specially liked Olga's story about the ghost boats. Did she make it up?'

'Definitely not. She saw them.'

'You don't believe in ghosts.'

'No. Olga doesn't either, does she? But she does believe that where there has been a tragedy, then something can get left unburied.'

'Well, that's ghosts. You both believe in ghosts then.'

'Maybe you're right.'

We're both looking forward to Susanna joining us tomorrow. At the end of the weekend she'll drive Max back home and my own father will then take over his bunk. From my son to my father: the bookends of my life.

We reach our designated meeting spot, Napton on the Hill, early in the afternoon and set off to explore on our bikes. At Napton Bottom Lock, a woman working the paddles looks up into the sky, shakes her head and says to Max, 'Looks like you'll get wet.' As we pass over the brick humpbacked bridge, we look back down to the lock in time to see her being helped out of the water. Max, naturally, finds it hugely amusing.

That night Susanna, Max and I eat at the lockside pub. The Folly is a splendidly unmodernised and eccentric pub warming itself round a large, brick fireplace. Decorating the walls and floor are weighing scales, flying pigs and a doll's house. Milk churns prop open the original wooden doors; and chairs for some inexplicable reason have been strung from the ceiling.

Above the bar itself, a splatter of photographs recall the pub's most memorable nights. The smell of burning firewood transports Susanna and me back to some of our own, sitting round fires in Himalayan lodges as we spent the first year of our relationship backpacking our way through Asia.

Max is a torrent of stories from his week afloat: the woman in the lock, the ghost boat, the games of *boules*, the number of fish he'd landed. Susanna tells me of news from home and work and passes on a message from Larne ('Dad you should be at home. Miss you. Love Larne xxx'). The pub is full of moored boaters who talk easily across tables as they tuck into the pub's speciality pies.

After our meal, I walk up through the Napton flight, while Susanna is sorting out bedding arrangements for the night. Midges are still zipping across the water, dragging ripples behind them like anchors. In the locks, the flowing hair and tangled beards of plant trolls undulate silently on the water under the moon. Crystal would have an apoplectic fit. In the fields cows graze, their uncomfortably bulging undercarriages the stretched parchment maps of the known world. Ahead of me is another arching brick bridge. Behind me, on a far hill, four silhouetted horses stand beneath the sails of a windmill. Beyond them the deep purple robe of sky has been edged with gold.

It is an idea of England we are still in love with, a composite of all that is good and unsullied. Like the man in the peaked cap sitting in the cabin of the battered old cruiser you just know he built himself, nodding as he slides by; like the sheep bleating, the skylark trilling, and the sun buttering the fields. Occasionally you may have to fast-forward over unexpected developments – the old guy suddenly venting his spleen with racist zeal; the majestic swan ramming its beak down on the duckling; the kids screaming 'wanker' as they leap off a bridge or crap on your boat – but these are mere blips. The Cut is still where we store our national treasures.

On my way back down to *Caroline* I chat with a guy in a Macclesfield Canal T-shirt walking his schnauzer. He's single-handed and has been afloat a couple of months on his current trip but has had to jump ship a few times for tests on his pacemaker which was recently deliberately stopped to shock his heart into a regular rhythm.

'Must have been weird. Like dying.'

'I'm used to it. I've told people all my life that my heart stops but they never believed me. I knew. I could tell even as a kid.'

We discuss the canals we've travelled. He's a member of the Caldon Canal Society.

'Great base for touring,' I enthuse.

'True. But our membership recently designated the Ashton Canal in Manchester a no-go area so that's one less that's accessible.'

'Why?'

'A couple of members were shot at and others had bad experiences too. It's had a bad reputation for a long time. Gangs jump on at one end of your boat, run through the cabin taking anything they fancy and then hop off at the other end. There's a boy called Tarzan who swung on a rope over one of our members' boat, snatching a bike as he went and dropping it on the far bank, before swinging back and taking the second. Last time I was on the Ashton, four years back, I was shot at with an air gun and had concrete lobbed off a bridge at me.' Fast-forward.

I go to bed that night untroubled by hooligans, bankruptcy, mourning, divorce, the sack, heart failure, or creeping disease. And with three-quarters of my family on board, I sleep like a lamb.

At 9 a.m. we continue on our 54-mile journey down the pathologically meandering Oxford. This is a contour canal to end all contour canals: a Brindley special winding so wildly and widely it almost makes circles in its attempt to never end. The Oxford Canal was the final flourish in Brindley's visionary Grand Cross, linking the birthplace of the industrial revolution (Birmingham and the Black Country) with London via the Thames.

In the meadow opposite us, two sand-coloured foals are rolling on their backs, cycling their legs in the air as if working out on gym mats. In another field around thirty Canada geese share their turf with a horned bull, assorted sheep and horses.

Today, with Susanna, I study a procession of pretty wild flowers – creamy clouds of meadowsweet, purple garlands of tufted vetch, flat-topped clusters of common ragwort, spindly beanstalks of pink rosebay willowherb, and handsomest of all, the humble meadow thistle. It's good to have time together. Max has slipped downstairs and is sleeping. Although we speak daily on the phone, it's not the same thing. Susanna looks forward to being aboard *Caroline* as much as Max does. We sit at the stern, gliding slowly, chatting with easy familiarity. Like all couples our marital journey has been a rollercoaster and there have been half a dozen times at least when one or the other of us could have walked out and given up. The children held us together. Now as they approach their own departure, we have no need of their cement, our shared history has bound us, our struggles for space or attention mostly consigned to the adolescence of our relationship. I cannot imagine ever being with anyone else.

Caroline's English journey will be remembered almost as fondly by other members of my family as it will by me. Susanna has a fantasy one day we'll buy a state-of-the art flash narrowboat, I meanwhile am keeping my eyes open for a cheap second-hand job that looks like *Caroline*.

Beside Claydon Top Lock, a gate in a picket fence leads through a pretty English garden to The Old Smithy Canalware shop. Decorating the walls and shelves inside are brightly painted pitchers, handbowls, dippers and cabin stools. Jane Selkirk is carefully painting a cream line between two dark banks of colour. She'll then leave the watering can for four hours to dry before adding the roses with enamel paint. It will be a total of eight hours before the paint job is complete and Jane can lean a card saying £85 against the can.

'The style's known as knobstick roses and it originated in Cheshire, unlike the Braunston rose associated with Frank Nurser.' The same Frank Nurser whose grave Hilda had shown me a couple of days earlier; the same Frank Nurser who painted Margaret's dipper. 'The best exponent of knobstick roses style is Bill Hodgson who was chief boat-painter at Anderton Dock in the 1930s.' She points to a large enamel handbowl on which she has copied the sailor (complete with 'Hero' emblazoned on his cap) from the Players cigarette pack. 'Bill first did that, as far as I know. It was one of the boatman's favourite non-canal-related images.' I decide to splash out £65 and buy it. I also buy a pretty baccy box for £5.50.

There are no roses on the bank today but plenty of dandelion clocks which Max and I blow, like my sister and I used to when we were kids growing up in Shropshire and Wiltshire. For a week, Max has had a glimpse of that childhood and would like more. Instead he now has to return to London and his nest of fears: fears of busy roads, fears of having his mobile phone jacked in the street, or being beaten up by marauding gangs in the parks (80 per cent of urban muggings happen to boys, not girls). And that's living in a pleasant, middle-class leafy London suburb. While my daughter Larne is in her element swinging through the jungle of a large London comprehensive, Max is a fish out of water, a fish made for country.

At Cropredy my father arrives in his red Micra and for the next couple of hours, he and Susanna zigzag across the countryside. First they head back to Napton on the Hill where Susanna picks up our Nissan, then they drive to Oxford, where Dad leaves his. Back at Cropredy, Max gathers his belongings and gives me a big hug before swapping places with my father in the car. I wave them both off as Dad settles in on board.

He's looking fit and his breathing seems easier than when I saw him briefly in Lymm. Nevertheless he comes with his nebuliser, pharmacy of medicines and aspirators. His doctor's orders are to do regular breathing exercises, go on the nebuliser three times a day, and bring up as much sputum as he can.

A shimmer of heat rises off a wheat field as we slip under a medley of lichen-stained humpback bridges. In shorts and polo shirt, Dad takes the tiller, and as we potter, we start planning his eightieth 'ramsammy' at our house in October. Dad's vocabulary remains peppered with Anglo-Indian words and phrases picked up during his 26 years in the army. He decides that for his celebratory meal he would like sucking pig (sucking pig – how the hell do you cook that?) to remind him of our time in Malaysia and Hong Kong. He has never been one for setting others easy tasks.

At Bourton Lock we chat with a woman I'd met in The Old Smithy

Canalware shop. She's locking down ahead of us. I'm curious about her accent. 'American diluted by thirty-nine years in this country.'

'Aha. It has almost an Irish lilt to it,' Dad comments.

'Now you can see how the American accent grew out of Ireland.'

'My father was Irish,' Dad says proudly.

The woman is from Ohio and lives year-round aboard her narrowboat, *Forward*. It's so highly polished I expect a genie to appear any moment. Dad is impressed. As a former drill sergeant he likes to see things well polished. The woman leaves and Dad steers *Caroline* into a lock for the first time while I nosey around the ramshackle lock keeper's house. Three plastic containers collect rainwater off a drainpipe, there's a wind generator, two solar panels, and a piece of string hanging from a window supporting a rose bush. I knock but the cupboard is bare.

I relieve Dad at the tiller so that he can descend to the cabin to nebulise. Afterwards, he lies on his side on the sofa-bed attempting to bring up sputum which he collects like gifts in a plastic container. A battered old boat called *Bo Diddley* passes. I've seen several Duchesses on my journey but doubt they were named after Bo's belle.

At Banbury we slip under Tom Rolt Bridge and moor where Tooley's Boatyard stood until very recently (it was here that Rolt had his narrowboat *Cressy* fitted out). Tooley's had been around as long as the canal and its dry dock is currently being rebuilt with glass walls so that it becomes a working adjunct to the relocating Banbury Museum. The old waterfront, like Tooley's, has already vanished, replaced by an anonymous shopping mall concealed behind a retro warehouse canal kit bought at B&Q.

Moored opposite me is *September Morn*. In the fore-end a ponytailed twentyish-year-old is throwing a ragdoll for his dog to chase along the towpath. On the far side of the hermetically sealed anywhere-mall, we walk through a jumble of Georgian and Elizabethan streets housing specialist shops, alfresco bars and cafés. Between a Chinese medicine emporium and the buckled frontage of a sixteenth-century coaching house, I buy a pair of cargo pants in a sale for £10. Dad and I then take a couple of pints and open baguettes and sit in the main square. Banbury Cross is still there but another rhyme comes to mind as we watch office workers eating on the hoof as they scurry back to their workplace. 'In Banbury he hung his cat on Monday for catching a rat on Sunday.' It alludes to the narrow-minded conservatism of the puritans who once ran the town. I doubt very much the Sabbath would get in the way of commerce today.

Leaving town, we pass a 40-footer called *Augusta* which has a subversive thought-for-the-day displayed in a port-side window: 'Life happens. Take time to live it.' It's moored next to a funereal-black boat named *Isis* – a timely reminder of the next waterway I'll be joining in a few days time.

Hogweed slows our progress, threatening to devour the boat. A cow, having sucked up half the canal, slobbers contentedly on the bank; a coot does the funky-chicken across the wet dance floor. Soon the grey steeple of Kings Sutton comes into view. Like virtually all the villages along the Oxford, the steeple stands a mile back from the Cut. The churches, built at villages' highest points, became the markers for the 'steeple-chase' when farmers raced on horseback between church steeples for wagers.

At 7.30 p.m. as arranged, Olga turns up in her battered Mazda with younger brother Raymond, a slight man with a white stubble beard. I introduce my father and we drive to the Butcher's Arms for an aperitif before supper.

Dad insists on getting the first round in and Olga helpfully waits with him at the bar. Raymond and I sit at a table on the terrace next to a group of babbling teenagers excitedly discussing an upcoming weekend party. I pump Raymond for more of the Olga story and discover that not only had she been Star Woman at the Royal Observatory, Bohemian Canal Woman, and Motorcycle Queen, she'd also twice appeared on *Mastermind*, winning the first time round with her chosen specialist subject 'Jenghis Khan and the movement of Asiatic tribes in the thirteenth century'.

Afterwards she rode a motorbike to East Germany on an international six-day trial when the Cold War was at its coldest, and went on to race in Poland, Czechoslovakia and Russia as well as throughout Western Europe. The last time she rode a motorbike, a vintage AJS, in a parade at Mallory Park in 1980, it inevitably degenerated into a race (which Olga made sure she won). I visualise Olga in her leathers, prone on the AJS tank doing a ton, staring down the track through misted goggles, knee scraping the tarmac as she swoops through a corner.

Olga's sense of the dramatic flourish did not desert her after she retired from the track and joined Raymond running the Three Tuns in the village. 'She used to pick herbs from the garden and add them to the cocktails. It didn't always work out. Occasionally she'd mistake ground elder for lovage and pop it in the Pimms!' When Olga and Dad finally join us at the table, I commend her on the multiplicity of her talents. Her sharp mind immediately alights on the church bells pealing in 8 p.m.

'It's not that unusual. If you were born Bill Ring, perhaps there's an inevitability about becoming the town bell-ringer, but in fact Bill – he was a real person and our last bell-ringer by the way – was also an expert welder of whom locals said, "The only thing Bill can't weld is the crack of doom."' Olga smiles. 'He was a barber too – fourpence an adult, tuppence a child. People can turn their hands to different things. Hopefully there's time left, too, to learn some new tricks. Our mother was one hundred and two when she died in the pub.'

When we finish our aperitifs, Olga drives us, surprisingly sedately, homewards. En route we pass the church (which dates back to the twelfth century), three thatched cottages that once housed monks, and a half-timbered courthouse. 'The village was once a Roman settlement,' Raymond informs us. 'Plenty of coins have been found, and one old boy who died recently – a member of the local Tew family – always claimed the Romans had a paint factory on the other side of the railway station.' Raymond doesn't laugh. 'He knew something. He wasn't daft.' No doubt Old Boy Tew will be vindicated one day.

'When I first moved here in 1966, there were just two privately owned cars on the council estate. The population then was six hundred; now it's two thousand but instead of the six grocers and three butchers, we're now served by one general shop; though we have managed to hang on to the post office.'

Olga pulls up outside a Victorian detached brick home. Doors slam and we walk up the garden path, passing two large bones lying on the front lawn ('Makes burglars think we've got big dogs inside'). Inside their home, Olga finishes off preparing the meal, and Raymond shows us a recent magazine article with glamorous pictures of Olga in her racing heyday accompanied by a fleet of admiring beaus and princes. One shows her holding her helmet, dressed in leathers, standing alongside an Ariel Four Square she'd just raced in Cannes. Behind her stands a line of broken hearts.

Olga serves our starters – an olive and feta salad that is a legacy of her

Greek forebears. She still has short auburn curly hair, and seemingly none of the creeping infirmities that spread like vines through the body of most 77-year-olds.

Turkey, mash and vegetables cooked in parsley and coriander – Greece's favourite herb – follow. The wine is Portuguese. As we eat and drink, Richard and Olga continue with their history lesson on Kings Sutton, which as it unravels, also tells the story of England.

To summarise: in the Domesday Book of 1086, the village was called Sudtone; in the mid-fourteenth century it suffered the Black Death; and during the Civil War was a Royalist outpost. A little later in the seventeenth century the village became a popular spa town. Unfortunately the greed of its inhabitants led to overhiked prices and its speedy eclipse by nearby Leamington Spa (though a local legend prefers to lay the blame for the spa's demise on a Dr Willis, who reputedly put a toad – and a curse of further infestation – in a well when locals decided to run him out of town for making a young girl pregnant).

As Olga and Raymond bat dates and stories back and forth, it's obvious they get on better than most married couples. Most of the trinkets decorating the home – the guns, jugs, steins, model boats, and seventeenth-century tavern glasses – belong to the male member of the house. 'I prefer to be out and about,' Olga confesses.

'She sits on various committees, meets friends, and travels on her quiz nights,' her brother expands a little. 'Our other brother – a pyrotechnical artist, and secretary of a fencing club in Birmingham – is the same.'

'He's seventy-five. We have a more prickly relationship, probably because we are more alike. Of the three of us, only Raymond married.'

'And only Raymond divorced! Publicans always do – too much temptation.'

'I've always been too selfish to walk down the aisle myself. My relationships have tended to last no more than a couple of years I'm afraid.'

One fling was with Murray Walker, the motor racing commentator. Raymond assures us there were many more. More drinks stir the mellow mood. We ease into comfortable chairs and talk about Raymond's garden, the Cut, my father's illness, and of course my trip. Although I've only met them twice, they already feel like old friends. It's midnight before we finally get home to *Caroline* who has sat up waiting for us.

*

Dad and I leave our mooring at Nell Bridge the next morning and are soon tracking the River Cherwell beneath a procession of half-cocked accommodation bridges shared with lonely farms. Dad does his exercises up on deck as I sit at the tiller watching him waving invisible planes in to land, attempting to transform his sodden tea-bag lungs into airy aircraft hangars.

Dad used to be a paratrooper in the First Forward Observation Unit of the Royal Artillery, part of the First Airborne Division. Once his chute didn't open till the very last minute. Getting a breath is now often like this. Sometimes he nearly blanks out. One day he will, and the chute won't open.

We pass white stethoscopes of bindweed and a boat called *The Great Escape*. We take a walk along the bank and into the village of Somerton. Dad's deep-breathing exercises don't seem to have helped this morning and he cannot make it up the hill. His pride is bruised. He's angry with his illness and this just constricts his breathing more. He stands in his burgundy fleece, blue shorts, and striped sports socks pulled up to meet his gnarled, veined legs. Two nine-inch zips run vertically down each where he had new knees fitted a couple of years back. That was just before the prostate cancer and emphysema took hold. I look at the leaves of the trees on the horizon, rippled by the strong breeze, and wish the wind could do the same for his dark lungs.

Dad insists we continue on the walk. I insist we return. Where once there would be no questioning, no arguing with the patriarch, now he's more than happy to let me take charge. Slowly we return to *Caroline* and he rests for a while in the cabin. When he reappears on deck, instead of talking about his problems, we discuss Mum's catalogue of illnesses. Since my brother died in a car crash in 1968, Mum has only enjoyed short bursts of good health between infirmities. 'It would be ironic if you finally went first,' I broach the subject that families prefer to avoid. 'Until a couple of years back you were always the fit one. If it does happen though and you go first, I have absolutely no doubt Mum will follow very quickly.'

'Really?'

'I reckon she's already quietly resolved to it, and there will be nothing Lynn or I will be able to say to shift her. You know what she's like.'

Dad nods his head. It's a candid discussion. Why tiptoe? As Epicurus said, 'Death is nothing to us, since when we are, death has not come, and when death has come, we are not.' I tell him how much I admire the way he's stayed positive throughout the past couple of years, how determined he's been to get back in the gym and get healthy again.

'All the doctors say that's the most important contribution I can make. I just pray you never get emphysema, Tid.' Tid, an affectionate abbreviated form of Tiddler, has been my nickname for close on half a century – the legacy of an early predilection for wetting the bed. Dad shakes his head. I don't tell him that like all hypochondriacs I believe emphysema has already stolen inside one night when the door was left ajar. Well, maybe not emphysema – after all I gave up smoking at 27 – but there is a weakness there. After bouts of bronchitis in Algeria, Spain and England, and altitude sickness on Mount Kenya, it seems to me everything goes straight to my lungs. Susanna says that's where I somatise my anxiety.

In the fields lining the canal bank, there's evidence someone has been out and about with something bigger than a scythe reaping the wheat. A mush of hay lies where eighteen-inch erect stalks once stood.

By 2 p.m. it's lashing down. I keep Dad downstairs. Every so often he appears in his waterproofs in the hatch asking if he can come up and help. I tell him he already has and that I've been managing alone for three months and it's really quite automatic now. It's not worth getting a chill or aggravating things. He returns downstairs with his tail between his legs. He puts Jazz FM on the radio and occasionally I edge back the hatch and see him down below beating his foot to Artie Shaw or Duke Ellington. In the locks, I watch him watching me through the fogged windows, as I go about my business. Throughout the afternoon he ferries me enough coffee to have me wired for a week.

The rain persists until we reach our mooring at the Rock of Gibraltar at Enslow Bridge. I'm soaked to the skin – all waterproofs eventually cave in. I take a hot shower and change and then we head up to the pub for a Thai curry. Inside the bar, three men on stools sway drunkenly and discuss mooring fees. Right now they're in need of mooring themselves, or seat belts. Like most pissheads in pubs, they have not got a good word for anybody.

The present landlord has only been ensconced a couple of months, having relocated from Newcastle. The first landlord, Henry Baker – who was also a contractor during the building of the canal more than two centuries ago – has a nearby lock named after him. I doubt the present incumbent will, but he's friendly enough. We sit beneath a canoe lashed to the ceiling and beside an aquarium where angel fish and other exotics live out their *Groundhog Day* lives. Dad keeps thanking me for the three days afloat. He's hardly slept, his chest has been a mess, and he's clearly worried about whether the deterioration is a blip or a new staging post in the illness. Despite all this, he says he's loved every minute on board.

As we leave the pub, as friendly and chatty as always, Dad throws a goodnight to nobody in particular (people always look at him peculiarly when he visits us in London as he can't pass anyone without a greeting). I hear one of the drunks meanly mouthing to another 'Captain Mainwaring.' Dad served 27 years in the army and made it finally to sergeant major. At the age of thirteen he was supporting his family. He is not Captain Mainwaring.

I go to bed agitated. It's not just because of the drunks, or Dad leaving. It's mainly to do with the fact that tomorrow I leave the Cut and *Caroline* for a month. Because my initial departure had to be delayed six weeks as a result of canal closures due to foot-and-mouth, the family holiday we'd booked ahead to coincide with the end of my journey now overlaps it. There is to be a hiatus and emotionally I feel as churned up as a ploughed field.

Fortunately our final morning is blessed with glorious weather. Today is what yesterday was for. If it had been a corker yesterday, would our spirits be soaring like they are now? Nature, too, seems bursting with renewal and enthusiasm after the deluge. Dad takes the tiller and sings 'That Old Black Magic', 'Amapola', and 'Mack the Knife'. The same songs I remember him singing around the kitchen when I was a kid growing up in a block of flats in Hong Kong's Kowloon Tong in 1960. On our arrival in the colony, Dad bought our first record player, and six records which we played over and over again. Bobby Darin's 'Mack the Knife' was his favourite, mine was Connie Francis's 'Lipstick on Your Collar'. I think Dad will miss music almost as much as he will Mum if the skiff-hirer in the sky does call his boat back first.

At Baker's Lock I chat with a bearded Dutchman and his young son on a rental narrowboat. At home in Holland he has a sixteen-metre sea-going barge that was built in 1911 and has never been modernised. Over the next month, away from *Caroline*, this is what I'll be missing. The chats, the tall grasses, raindrops glistening on reeds, yellow irises sashaying in the breeze.

As the Cut joins the River Cherwell, canal walls dissolve and the waterway broadens and deepens in a series of sweeping bends. Beside the bridge at Shipton-on-Cherwell, I moor beside church walls, leave Dad on board, and wander up into the hamlet. Every town and village has its moment. Shipton-on-Cherwell's came on Christmas Eve 1874 when, a quarter of a mile away at Shipton Bridge, nine railway carriages plunged into the frozen canal killing 34 people. Rail crashes are not a modern invention. There was another moment just a dozen years earlier when pulses also ran a little faster as the thigh-bones of an immense dinosaur, *cetiosaurus oxoniensis*, were found in a nearby quarry (they're now in the University Museum in Oxford).

In the cemetery of Holy Cross Church I walk down a procession of ancient lime trees that are sprouting new fifteen-foot limbs from pollarded trunks. 'Gone But Not Forgotten'. 'Safe in God's Keeping'. 'Rest in Peace'. 'Rest after Weariness'. 'Peace after Pain'. The English names beneath the mossy arêtes of tombstones – Dyer, Broome, Foster, Thorpe, Howell, Roberts, Adcock, Robbins – keep reappearing like reflections on the canal. A rabbit hops by. A dog sees me moving in the dark shadows and shifts into attack mode before a woman in an M.G.M. Studios T-shirt, who's out walking the brute, gets it under control and apologises.

As I return along the towpath to *Caroline*, the sun disappears behind a cloud the shape of Britain before Wales and Scotland sailed off. We burrow through Oxford's hinterland, the average age of boaters dropping dramatically as we do so: ponytails replace beards, and names of long-term moored boats appear to have been dreamed up on drug binges by poor spellers: *Andian Hazelnut*, *John Hustler's Dream*, *Hieronymous Peyps*. The narrowboat *Tuna* has a female bust wearing a motorbike helmet as masthead, another has an Aboriginal design and is called *Song Lines*. We negotiate several lift bridges before Dad returns to his Micra and I bring *Caroline* into College Cruisers' boatyard.

8

Hiatus

It takes several hours to clean the boat and pack up. Peter from Adelaide Marine has done a deal with Charlie at College Cruisers to hire *Caroline* out in my absence. Peter makes money, Charlie makes money, and I avoid paying out around £150-worth of mooring fees. I did discuss it with *Caroline* before agreeing to the deal. The old slapper said she'd prefer a stranger on board to nobody.

I have a shave and take a look in the mirror: no parrot on my shoulder, no discernible sextant squint, and the stubble that was a whisker away from beard has gone (even the children come with beards on the Cut). Alison, my niece who's been staying at our house following a stint guiding Explore groups in the Sierra Nevada, picks me up in our Nissan. I have the bags waiting on the quayside and insist on driving on the way back to London.

It's been a while and it feels both weird to be on the left instead of the right and decidedly dangerous to be propelling myself, albeit at a snail's pace, through clotted traffic. Out on the dual carriageway, competition replaces cooperation and soon, unconsciously, I'm sucked into the pressure to speed, whipping lanes to save ten feet and get ahead. Once we reach the M40, pregnant clouds burst and thunder crashes. The rain hammers down apocalyptically and I can barely see through the windscreen. I slow to 35 m.p.h. Lorries try to climb through the rear window, cars flash their lights and then aquaplane past. It's a nightmare.

London is almost sedate by comparison and Muswell Avenue comfortingly calm and familiar, though there's the usual difficulty finding anywhere to park. As I turn the key in the door latch there's a scream from inside and Larne hurtles towards me. A family scrummage ensues with Larne getting the biggest hug of all. It's been a long while.

The cat runs – as much as a cat dragging her stomach along the floor can ever run – from the stranger in her home. I look round the house: to cabin-trained eyes it resembles a palace rather than an early Edwardian terraced home. The garden looks particularly lovely and I make a silent promise to get more involved with it.

As the dust settles I start putting away my clobber from the boat; attempting first to hang a couple of shirts and a pair of jeans in my cupboard. There is no room for expansion: over the years I've already expanded too much and there is no longer room to slide so much as a tie between the crammed shirts, jackets and trousers. I could clothe the Isle of Man from this one cupboard. T-shirt and socks drawers pose similar problems and I make a second resolution to have a major clear-out and to adopt a new frugality.

The night of my return coincides with the final meeting of Susanna's monthly book group. The circle is breaking up because several of the members have new pressing this-and-thats. No time for books and discussion. Tonight they're discussing *The Songlines* by Bruce Chatwin. I tell them I passed a boat with that name just yesterday. They stare at me indulgently and comment on my weather-beaten face. Their looks tell me I'm older but no wiser. Politely one asks about my trip before her eyes put up a defensive glaze which is the equivalent of the note on the front door saying 'Sorry I'm out'. I mumble a few limp clichés that measure nothing: Great, really interesting...I mention Olga briefly and eyes flicker back to life. I mention the fact I've been shocked by the level of racism; that living in the middle-class ghetto of Muswell Hill I'd been under an illusion that things had been improving. One of the women nods her head, showing understanding. But she has had her learning in sanitised packages by radio, television and newspapers just as mine had been. My new take on it is more visceral. I remember the people I'd warmed to on the Cut who had suddenly started spouting about Pakis and Powell's Rivers of Blood. I remember the crumbling face of the farmers in Church, the boat people, the easy meetings over a paddle or a pint, and conclude the epic will have to remain in my heart. I hope that it will also reside in this book.

On board *Caroline* over the past three months, the only comfort I'd missed, apart from family, has been a bath. And I didn't really miss that much. More like I felt there had to be something I missed and so that would do. The cassette player never worked for more than twenty seconds, the TV was a waste of time, and I didn't miss either. So I have a bath. It isn't up to much.

The spell of the water road is suspended. The consolation is that my departure from the boat is not the end but a rehearsal. Just as the purpose of yesterday's rain was to appreciate this morning's sun, so this hiatus will make the return to *Caroline* even more valued. I'm excited already at the prospect this first night back on dry land.

In bed I close my eyes and see the simple brick arch of a bridge and fields sloping off either side. Telescoping through its tunnel, I spy a deserted lockgate with its black paddles, black and white balance beam, and white steps tracking up to a grassy knoll.

That night I sleep badly. I dream I'm in a large temple complex in Nepal, landlocked and surrounded by the world's highest mountains. I am in an underground network of temple ponds. A pilot has crashed through the roof of the one I'm in and his plane is in the water. When I report it, I'm told to ascertain whether his body has already stiffened. I make my way to the plane and shout back to the party on the bank that I've found the pilot, that he's upright but his legs have disappeared. At that moment he falls from the plane into the water and is washed into the shallows, head down and legless. I am told it's OK to leave him there. When I reach land, the group has gone. I wander the temple grounds. Skirting the bank of the next temple pond I find the dead bodies of a bear, a panda, a dog and a fish that have all drunk from the water.

In the morning my psychotherapist wife tells me being in water is being in feelings and being overtaken by feelings. I feel like I've been flushed out of a lockgate. In the classical Greek world, the dead drank from the waters of Lethe, the River of Oblivion in the Underworld, to forget their earthly life. It was also believed, however, that spirits drank from this same water before being reincarnated, removing any trace memory of what they had seen in the Underworld. It's hard to know which direction I'm heading in.

And so, after two 24-hour ferry trips down and up the Atlantic to Santander, 2,000 miles of tarmac and a wonderful month spent back inside the bosom of the family, I drive back through the Victorian neighbourhood of Jericho to College Cruisers' boatyard. I unload my bags once more and Susanna reluctantly returns to London, as envious as hell of the water, the weather, the slow life, the escape.

Alan, a boat engineer with 25 years' experience, has fixed several of the glitches including the bilge and shower pumps. *Caroline* looks considerably cleaner than I remember her as if she's made a special effort for my homecoming. Charlie, who has owned the yard four years ('I spent ten years working in Saudi so that I could buy one'), tops up the diesel for free and I'm off.

My intention had been to spend a day exploring Oxford but the pull of the water is too strong. *Caroline* is equally eager to be off. We potter along a line of boats on permanent moorings. Most of their owners live and work year-round here, preferring a floating home in Oxford to a terraced street or apartment block. The tiller feels warm in the hand and Alan has tuned the engine to a purr. I enter Isis Lock, once a favourite with suicides and drowning drunks. Thankfully lock operation is grooved into me now and there is no need for memory. I ask a twenty-year-old sitting on the bank to shut the gate for me and swing hard to starboard into the 200-yard channel linking the canal with the Thames.

9

Epiphany

Isis and the Thames

With Isis Lock behind me, I sail on to the River Thames three months after I left it at Brentford. When I say the Thames, I'm not being pedantically accurate, for although many use the name for the entire stretch of the river, hair-splitters will tell you otherwise. In Julius Caesar's time, the river was known as *Tameses* but at some point the name unhitched itself and became the *Thames* east of Abingdon, and *Isis* west to the source.

The goddess Isis in early Egyptian mythology ruled magic, the water and the fruits of the earth. She was also the sister-bride of Osiris, lord of the dead and resurrection. According to Sir James Frazer in *The Golden Bough*, 'At his nativity a voice rang out proclaiming that "The Lord of All had come into the world".'

A hard act to follow for younger siblings then. Seth, the runt of the family, grew increasingly irritated at his big brother's power and popularity. So, as younger brothers do, he plotted to get even. One night, when it got round to party-games time, Seth offered a fabulous cedar coffer to the reveller who fitted into it most snugly. When Osiris was eventually persuaded to try it for size, Seth leapt on top, nailed it down and jettisoned the coffin into the Nile. The cruel brother had murdered the king. The old story. Deed done. Only it wasn't, for Isis tracked the coffin all the way across the Mediterranean to Byblos, jemmied it open and brought Osiris back to Egypt where she hid him in river rushes.

Unfortunately for the incestuous lovers, Seth discovered Osiris and, before Isis could breathe life back into him, ripped the body apart, scattering the fourteen pieces across the land. Indefatigable Isis gathered every last bit, apart from the genitals which had been eaten by a fish. She

271

embalmed the body (thus inventing mummification), ensuring Osiris could live on in the Underworld and breathed life back into his lungs through his nostrils, thus raising him up: the dead and risen Lord.

The love story appears in different guises in pagan fertility myths throughout the Middle East – for Osiris, read Tammuz, Mithras, Attis and Adonis. And it is this same mystical raising-of-the-dead that symbolises nature's annual death, rebirth and regeneration that underpins Christianity. Every year loyal Ishtar visits Tammuz in 'the house of darkness, where dust lies on door and bolt'. When she returns to the earth, she draws back the curtain of winter and brings life back to the land.

Lazarus, Osiris, Jesus, Tammuz – even the names slip comfortably into each other – offer the same message of hope for our own, and our loved ones' resurrection. The prayers in stone I've been reading in graveyards tell the story: 'Till We Meet Again', 'Gone to a Better Place', 'Reunited in Heaven', 'Life Everlasting'.

The central faith of the major religions, be it Hinduism, Islam or Christianity, has it that just as births are the lockgates to our deaths, deaths are the lockgates to the afterlife. We can travel both ways through water and all waterways eventually join up. Even atheists like me who have difficulty with the idea of a *literal* human afterlife (I follow Epicurus's succinct epitaph 'I was not, I was, I am not, I do not care'), would happily invest in a new pair of sandals if invited to the local pagan spring fertility party. As a matter of choice, however, I'd rather spend the winter in the Caribbean than the Underworld.

We may prefer it otherwise but beneath our skin we're all related to the devotees of Osiris, Adonis, Mithras, Attis and Tammuz. The fertility myth and the rising river (it was the overflowing of the Nile that gave life to the crops and created the first farmers) is the root of all the faiths. It is also the world's first great love story: a love so strong it cannot be stopped by death.

The scattered parts, or canals, have led me back to the Royal River which will carry me home. I have a month before I descend into London's warren and winter hibernation and I'm going to make the most of it. The sun's out, I'm back on the Cut. In terms of my journey, the Thames was the beginning and so it will be the end, the alpha and the omega. In a few days I'll follow it eastwards from Oxford past Windsor

Palace, Hampton Court, Kew Palace, Fulham Palace, Lambeth Palace, the Palace of Westminster, and all the way to the Tower of London where royalty discovered at their last gasp they were indeed human. First, however, I'm heading westwards, travelling as far upstream as I can before seeking out the source of the river on the back of Rosinante.

Caroline purrs and whispers as we slip beneath languorous willows bathing in their green liquid reflection like Narcissus (the ancient Egyptians believed each person had a double, known as 'Ka', that was released in dreams and in death). Maybe it has something to do with spending the past month hurtling at 90 m.p.h. on Spanish and English motorways, but I feel all the cogs and wheels inside slowing right down. I'm no longer speeding to get somewhere. I'm here; back where the journey is the end in itself.

In *FASTER*, James Gleick writes:

You are bored doing nothing so you go for a drive. You are bored just driving so you turn on the radio. You are bored just driving and listening to the radio, so you make a call on the cellular phone. You realise that you are now driving, listening to the radio, and talking on the phone, and you are still bored. Then you reflect that it would be nice if you had time, occasionally, just to do nothing. Perhaps you have a kind of sense organ that can adjust to the slowness, after being blinded by the speed. The void is not so dark after all. With the phone not ringing, the television switched off, the computer rebooting, the newspaper out of reach, even the window shade down, you are alone with yourself. The neurons don't stop firing. Your thoughts come through like distant radio signals finding a hole in the static. Maybe they surprise you; maybe they disturb you; maybe they assemble themselves into longer strands – ideas, or knowledge, that might not have formed in the usual multitasking hurly-burly.

Recently 33 Italian towns formed an association of 'slow cities' vowing to promote the joys of quiet, reflective living. To be permitted to display the movement's snail logo, cities must meet a range of requirements, from banning car alarms, to setting up centres where visitors can idle over slow-cooked traditional foods.

My neck muscles relax, shoulders melt. It's 4.15 p.m., I'm on the

Thames and as happy as Larry, whoever Larry is. I take a glass of chilled wine up on deck, breaking with the usual tea and Dundee cake tradition, and drink a toast to Edmund Spenser – 'Sweet Thames, run softly, till I end my song'. There's a tangle of white beard in the sky. Old Father Thames or Old Father Time timing me in and timing me out.

I have the throttle down low so that I can stretch and savour the couple of miles I plan doing before mooring up for another hopefully long summer evening. A sign warns to keep to the centre of the channel and avoid the shallows. Sound advice. The river broadens, slipping beside a meadow grazed for 4,000 years without the rude intrusions of plough, pesticide or fertiliser, thus allowing a profusion of rare wild flowers such as devil's bit scabious and greater burnet to flourish.

A single skuller powers by, nodding a greeting. A topless father and son throw a Frisbee between them, chased by a manic dog. A forty-year-old ponytailed male, dressed in muddied swimming trunks, enters the water in a side creek followed by seven young children. It reminds me of a mass river baptism I once stumbled on in Florida. Cyclists dawdle, butterflies flicker, kids dangle feet. A herd of maybe sixty horses wander down to the water's edge to drink and hang out, the ponies swishing their tails like teenage girls swinging their ponytails down the high street. I've just spent a wonderful month in Spain but there is simply no contest: riverside England in sunshine wins hands down. And the race memory of English summers has taken up permanent residence on the Thames.

Godstow Lock, like all the other locks on the Thames, is manned by Environment Agency rather than B.W. employees. Green and white uniforms are traded for blue and white, and lock-keepers' cottages and gardens walk straight off the pages of *Country Living*. I pass the crumbling walls of Godstow Nunnery. Physically today it's on its knees but scandal suggests its occupants were similarly positioned considerably earlier: it was at the Godstow nunnery that Rosamund Clifford met Henry II and became his mistress, bearing him two sons; the nunnery then became notorious for its 'hospitality' towards the young monks of Oxford.

On an adjacent bridge boys are shivering between plunges into the water thirteen feet below. I moor up beyond them and finish stashing my gear to a percussion of splashes and hollers. There is a delicious feeling of escape and summers stretching like the 215-mile river to eternity. I

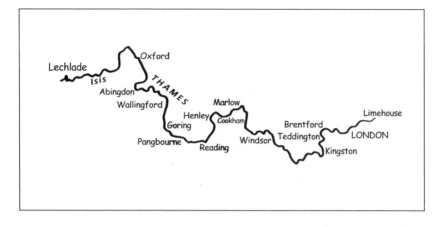

The Thames from the Source

take a stroll. Up ahead I see a woman towelling herself down beside bulrushes. She has her back towards me and is vigorously rubbing a towel through her hair, accentuating the nakedness of the rest of her body. When I get closer, I notice there is no bikini mark, just the overall light tanning of the naturist. Her face, when she finally turns towards me, has the life-lines of middle-age but the rest of her torso is that of a thirty-year-old artist's model. Her buttocks have not even begun to sink yet. Her companion is now fully clothed and running a comb through his white hair. 'Lovely day,' he smiles. I smile too, not too lasciviously I hope. 'It certainly is.' I leave Osiris and Isis to it. As I continue on my way I hear the two chuckling like school kids who've outraged the vicar.

Light shivers on steel and fibreglass hulls. Beside a blue and red narrowboat moored to the far bank, kids are chasing their dog through the fields. 'Don't go far,' their father shouts across, interrupting his mechanical fiddling on board. Emblazoned on the side in upper case is the word *HOPE*. A long-haired dad with cut-off shirt revealing a Japanese geisha tattoo hammers by in his cruiser puffing on a fag. His young chubby daughter sits beside him. A yellow and green narrowboat called *Beauty* follows steered by a buxom woman with blonde hair stacked into a pyramid. My guess is they've all rushed home from work to steal a few hours resurrection on the water. We are an island race who created the most concentrated canal network dug anywhere on earth for navigation. Is it any wonder that our element is water?

Later that evening, while chopping onions and listening to Susan Maughan's 'Bobby's Girl' on the radio, I notice *HOPE* is now moored opposite me for the night. I talk to Larne on the phone. She tells me about the Voodoo Glow Skulls and Capdown gigs she's going to at the weekend. We then discuss whether the Red Hot Chilli Peppers or the Smashing Pumpkins will be the band most remembered from the nineties. She passes me on to Max and I nimbly switch register, bemoaning Chelsea's feeble 1–1 season-opener against a crap Newcastle side that had their four best players out injured. Finally I debate the pros and cons of adding a dog to our household with my wife. When all the talking is over, I switch off the mobile and dance to Carlos Santana's 'Leave Your Light On' as I stir the mushrooms, onions and chicken in the pan and the sun scorches its farewell message along the skyline. All hatches, windows, skylights and doors are flung wide.

At around 9 p.m. I slip over the bridge the boys are still shivering on and jumping from into darkening eddying pools. The creekside terrace of the Trout Inn is heaving. It's a summer evening, work's over and no one wants to be indoors.

What is it about waterside pubs that makes them call like sirens to me each night? Samuel Johnson pretty much summed it up when he said, 'There is nothing which has yet been contrived by man, by which so much happiness is produced as by a good tavern or inn.'

In the newspapers and on the radio over the past week, Prince Charles has been touting the idea that rural pubs should take a leaf out of Ireland's book and assume the additional roles of grocer and post office. He reckons it's the only way to staunch the haemorrhaging of country services and put the brake on an average six pub closures per week. But pubs are meant to be a retreat from business, a place for slacking, not doing jobs. And I, for one, would strongly object to having my pint interrupted for a robbery or someone asking where the fish fingers are.

This seems to me to be potentially a greater curse than even the abbreviated opening hours which were introduced in 1914 by Lloyd George as an emergency measure and then forgotten about until 1988 when all-day opening became legal again from Monday to Saturday. If our pubs were like Ireland's and our grocers like Ireland's then I'd be the first to sign up, but I fear in England we'd end up with Spar moving in selling Heineken lager and nothing else (at present over 250 different beers are still produced in England).

Whereas urban pubs get christened the Junction Arms and the Railway, out in the country today I've already passed the Perch a few hundred yards back and will pass two more Trouts before reaching Lechlade. The most popular pub name in England is the Red Lion and the oldest is the Olde Trip to Jerusalem in Nottingham (dating back to the twelfth century). Talking of Jerusalem, William Blake was inspired to write much of it while staring from a pub window in West Sussex.

Tonight a noisy peacock is entertaining the hundred-strong crowd out on the terrace as they tuck into prawn teriyaki, trout in garlic, rack of lamb and Mediterranean chicken soup with bacon and coriander. The Trout Inn is one of those pubs that make you dream of upping sticks and moving to the country. Ivy has consumed rather than clad the place, water is cascading over a weir, and kids are throwing leftovers to shoals of obese fish from the wall.

When I return at 10.30 p.m. the boys are still there shivering, but huddled this time under their towels on the bank. When they draw their last breaths, it will be today they rerun: walking out their doors first thing in the morning and still playing in the river under the stars at night.

Under the bone-warming rays of the early sun, I sit on deck feeding squawking ducks the tails of my Marmite toast. The sense of the day being an open book is back and I decide to cycle a couple of miles back into Oxford to visit the Ashmolean.

I untether Rosinante and start bouncing along the rutted grass bank of the river. Soon, however, I discover the gate to an adjacent field is blocked by a herd of cows that are being mounted one after the other by a rather large bull. I decide that asking a bull with sex on its mind to move is not a great idea. I turn back and cross the bridge beside the Trout. Beneath a second bridge, young children are already splashing and swimming, one small boy threaded through a tyre on a rope like a finger through a wedding ring. A man with a distant memory of a Scottish accent smiles and applauds the glorious day, pointing to a swan and her two adolescents sailing under the bridge. 'That's her second brood this year. The first two she had just after Easter and they were had by the mink on their first day.'

A couple of hundred yards further on I descend to the Oxford Canal towpath and re-enter the city the way I had a month earlier, passing narrowboats *Andian Hazelnut*, *John Hustler's Dream*, and *Hieronymous Peyps*. A man is sitting with a parrot on his shoulder. A young mother is rucking her six-year-old for not being ready to go to town.

I join the army of cyclists weaving through Oxford and lock Rosinante to a drainpipe outside the Ashmolean. The Leeds Room, upstairs, is a rich seam of Roman and Anglo-Saxon treasure from burial sites excavated along the Thames. Displayed in one cabinet are the contents of relic boxes discovered in female seventh-century Anglo-Saxon graves – amulets, necklaces, Roman glass keepsakes, toilet implements, chatelaine, jewelled gold pendants, and the iron frames of handbags. The dead clearly thought they'd need their best togs and jewellery where they were heading. Among the Roman exhibits is an exquisite small-breasted Venus (the Roman Aphrodite) admiring herself in a hand mirror, and pottery lamps found under the Thames shaped as the Indian yoni female genitalia. In the Cyprus Room there's a thirteenth-century BC statue of Aphrodite, Goddess of Love, and earthly consort of Adonis; in the Tradesant Room is a statue of Mithras (altars to the Indo-Iranian saviour god who offered rebirth into an immortal life have been found at several Roman sites in England). All the gods and goddesses swim in the same pool as the boys jumping off the bridge. All the waters run back to the overflowing Nile.

In *The Masks of God: Occidental Mythology*, Joseph Campbell points out that, while we prefer to highlight our differences, our gods suggest otherwise:

> No good Catholic would kneel before an image of Isis if he knew it was she. Yet every one of the mythic motifs now dogmatically attributed to Mary as a historic human being belongs also – and belonged in the period and place of the development of her cult – to the goddess mother of all things, of whom both Mary and Isis were local manifestations: the mother-bride of the dead and resurrected god, whose earliest known representations now must be assigned to a date as early, at least, as *c*. 5,000 BC.

The nativity of Mary's son, the Christian Saviour, climaxes on 6 January with the Epiphany. The Feast of the Visit of the Magi marks the arrival of the Wise Men who followed the star to Bethlehem. This was in fact the date of a much earlier festival in Egyptian Alexandria of the birth of the new 'Aion', the Saviour-Son:

> of whom the bright star Sirius (Sothis) rising on the horizon had been for millenniums the watched-for sign. The rising of the star announced the rising of the flood waters of the Nile, through which the world-renewing grace of the dead and resurrected lord Osiris was to be poured over the land. Saint Epiphanius (*c.* AD 315–402) states that 'on the eve of that day it was the custom to spend the night in singing and attending to the images of the gods. At dawn a descent was made to a crypt, and a wooden image was brought up, which had the sign of a cross and a star of gold marked on hands, knees and head. This was carried round in procession, and then taken back to the crypt; and it was said that this was done because "the Maiden" had given birth to "the Aion".'

That afternoon, as the river twists and turns, shaking off towns and villages, *Caroline* passes more long-forgotten meadows that somehow slipped the attention of industrial-estate makers, intensive farming and Barrattland stalags. I moor in nowhere about a mile's walk from Newbridge, one of the river's oldest and most beautiful bridges. Mirrored darkly and tremulously in the water are a succession of graceful thirteenth-century stone arches that would not be out of place in a Romanesque cloister. Once the river was forded here by Romans and Roundheads; now the only battle is between two pubs for my custom – the Rose Revived on one side and the Maybush on the other.

I decide to dispense my cash evenly, taking a pint first in the garden of the latter before switching sides to the Rose Revived. The name alone, let alone its tree-shaded river terrace, its mellow Cotswold stone exterior and inglenook interior, wins hands down. For a moment I waver again on the title of the book – 'The Rose Revived' – maybe.

I ask a barman how the pub got its name. 'I've only been here a week but the people at the table by the door will know – they manage it.' The six-strong party are hurling banter like rugby balls across a round table. As I interrupt, eyes turn to a large man probably not yet out of his thirties who is either the manager or spokesman. 'The pub used to be called the Rose and it was doing badly. Then about a century ago, or it may be a bit more, a larger company bought it and the Rose was revived.' The disappointingly prosaic tale is thankfully upstaged by a second man with bulbous eyes and longish hair that only kicks off from the halfway line. 'It's not been proved yet but there's also a story that Oliver Cromwell himself came to the pub one day wearing a rose in his lapel or whatever they had for lapels in those days. The rose was as good as dead so the barman stuck it in the beer and it revived.' That's more like it.

Walking back towards *Caroline* through the quiet of dusk, I notice an orange fireball is ablaze on the horizon. I am on a Permissive Footpath. Does this mean it is inhabited by licentious, debauched villagers? On a small plaque beside a hawthorn bush where a man is missing I read 'Witney Angling Society August 2000. In Memory Bill Kirk. Angler, long time member, and our President. We will all miss him.' One gone under. Three anglers are fishing nearby, hoping to magic things up from the deep.

That night the dreams come thick and fast and at one point become so oppressive that I'm moved to let the sky into the cabin to help me breathe. I slide back the large hatch above the bed and stare up into the stars. From somewhere unseen a similarly insomniac bird tries to strike up a conversation. Bankside existences continue to be guided and lived vicariously through *Big Brother*, *Survivor* and *Blind Date*. Up above the hatch, stretching across the night sky are the spiralling arms of the Milky Way with its billions of suns disguised as cloud. The three bright stars that form the Summer Triangle – Vega, Deneb and Altair – are easy to spot but brightest of all is the Dog Star, Sirius. Twenty-two times more luminous than the sun, it hangs from Orion's Belt closer now to the earth than at any other time of year (hence the saying 'the dog days of summer'). It is the only star that appears to be coloured, its reddish hue created through its rush to get away (if it were approaching, it would be blue), proving the universe, like our knowledge, is constantly expanding.

I return to the dreams. In one, a child is trying to hunt down a picture of her mother. She doesn't know what her mother looks like but knows she will recognise her straight away if she finds a picture. Someone allows the girl to rifle through a pile of official photographs. She finds one that was taken of her mother immediately after her death: a beatific smile, hands folded on chest, eyes open as if still alive. The child is now uncertain whether the mother is living or dead.

The dream fast-forwards to a packed flotilla of coffins bumping their way along the river. It's like the caskets in New Orleans that rise up and collide like dodgems each time the Mississippi floods. I notice each coffin, like each narrowboat on the river, has a different name, but inside each identical bodies are entombed. The congestion eventually eases. Bringing up the rear is a cross of five caskets. My brother heads the cross, followed by my father, mother and sister. I think I can guess the name of the one at the foot of the cross.

Finding it impossible to sleep, I pick up a book – *Romans and Barbarians* by Derek Williams – which I'd been reading earlier. 'There were superstitions regarding Britain herself; for example the eerie story that she was the abode of the dead and that souls were rowed across in unmanned boats, which left the coast of Gaul at nightfall and returned before dawn.' Williams claims Thanet, the last landfall in the Thames estuary before open sea, is the legacy of this legend (in Greek 'Thanatos' is the personification of death; and Freud usurped the name to identify a universal death instinct as bride to the procreative instinct Eros). Maybe the Roman Tameses is a syncretic union of Tammuz (the Babylonian God of the Dead) and Isis?

I listen to the breeze rustling the dry rushes. Why am I wading through death and darkness amid such beauty? Thanatos was the son of night (Nyx), and night is where the world tilts. I'm out on the river, the hatch to the unconscious is open and the metaphors are raining down. The dreams are all about ending of course. I'm kidding myself that I still have an eternity ahead of me on the boat. The finishing line and Adelaide Marine is probably no more than seventy miles away.

With daylight, I enjoy another fertility resurrection in Shifford Lock, rising up before the doors swing back on another bit of paradise. If, as David Byrne (following G.B. Shaw) believes, 'Heaven is a place where nothing, nothing ever happens', heaven does not reside on the river. The river, like a puppy, is forever moving and exploring, suggesting an eternity of play.

People only play when they feel safe and have a secure home to explore from. The narrowboat is exactly this: the safest of homes and, even better, it explores with you. A couple of months before I set off from

Adelaide Marine, Ellen MacArthur made port on the French Atlantic coast having spent three unprotected months sailing the savage sea alone with no locks and no waterside pub to pull into for company and a chat. Her journey is an epic in the mainstream adventure tradition: risking everything, pitting herself against the elements, stretching herself daily to the limits. My own three months afloat in comparison is the equivalent of a remote-control model boat skimming across a pond. This small Everyman adventure is more akin to the child's exploration of the garden and adjacent woods.

Habitation has now vanished altogether from the landscape. Large cotton-reels of hay have been left abandoned where the giants finished playing with them. On the water, swans are fishing head down, their rear ends rising out the murky water like Klan hoods. I pass yet another boat called *Noah's Ark*.

The Ark was also the name of the hired houseboat that William Morris, English craftsman, poet and staunch socialist, made this same journey on in August 1880, and again the following year (his daughter May said it resembled 'a sort of insane gondola'). It took six days to row upstream from their Hammersmith home, Kelmscott House, to their Cotswold retreat Kelmscott Manor, and the journeys became the basis of Morris's prose romance *News from Nowhere*, in which he travels up the Thames in a utopian time when the river has been cleared of pollution. Pretty much as it is today then. In the mid-nineteenth century the river was an organic waste tip that reeked so badly that in 1858 ('The Year of the Great Stink'), parliament had to be suspended. In the early twentieth century bacteria were still the only organisms that managed to survive its polluted waters. Chlorination slowly improved things and by the end of the century, with the introduction of U.V. treatment and carbon filtration, the Thames had been recolonised by marine life.

I pull in at Kelmscott and walk half a mile to the home – now a museum – of the father of the arts-and-crafts movement. It was in the garden at Kelmscott Manor, by careful observation of acanthus and vine, and watching a thrush pecking wild strawberries growing between paving stones, that Morris got the inspiration for his hallmark designs. The 'good life' that Kelmscott Manor symbolised became his socialist clarion call, an ideal that all men should have access to. In 1880 he wrote 'I have more than ever at my heart the importance for people of living in beautiful places; I mean the sort of beauty which would be attainable by all, if people could but begin to long for it.'

In the garden of his 'Earthly Paradise' today are the same luscious acanthus, hollyhocks, roses and mulberry tree that inspired him. Inside the home, the same equally vibrant motifs run riot over wallpaper and fabrics. For Morris, art and the art of living were one and the same thing. Like Ruskin, he believed the quality of life was reflected in the quality of work, and he showered scorn equally on impersonal, mass-produced factory goods, and the clutter of the Victorian home. His experience as a master craftsman, and his passion for the Gothic, led him to conclude that the excellence of medieval arts and crafts was a result of the freedom of its practitioners; a freedom that had been destroyed by Victorian mass production and rampant capitalism. Find a job you love and you'll never have to work.

Morris transformed crafts into art, and decorative art into high art, neatly summing up his philosophy by admonishing us to 'Have nothing in your houses which you do not believe to be either beautiful or useful.' Ironically the design on the handmade tapestry that's on display inside the house and which took him 516 hours to weave, is now mass-produced and plastered on the wallpaper and curtains of virtually every country-house hotel in England. Morris's artisan movement has been reduced to literal wallpaper no less than Che Guevara's revolution has become a poster.

Hanging from his four-poster bed on the upper floor is a self-penned poem embroidered by his daughter May. It starts:

> The winds on the wold and the night is a-cold
> And the Thames runs chill twixt mead and hill.

And it concludes:

> No tale I tell of ill or well
> But this I say night treadeth on day
> And for worst and best right good is rest.

When the man actually managed this – rest – however is a mystery. He had busy hands and a busier mind. When he wasn't writing poetry, novels, or socialist tracts, he was painting, designing and producing textiles, travelling extensively, translating Icelandic sagas, and campaigning for the Society for the Protection of Ancient Buildings (which he founded and which is still running today). He even managed to squeeze in time to get arrested for incitement to riot when soapboxing his socialist ideas.

In the Elizabethan trinket box of romanticism and romance that is Kelmscott Manor, Morris's wife Jane is as prominent as Morris himself. Seemingly at every turn there is either one of her infinitely superior embroideries or tapestries on display, or a Pre-Raphaelite portrait of her painted by Edward Burne-Jones or by her lover Dante Gabriel Rossetti.

Morris first spotted Jane, the daughter of an ostler, at a concert in Oxford. When she agreed to marry him and join his bohemian world, she was 18 and he 24. Several years later and the mother of two, she became infatuated by Morris's friend Rossetti. Her husband obligingly set them up in Kelmscott to escape London's rumour-mongers. Rossetti, however, soon started losing the plot. Already addicted to chloral, he exhumed his first wife to retrieve poems he'd buried alongside her. When his next published collection of verse was panned, laudanum failed to hold him up and he disappeared behind a fog of paranoia and mistrust, eventually fleeing back to London and thus leaving the field clear for Morris to return and make Jane fall in love with him again.

> What better place than this then could we find
> By this sweet stream that knows not the sea,
> That guesses not the city's misery,
> This little stream whose hamlets scarce have names,
> This far-off, lonely mother of the Thames?

After his death, Jane continued to live in the house, and daughter May occupied it until her own death in 1938. The latter set up the village school and was a founding member of the Women's Institute. She also ran a soup kitchen but rumour has it cooking was not her strong suit and the villagers threw it in the ditch. May loved the manor as much as Morris did. Instead of writers and artists, however, she took in Miss Lobb, a member of the First World War Land Army who had been thrown off a farm for foul and abusive language. Six foot tall, sixteen stone, pipe-smoking, dressed in tweeds and breeches, and said to go to bed with a pistol and a bottle of brandy, she terrified everyone (including George Bernard Shaw) except May, who became her companion for eighteen years.

I leave the sixteenth-century home and its knowledgeable, white-haired fleet of enthusiastic volunteers and sit out with a cream tea in the garden. On the next table an elderly man in baggy needlecords is holding forth on a little-known vein of Thatcherism: 'Symptomatic was her loathing for Egyptology. Thank God she never knew about Assyria. Boy, she would have hated them with a vengeance.' Where it's going is uncertain but I don't care – it's elliptical, passionate and as English as my cream tea. Someone on another table in this heart-of-England garden is discussing the source of the Thames. He pores over a book with a companion and then marks a spot on an unwieldy Ordnance Survey map spread out before them on the table. Nearby, an elderly man with raised sunshade attachments to his spectacles, and dressed in a viciously starched blue shirt and trousers pulled too high, ferries a tray of cake and tea to his ancient mother who sits like Buddha on the grass in the centre of the garden. Baggy Cords has meanwhile moved from Thatcher to Mendelssohn to Mugabe and the terrible legacy white Zimbabwean farmers are now reaping.

I wander up to the Church of St George the Martyr, where William Morris's coped stone tomb (the work of Philip Webb) lies like a dolmen in the garden. The studded church door opens into a time capsule of musty smells and history. St George's is virtually unchanged since the sixteenth century (Victorianisation was valiantly resisted with the assistance of the Morrises). The stained glass has decorated the east window since 1430, one of the church's two bells has been ringing since the early thirteenth century, and the font and piscina are at least a hundred years older. The only recent incursions are a seventeenth-century Turkish brocaded altar-front donated by Jane Morris in 1893, needlework kneelers produced by fellow parishioners (some in Morris's 1893 strawberry-thief design), and four black-and-white photographs of the church and the Morris tomb. The latter are pinned to the wall by the entrance alongside a sign that declares, 'These photographs can be obtained at Kelmscott Post Office Price 2/6 each in aid of Church Repairs.'

As a service is only held once a month, the angels, saints and green man sculpted by medieval arts-and-crafts masters are mostly left to their own devices. Littered across the floors and walls are loose-leaves from the book of England. On one garlanded marble memorial slab, complete with family crest, is a single circumlocutory sentence that weaves, like a river, the totality of a life:

Edward Dore late of London Born the 5th July 1724 who After
having surmounted the Dangers of Five Voyages to the East Indies,
and the Hazards incident to his Services under George Brydges
Rodney Esq: Rear Admiral of the Blue Squadron at the Reduction
of the island of Marrinico; and also at the taking of the Moro Castle
and Town of Havannah in the island of Cuba in the West Indies
under Sir George Pocock Knight of the Bath and Admiral of the
Blue Squadron in the year 1762; at length came to this Hamlet, and
at the Residence of his Relation and worthy Friend Mr Edward
Bradshaw, finished this Life, in the hopes of a better, on the
eighteenth Day of May 1773.

On a brass plaque on the floor is a shocked, less loquacious lament for
a 30-year-old: 'Death must be long.'

I squeeze past an abandoned organ and into the original chapel
where I discover red-ochre figurative biblical scenes dating from 1280,
depicting Cain and Abel, the Massacre of the Innocents, and the Last
Judgement. Pinned to the side of a cupboard is 'A Table of Kindred
And Affinity' to remind the visiting vicar whom he should continue to
exclude from marriage vows. The list starts with man's forbidden
fruit: 'Mother, Daughter, Father's Mother, Mother's Mother, Son's
Daughter, Daughter's Daughter' and sinks further and further into
the gene pool to Wife's Daughter's Daughter, and Daughter's Son's
Wife. Isis and her husband-brother Osiris would not be welcome in
this parish. A similar long list awaits the female of the species.

On a table, a pile of leaflets has been left headed 'Poverty is bad for
your health.' I have spent an inordinate amount of time in churches
and with the graveyard dead over the past months and it suddenly
dawns on me that this bastion of conservatism is also the most think-
small, think-globally, radical community experiment so far attempted
on our shores. The Church manages to commemorate those who have
gone, care for those still here, and provide weekly (or monthly in this
case) ethical sermons to ponder concerning human values, humility,
the quality and meaning of life, responsibility to others and equal
respect for all. Where else is ethics given any time and space? Where
else is there time for quiet meditation? I may not believe in God but I
do believe in the generosity of his spirit that passes through the

congregation. The church and the pub remain the hub of the community in every village.

That night, back aboard *Caroline*, a fog creeps across the meadow and hovers on the bank as if scared to go further. Luminous shapes rise up indistinguishable in the darkness, a pointillist canvas of suggestion. Shadows tremble in the water, cathedrals rise from tree trunks. Sounds are muffled. A croak. A scream. I am the only boat on the river.

From Kelmscott to Lechlade, my route is tracked by the mean slits of pill-boxes keeping an eye out for any German Second World War soldiers still wandering the countryside. At Buscot Lock, I ask the man on duty if the reason for the pillboxes' construction was fear of a German invasion up the Thames. 'More like that their armies would have to cross the river at some point wherever they landed on the south coast. Either way, they'd have hammered the shit out of 'em whichever way they came.'

This western stretch of the Thames, or Isis, was considered so remote sixty years back – when many locals were still employed in fishing, hunting, and basket-making – that journeys upriver were referred to as 'Up the Amazon'.

At the second lock, St John's, a statue of Old Father Thames lies out on the bank sunning himself, his beard shimmying like eels, his only covering a small cloth casually draped across his procreative organ. With his muscular torso and long, swept-back hair he is a cross between Neptune and David Ginola.

As I sail into the final navigable bends of the Thames, the river composes itself into Constable's brush strokes and Shelley's stanzas. Across a meadow I see the Cotswold wool church that inspired the latter's 'A Summer Evening Churchyard, Lechlade', after he rowed upriver from Windsor in 1815.

> Clothing in hues of heaven thy dim and distant spire
> Around whose lessening and invisible height
> Gather among the stars the clouds of night.

Shelley spent the whole of that summer on the Thames with his mistress, Mary Godwin, staying with friends. After his first wife committed suicide the following year, Shelley and Godwin married and bought their own home on the Thames at Marlow. Over the next summer Shelley composed the long poem *The Revolt of Islam* while sailing his boat under the beech groves of Bisham. Mary meanwhile worked on a book that was to make a considerably bigger impact worldwide – *Frankenstein or the Modern Prometheus*. The mummified monster rising from the dead may owe much to the Titan who made man out of clay, but it is equally indebted, in my view, to Osiris and Isis.

Written when she was still in her teens, it became an instant success despite being panned by reviewers. Interestingly, Mary's own mother died ten days after she was born, and in February 1815 Mary's own first child died a few days after it was born. She wrote in her diary 'Dream that my little baby came to life again; that it had only been cold, and that we rubbed it before the fire, and it lived.' The following year she started on *Frankenstein*, a creature raised by science from the parts of the dead; 'A mummy again endued with animation.'

Having reached the end of the navigable river, I cycle from Lechlade sixteen miles to the source, south-east of Cirencester. Having concealed Rosinante in a bush, I scramble across a meadow called Trewbury Mead to a limestone block on which a barely legible message is written beneath a large ash tree. 'The Conservators of the River Thames 1857–1974. This stone was placed here to mark the source of the River Thames.' I place my hand in the small hollow in the earth. It's high summer and the source of the Thames is damp rather than wet. A jumble of stones, like Osiris's gathered bones, rests in the earth's omphalos a couple of hundred yards from Foss Way (A433). I imagine Old Father Thames, a patriarchal British figure reaching back into the beginnings of time, leaning against the elephant hide of the trunk, throwing sticks, catching pebbles on the back of his hand, making a daisy chain, wondering aloud whether Aphrodite, Isis, Boadicea, Jane Morris loves him or not. When the nights draw in, he'd have a fair bit of digging to do here to reach the Underworld. What did I expect to find? A Sumerian snake with the world's first written word in its mouth? A legion of Roman soldiers making an orderly descent? Arthur and Sir Gawain stealing like thieves downstairs with the Grail to Avalon? Osiris himself packing his thermals for the winter?

I start walking, seeking water. In the middle of the field a hunched heron, the aquatic grim reaper, stands guard. Water cannot be far away. It isn't. The heavens open, and a soft summer rain envelops the field. I shelter under an oak tree on a rise in the meadow and look back. The source has disappeared beneath the veil.

When the rain stops, as suddenly as it began, I cross Foss Way into the next field. The mons veneris tucks into a trickling uterine channel dotted with forget-me-nots. I crouch down and feel the earth with the back of my hand. Damper, but hardly flowing. One hundred yards later, the channel has already widened to a fifteen-foot-wide U-shaped conduit. The dampness burrows under the A429 into a stream of watercress. In winter it would be a torrent of white water.

The following morning, with my bum aching from the cycling, I listen in bed with curtains still drawn to an asthmatic pigeon – her her her her her her. Her her her her her. Who? I hear the manic slap of wing on water, the precursor to flight. I listen to the low groan of a plane, the hysterical caw caw of the crow, the splintering of water as fish imagines life as a bird.

Back at St John's Lock, on the return journey to Oxford, a couple of canoeists thumb a lift and cradle their yellow plastic boats on the cabin top before joining me at the stern. They are on their pre-honeymoon holiday and have fallen behind schedule. The problem is Jacqui's boat, a snip at £40: unfortunately it has no keel and prefers going in circles rather than in a straight line. I'm sure Jacqui's slight 33-year-old frame will be better suited to the couple's three-week Caribbean honeymoon cruise in December. 'We'll be married by the ship's captain under Bermudan law at sea,' she tells me a little dreamily.

'It will be lovely,' Alan Massenhoveon, aged 38, adds. A strange adjective for a man who closely resembles Neil Morrissey from *Men Behaving Badly*. A few years back Alan canoed all the way from Lechlade to the Thames Barrier in ten days. But with Jacqui, he's going to be hard-pressed to make Staines in five. I drop the pair at Newbridge where I moor for the night for a second time. The sky is so sensational I sleep with the hatch open and drift off to sleep keeping tabs on Sirius and Orion.

Entering Oxford for what feels like the umpteenth time – was it Heraclitus who said you couldn't pass through the same stream twice? –

I slip off the Thames to moor beside the winding hole at Isis lock. My plan is to nip to the nearest supermarket, Sainsbury's, to restock the galley. As usual I get distracted. On the far side of a six-foot-wide isthmus of grass separating me from the permanent moorings of the Hythe Bridge Arm, Mark Davies is bending, clearing some weeds from his small canalside garden. Apart from rubber gardening gloves, he's wearing a blue T-shirt, and thin green cotton shorts decorated with stick-men copied from the wall of some Tassili or Cantabrian cave. Alongside him is *Bill the Lizard*, one of thirteen narrowboats moored here with individual post boxes, and electric and water hook-ups.

Mark works just a five-minute walk away in the city centre as a fundraiser for Oxfam. A sign leaning up against the boat also identifies him as the author of the local historical guide, *Our Canal in Oxford*, and that it can be purchased from the boat. There are two other published authors living on neighbouring narrowboats but Mark is the only one actually publishing his books on board. Other neighbours include an Oxford University lecturer, a social worker, a health worker, and an African Grey parrot which at this moment is whistling what sounds like Dusty Springfield's 'I Just Don't Know What To Do With Myself'.

'A while back Holly – that's the parrot's name – was stolen,' Mark fills me in on recent history. 'Des – you've probably seen him on the *Gremlin Castle* a couple of boats down – he's got curly hair and wears a waistcoat with nothing underneath – was devastated. Des put up notices everywhere and even managed to get the local TV station to put out a plea for Holly's return. As a result a man contacted Des, saying someone had sold him the parrot but that he was willing to give it up if he was reimbursed. Unfortunately Des is always broke and so members of the boating community's Catweazle Club – we meet weekly in a bar to recite poetry and song – had a whip-round and bought Holly back.'

Between Des's *Gremlin* and Mark's *Bill the Lizard* is the *Phoenix* and the *Wigwam*. The former is home to Catou – a hippy version of Catherine – Mason, now in her mid-forties, who's lived twenty years on the site and has been a tireless campaigner for official mooring rights for 'squatter' boats like *Hieronymous Peyps* further up the Cut. 'Until recently the boats had been allowed there under sufferance,' Mark continues his potted community history. 'It was basically a case of turning a blind eye, but people like Catou kept on campaigning and finally, just a week or so back, we heard the boats have been recognised by the local council and B.W. under the 1992 Rio Earth Summit's Initiative 21 which encourages everyone to do their own bit for a greener world – you know, the Friends of the Earth maxim, "Think Globally, Act Locally".' Mark beams with pleasure over the hard-fought victory. 'Oxford is home to Oxfam and the green lobby's strong here too. The council and B.W. have also recognised the right of the large colonies of water vole inhabiting the canal through Oxford to be official residents. They've got their own designated undisturbed sites now.' Kenneth Grahame, who wrote *The Wind in the Willows* on the Thames at Pangbourne, would have approved.

Catou's boyfriend, John, lives on *Wigwam* next door to her. 'That's the one with the roses-and-castles postbox,' Mark helps with my identification.

'Do you really get post delivered here?' I ask.

'Of course; it's not just decoration. That was the problem with getting the other sites recognised – the emergency services had to be happy with access and safety. Once you're recognised, you're entitled to the normal services.'

Mark is clearly hooked up to more than electricity and water for it's the fellowship of the Cut that is the real draw for the floating community. 'You may not know someone's surname, but you will know their first name and the name of their boat. We understand each other's problems and also share the same pleasures so there's a lot of empathy. It's particularly good for single women who can have their own independence but at the same time, have neighbours next door they can depend on. Boaters are always willing to help out – you must have found that, Paul.'

I nod emphatically.

The process of living on a narrowboat for a decade has kept Mark lean. He has shed excess baggage, unconsciously following Morris's maxim to keep nothing on board that he does 'not believe to be either beautiful or useful'. The narrowboat is the high-water mark of this injunction and Mark's frugality and refusal to waste is easily transferred from work at Oxfam to home. The art of living. He may not have actually travelled far on his narrowboat, preferring to dredge the social history of his own patch rather than exploring further afield, but every six months he turns his boat round to get a different view. Soon, as leaves start to fall, he will once again rotate 180 degrees to face north and allow the winter light to stream through the bare branches into his kitchen.

I leave Mark to his weeding and make a quick snatch and grab at Sainsbury's before exiting the city on the Thames, tiptoeing round punts and rowing boats and slipping under a bridge on which is written 'When freedom is outlawed only the outlaws are free. One world without borders. Close Campsfield.' The hullabaloo about refugees flooding the country would appear to have been resisted in Oxford where there has been a long-running vigorous campaign to close Campsfield detention centre. The fight has gained new momentum with Judge Collins' recent ruling that the locking up of four asylum seekers who had committed no crime was unlawful.

I leave alternative lifestyles, detention centres and frugality behind to sail into the acquisitively aspirational *Tatler/Harpers and Queen* world of excess. Between Oxford and Windsor the obese homes of the blessed rise up like fairytale palaces, their rolling lawns heaving with marquees and wicker hampers as the Season passes down the Thames Valley in a succession of summer parties and regattas (Henley Royal Regatta may be the most famous, but Goring and Streatley Regatta, Wargrave and Shiplake Regatta, Hurley Regatta, and Cookham Regatta are not far behind). The Thames is where commoners and royals played in the past and continue to play. The new royalty in situ, however, are pop and media stars, and the new aristocracy play football for a living.

In the tiny village of Goring, George Michael, Geri Halliwell and Michael Caine have their country seats. On the banks of Pangbourne is the large Victorian home that guitar legend Jimmy Page lived in until recently. Bray has a greyer hagiology: Michael Parkinson and Rolf Harris.

The Royal River slips along its green corridor, enjoying its unchanging pageant, while just miles away, strung out along the M4 corridor, thankfully shielded from its gaze, are the high-technology industries of the country's Silicon Valley, a nuclear power station, one of the U.K.'s largest power stations, and laboratories genetically modifying whatever they can get their hands on.

At Abingdon, plummy announcements over a tannoy set the Thames Valley tone and introduce me to the mother-and-daughter jodhpur world of the gymkhana. The crowd has thinned to family and friends for the final horses-and-hounds competition in which a competitor completes an equestrian jumps course then hands the whip over to younger sister or mother who sprints round a smaller course with the family mutt negotiating a series of miniature jumps.

Next week antique Rollers, Jags and Bentleys will be buffed before gathering on the same site for a veterans' rally. Any excuse for a get-together. A month ago it was the turn of the Queen's Swan Marker, David Barber, dressed in all his finery and pursued by a flotilla of small craft, to be greeted by the riverside crowd (the annual ceremony, known as Swan Upping, ends on Abingdon's willowy banks, having commenced at Sunbury-on-Thames five days earlier). The swans that are counted and marked are thought to have been introduced to England, like so much else, from the Levant by Richard I. Later, in the same twelfth century, the Crown claimed ownership of all male swans in the country and Swan Upping became an annual ritual.

The gymkhana stragglers pack their horses into boxes and bump their Land Rovers across the rutted field. Across on the opposite bank, an attendant is spinning a pole in the air, playing majorette to an equally thin crowd that has gathered in the open-air pool this dank August Bank Holiday Sunday.

I spend the rest of the late afternoon and evening sweeping and mopping the cabin and deck, clearing away assorted debris, sorting books and clothes and generally tidying up. I wouldn't put myself quite in the Nigella Lawson Domestic Goddess league but find the chores strangely enjoyable.

Bank Holiday Monday appears, equally sprung-clean and glowing. The dankness has been slid back like a roof hatch and it's already a baking day. No doubt, at this very moment the Met. Office is announcing it is the hottest August Bank Holiday Monday since Jesus bathed in Galilee. For most of England, today will be the last chance to dip toes into summer. Launches appear to have had their wooden hulls lacquered this very morning; and the white reflection off fibreglass cruisers is threatening me with blindness.

Reluctant to risk breaking the spell, I decide not to travel far today. *Caroline*, unfortunately, is not smelling as rosy as the countryside. She is having problems with her bowels and is decidedly whiffy. From maybe the second sewerage pump-out of our trip, she has never been quite herself, unless of course this is herself. I should not complain too much of my companion's affliction as it appears to be the only infirmity of her advanced age. I get a pump-out at Kingscraft, and bung in half a litre of concentrated Blue on top. *Caroline* still remains a tad whiffy so I buy another concentrate the owner recommends. 'Stick half a cupful a day down the loo and it works like worm syrup – she'll be right as rain in a couple of days.' I think back to the worm syrup that used to make me retch as a boy. It's hard to inflict it on *Caroline* but I remind myself she is just an animal not a human.

The river broadens yet again and as it does, so do the houses; even boathouses are the size of the streets up in Stoke. Sit-up-and-beg motorboats hammer past the zillion-pound summer retreats of the rich and famous with their Hyde Park-sized gardens and lake-sized pools. We proles fortunately still get a slice of riverfront to share with friends, lovers and grazing horses, from which to toast the olive-coloured Thames.

Southern England has come down to mess about on the river on the final day of the long weekend that flags the end of summer. Rowing boats are tied to tree trunks, rugs are spread on banks, and wine corks being pulled.

At Clifton Hampden Bridge I join the revellers, stopping for lunch and a swim. A little further on, at Shillingford Bridge Hotel, bodies have been tipped down the grass bank and into the open-air pool. Beyond it is a caravan site and a camp site, providing further Thames-side options for the less well flush. A fuck-off cruiser called *Anti-Depressant* hammers by.

I moor just before Wallingford, under the shade of a willow. On the opposite bank a pair of legs suddenly appear on a rope swing from the petticoat of another giant willow. Suddenly the swing and the swinger disconnect, the latter squealing with delight before disappearing beneath the water.

At the back of a moored white cruiser, a sun-ripened woman has crashed out on the white imitation leather banquette beside three white terriers and an empty wine bottle. The gorgeous grey-stone town follows suit, stretching itself out beside the swans and ducks.

This Bank Holiday Monday, Castle Gardens – a secret oasis of ponds, cypresses, flower beds, and the crumbling towers left by Cromwell – is abandoned. Market Square, too, is quiet: the Georgian banks and solicitors' offices snooze through their bonus day off, red telephone boxes are empty, the war memorial unvisited, and no one is queuing at the cinema/theatre/concert hall that has set up home in the old Corn Exchange beside Mousey Lane. The world and its dog is down at the river.

I wander the Georgian High Street passing the Wallingford Model Centre whose broad shop front is guarded by the construction kits of an invincible England – Tornados, Super Hornets, H.M.S. *Belfast*. On several of the model boxes, the fantasy of German subjugation is fully played out with German Albatross or Messerschmitt in fiery tailspin pursued by bullet-spitting British Camel or Spitfire. Tucked in a corner of the window, however, is a Mercedes-Benz Wrecker Truck, a reminder of the usual outcome of our post-war encounters with the old foe. In five days time England will attempt yet again to reverse the pattern and put the German national football team into a tailspin in a crucial World Cup qualifier. I have to make sure I'm parked by a television that receives more than snowstorms for that one. It's a shame I shan't be watching it with Max.

Beyond a line of black railings, a Hampton's 'For Sale' sign stands beside the portico of a plain Georgian home. I punch the phone number into my mobile and am told the property, St Nicholas House, has five bedrooms and 1.3 acres. 'It's on the market for £1.5 million. Can I send you details?'

'No, I don't think so.'

'It has got a cinema in the cellar and a detached gym and all-weather tennis court.'

'Hmmm... I'm tempted but I think I'd better say no.'

In the window of the fourth and final antique shop on the High Street is the clever marketing boast, 'Hundreds of previous owners', and 'Quality is a thing of the past'. I wonder if it really is, or whether this is just laziness – a sloppy faith that the best is always in the past and that anything that survives must, by its very nature, be suffused with 'quality'.

Back aboard the whiffy *Caroline*, I watch a swan and her seven dusky juveniles sailing through a heat haze. Dabchicks, coots, mallard ducks, a Canada goose and common gulls are cooling their feet in the shallow water beside a beach on the far bank. A shower of white feathers slips by like a flotilla of yachts.

Most people would probably agree that the natural world is an awesome, beautiful one. Whether its beauty is down to a subjective conative connection to perception, or is simply a priori objectively gorgeous, is immaterial. The real dilemma is the human condition that often prevents us taking pleasure in it. Just as there is no *quality* in a life filled with hatred, bigotry or meanness of spirit, neither is *quality* the exclusive preserve of the past, it is timeless like the gods. As William Morris observed, the human spirit soars if it is free of its own shackles. One farmer/office worker/tug captain/artisan is consumed by resentment and is absent from the job in hand. Another paces his heart to the work, is attentive, and has life breathed into him. Attention and quality are the key in life as in workmanship. It's no wonder the two – quality and time – merge into the modern holy grail of 'quality time' in our own age and are sought as the all-purpose panacea against the distemper of hurrysickness.

As I leave my mooring, an open-top Mazda sports car booms its way across the seventeen arches of Wallingford Bridge, its sound-system at full wallop ripping up the peace with the remix of Eddy Grant's 'Electric Avenue'. There are no electric avenues now until Reading.

The six-mile stretch from Wallingford and the flat Oxford Plain to Goring Gap (the steep-sided valley that separates the Chilterns from the Berkshire Downs) is an absolute peach. Four hundred thousand years ago a giant glacier scooped out the valley forcing the river south. Two hundred thousand years later the Thames Valley was inhabited by bear, mammoth, lion, elephant and bison. Another two hundred thousand years on, it's my turn.

Together *Caroline* and I slip through Goring Lock and a gauntlet of gongoozlers and moor. On the equivalent of a private water drive, a small thatched cottage stands, doubling up as a boathouse. Unfortunately the sleek chrome and highly varnished wood Amoreena cruiser is three times too long to fit into it and has been parked outside. I wonder if it belongs to Kevin Maxwell, Michael Caine, George Michael or Geri Halliwell. My money's on George Michael.

The old rivals, Goring and Streatley, face each other across a stretch of water that's possibly the oldest fording point on the entire river. The Ridgeway and Icknield Way provided a prehistoric trade route from Dorset to East Anglia, crossing at Ferry Lane. The Romans raised it into a causeway. The grassy knoll that remains today is littered with bodies soaking up the sun.

Goring's flint-and-brick homes nestle in their wooded valley, slumbering through the chimes of one of the oldest church bells in the country. The village manages three churches and three pubs in all: a draw in other words. From beyond the cemetery wall of the Parish Church of St Thomas of Canterbury (built by the oily Robert d'Oilly, the Norman baron who supported William the Conqueror and was rewarded with the manor of 'Garinges' as well as 59 other manors), laughter and swimming-pool mayhem waft across, as if from another world. I wonder if it's George or Geri.

Inside the cosy Catherine Wheel, four local males are already dug in for the night. Behind the barman's right shoulder are a dictionary, a thesaurus and a Brewer's Phrase Book to settle any disputes that arise and threaten to get out of hand when alcohol and male egos mix.

I quickly ascertain that one of the men is divorced, and one – a pilot – widowed. The other two offer no signals as to their domestic arrangements. One of the pair, probably in his early sixties, is boasting that he, along with George Michael and Geri Halliwell, is lowering the average age of the village, 'Compensating for old farts like you and Michael Caine and that born-old fart Kevin Maxwell.' The object of his badinage is at least twenty years his junior but this doesn't seem to faze him. 'You relics have had your day. We're the future. We are Goring's youth policy.'

A visitor temporarily quells the banter as he orders a drink. Foolishly he then transports his pint outside, missing out on a deep seam of heaven-and-hell jokes that the imbibers start upstaging each other with. The most memorable one concerns Bill Gates.

'The world's richest man finally dies, his work done, having flogged computers to Outer Mongolia and the Arctic,' the pilot sets the scene. 'Bill stands in a queue in front of God waiting to learn his fate. When it's his turn, God rubs his beard and says, "This is the first time in my long career that I've been stumped, Bill. I just can't decide whether to send you upstairs or down. You've put computer monitors in the homes of every living being on the planet but is that a good thing or bad?" God rubs his beard. Bill shrugs his shoulders and offers no comment. "I tell you what," God continues. "I'll leave it up to you to choose." Bill says OK but he's not signing up to anything without first taking a look at his options so God takes him down to hell first to sneak a preview. Down below, Bill sees hundreds of naked women on a beach in paradise sipping cocktails. He's impressed. No men. No competition. Then God takes him up to heaven where he finds just old tossers like you lot playing harps. "No contest," Bill says, "I'll take hell."'

'Wise choice,' the divorcee interjects.

'A week later God visits him and finds Bill screaming in agony staked out naked over a pit of flames. "This isn't what I signed up for," screams Bill. "Where's the beauties and the beach?" "Oh, that," answers God. "That was the screensaver."'

The laughter has barely subsided before Goring's Youth Policy tries to mop up a little of the glory himself. 'That's like the one where God asks the man burning in hell how he's feeling and does he see the foolishness of his ways now. "No," the man replies, "This is nothing, I come from Basildon."'

By the time the jokes dry up and conversation moves on to golf, I've downed my pint of mild and am into my second. Golf for me is Basildon, so I strike up a conversation with a man who's just perched himself at the bar next to me. His name's Chris Coburn and he's just got in from a narrowboat jaunt up to the I.W.A. Festival at Milton Keynes. He's exhausted.

'Took so long coming back down because of all the inexperienced boaters clogging up the locks. I got pissed off and started moaning about how slow they were to my companion Ian Smith, he's the locksman from Cleve Lock – that's the one just before Goring Lock. He quickly shut me up. Said *I* was the problem not them. "Shouldn't be on the Cut if you're in a hurry," he said. "They're enjoying themselves. Are you?" He had me there.'

It transpires Chris is in toilets, or rather 'marine sanitation'. His father ran a passenger boat along the south coast between Eastbourne and Hastings after the Second World War and Chris followed in his maritime footsteps starting a boat-equipment business nearby. Occasionally he took trips on the inland waterways and noticed how narrowboat owners constantly complained about not being able to get spare parts – 'At the time, most of the marinas were on the coast. So I moved the business inland and eventually specialised in sanitation because that appeared to be a big hole in the market if you get my meaning.'

Now here was the man to ask about *Caroline*'s own bowel problems.

'Basically if you have a flush-through unit like yours, you have to pump out every week. They were designed for hire boats and hire boats aren't usually out for more than a week at a time so they can keep them regularly pumped and fresh. If you were to buy your own boat, you should go for the sealed vacuum system which uses air more than water and the tank is sealed off from the toilet.' I file this crucial piece of advice away for the day I can finally afford my own narrowboat – *Caroline* preferably. 'Twenty years back, all the boats being built were hire boats, now they're mostly private and they need a different system.'

At around nine the next morning I'm checking engine oil just before casting off, when I notice two young mothers who've abandoned their buggies to peer through the towpath hedge. 'What are you looking at?' I ask, maybe a little too officiously as they jump back guiltily.

'George,' one answers meekly.

'George Michael,' the second adds helpfully.

George's home is slap-bang alongside *Caroline*. Last night I slept next to a god. I join the women at the fence and peer through. All I can see is the statue of a dog on a lawn. 'You get a better view from the bridge,' one encourages me. 'Up there,' she points ahead. I walk twenty yards up. She's right. Beyond another moored narrowboat named *Brindley* is Michael's medium-sized Victorian gabled home. Quite modest really.

I quit my moorings and fifteen minutes later slip into new ones belonging to Chris Coburn, the toilet man, who's invited me in for coffee on the way down. Next to me is *Catspaw*, a narrowboat owned by Emrhys Barrell, editor of *Canal Boat* magazine, who lives close by and keeps his boat here. *Canal Boat*, *Waterways World*, and *Canal & Riverboat* are all monthly, dedicated inland-waterways U.K. magazines with a joint circulation of around 60,000. There are more of us than you think.

Chris must have seen *Caroline* approaching and comes down to meet me on the pontoon. He's dressed in a short-sleeved shirt and looks tanned and rested after a good night's sleep. Nosily I pry into the sensitive details of his mini-estate. Chris paid £500,000 for it about six years ago ('My father would have turned in his grave'). The one and a half acres is now worth three times that.

We walk up through the walled garden past his private residence and office. On the river-side of the estate is his mother's house. She is just pulling into the drive in her car, slowly negotiating the newly laid cut-granite stones that were shipped from Portugal to London and then ferried by Chris and a friend aboard a working boat up to Goring. 'She's ninety-one and takes the boat out alone too,' Chris whispers proudly in my ear.

In the garden of the nearby White Swan I discover that my new friend, the marine sanitation businessman, is in fact an inland-waterways guerrilla. Over a pint and cod and chips I listen to some of the narrowboat epics he's made over recent years to publicise various canal restoration campaigns round the country. In 1998 he took his narrowboat *Progress* up to Chester, onto the River Dee, round the North Wales coast, and through the Menai Straits to Caernarfon. 'I was steered out by Jo Hollinshead, an old boatman employed by B.W. He had to count off stations on the train to Chester to meet me as he couldn't read or write.' Jo and Chris then sailed back with a cargo of roof slates, retracing the journey that boatmen took every week earlier in the century via the Lichfield & Hatherton Canals to Birmingham.

'It was a stunt to draw attention to the Lichfield & Hatherton Canal Restoration Trust. We got a lot of support and media coverage. In 2000 I crossed the Channel in *Progress* and sailed the canal to Brussels to attract more attention.' Having personally negotiated the tidal Thames and the Trent I cannot quite believe anyone would be foolish enough to take a narrowboat onto open sea. But it wasn't the first time a narrowboat had sailed across the Channel. Chris had been the first in 1990.

'It was all worth it,' Chris dismisses my horror. 'We kept up the pressure and got a major result a couple of weeks back when the government reversed its earlier decision to allow the new Birmingham Relief Road to be constructed without any consideration for the Lichfield & Hatherton. All along they'd argued that as the canals were closed there could be no provision made. We kept stressing the fact that the Canal Restoration Trust, headed by the chairman David Suchet – he's the actor, you know, and he's a keen narrowboater – was raising funds to reopen them. They finally saw sense and have insisted the road contractors now allow for a bridge and an aqueduct. Soooo,' he stretches the word out to climax resolution, 'there will be no new financial hurdles to negotiate when we do finally raise the funds to open the canals.'

Though hugely time-consuming, the Lichfield & Hatherton has

not been Chris's only crusade. In 1995 he sailed from Liverpool up the coast to Lancaster to highlight the need for the Ribble Link. It opens in autumn 2002.

Then there was the jaunt to Ipswich which resulted in one of the most famous inland waterways photographs ever to appear in the national papers, of a narrowboat see-sawing on a weir. 'Didn't know there was a weir there,' Chris explains matter-of-factly. 'It was seven in the morning and I suppose I was a bit bleary. We were trying to publicise a little-known route known as the Ipswich & Stowmarket.'

That was in 1997. That year he also made a 276-hour coastal voyage from the Wash to Great Yarmouth. 'The only one to do it so far. That was a bit of a hairy one.'

Chris speaks highly of both John Prescott, the Deputy Prime Minister who in 2000 launched the first Inland Waterways Charter for thirty years, and British Waterways who he believes has shown 'nerve and vision' in transforming many backwaters into thriving concerns. 'There have been really important openings in recent years such as the Rochdale and the Huddersfield which were both considered impossible restorations. It's a fantastic time right now to be involved in the waterways.'

In his personal life Chris admits to having been less successful at conservation. Last year his second wife ran off with someone 25 years her junior. What made it even more humiliating was that her new man was one of his employees. 'I think all the campaigning and journeying kept me away too much and when we were on the boat together at weekends, there were invariably friends I'd invited along.' Fortunately he still has the friends. 'They call in from every corner of the network and stay a night or several. That's the beauty of the linear village. They come from Blackburn and Boston, from Liverpool and York. The great thing is that boating, unlike yachting, is classless and people from all walks turn up.'

Chris's mum is pruning the roses when we return. We have the cup of coffee Chris had promised me, before he waves me off. I notice, as I pootle out onto the main thoroughfare, his mother waving to me from the conservatory window of her highly desirable old people's home.

I sail past the Twelve Deadly Sins, a line of vast Victorian houses, and into Pangbourne (where *Three Men In a Boat* threw in the towel and returned to London after one disaster too many). The river is quiet, the riverbank carpeted in goldenrod, wild celery and willowherb, and the stashed boats of Pangbourne College enjoying a siesta. I moor up for the day on the National Trust's Pangbourne Meadow. I have taken three hours in three days to cover ten miles from Wallingford! Do I note a slight reluctance to end?

After a swim in the river and a bite to eat, I wander up into Pangbourne and pass Church Cottage where Kenneth Grahame, Secretary to the Bank of England, first told the story of *The Wind in the Willows* to his four-year-old son Alastair in 1904. The road loops back to the Whitchurch Bridge where kids in canoes from the Adventure Dolphin Centre are twirling and dipping under water like coots. One boy manages to turn continuously, nose down, on the spot, swirling like a dervish, creating a whirlpool which he looks in danger of being sucked down.

On the Whitchurch side of the bridge, car drivers are being requested to pay 10p at one of only two toll bridges along the entire river. In the church-yard of St Mary the Virgin, two boys sit on a bench sharing a can of Carlsberg Export beneath the shingle spire. I seem to have spent an inordinate amount of time in graveyards of late. Like Osiris, trying the coffin for size. By the door, a plot commemorates someone called Fred Maggs – 'In Loving Memory 1918–1999. Lock keeper, Riverman, and Rascal 1956–1983.' Eight bell-ringers walk down the path and into the flint building. Soon they're sprinkling chimes over the village like pixie dust.

Although the village cannot muster a single shop, it does manage two pubs. I carry out my usual field research and plump for the Ferryboat, a cosy shoebox stuffed with antlers, a fox wearing shades, and a witch on a broomstick. A sign at the bar declares 'There are no strangers here, only friends we haven't met.' Just like the Cut then. Fortunately the Ferryboat lives up to the PR and I'm soon chatting to Catherine who's working her last night behind the bar. When Catherine's father, a science teacher and skulling coach, became a master at Pangbourne College, he wanted her to be the school's first female pupil but she refused and went to another school. She's now at Chichester University and earning spare cash in the pub. Tomorrow she takes off on a fortnight's holiday in Bangkok and Penang before returning in time for her final year of a religious–studies degree (her own faith she describes as 'nonspecific, but spiritual').

Catherine gets busy so I listen in on six men in their late twenties sitting behind me discussing the hold-ups on their morning commuter train to London that doubled the one-hour journey time. Late or not, the train that rattles along the Thames Valley ensures it is London's most coveted dormitory with house prices to match.

I look along the bar and decide the man puffing on a pipe on a bar stool might know something of Fred Maggs, the 'rascal' now inhabiting the churchyard. I'm right. Mike tells me 'Fred used to carouse in here of a night on his squeeze-box once he'd finished working the locks or poaching. His favourite trick was shooting the candles out with a Cromwellian smoothbore muzzle-loader. He only used the priming cap, mind, so the plosion of air blew the candle out not live ammo! Having said that, sometimes he did sprinkle a little black powder in the ashtrays so when someone stubbed a cigarette out they'd get a bit of a shock!'

Pam Freeman squeezes in between us at the bar to buy another round. She says she's bored with sitting between the other two men in her life, indicating with a motion of her head what I take to be her husband and son. Recognising a stranger in her midst, Pam asks where I'm staying.

'On a boat on the river.'

'Shame you didn't meet my mother-in-law Jenny, she was the river warden but died recently. We've got a commemorative bench ready and a plaque with a cigarette on one side and a wine glass on the other. We're going to place it down on the bank and scatter her ashes along with the stub of a fag and some whisky. She'll like that. How long have you been out on the water?'

'Just coming up for four months.'

She does a little more digging and discovers I'm writing a book of the journey. She then excuses herself and disappears with her drinks. Mike suggests, as I'm writing a book, I might like to talk to the four men at another table who knew Fred Maggs better than he did. I look across at the four oddballs. 'It's the regular monthly meeting of the local Muzzle Club. Fred was a member.' Just as I slide off my stool to follow Mike over, Pam reappears with a man in tow. She introduces me and suggests we sit together. It's one of those awkward moments. Mike is already talking to the muzzle-loaders and they're pulling up a chair for me. I explain the situation to Pam and her friend and apologise.

The four at the table become five.

After our introductions, the first to speak is Bob Staunton who joined the association in 1952, the year it was formed, and appears to have been growing his beard ever since. 'My membership number is eight. There are two thousand four hundred current members nationwide though we only have around seventy or eighty locally.' Bob has wild eyes, is pencil-thin and reminds me of Ginger Baker of Cream. 'In the early days I used to rent an Edwardian boat on the river for £5 a year – it had been cut in half and had a Triumph Herald car engine.' Bob, now 70 but as wiry and athletic as a Manchester terrier, has spent his lifetime boating and shooting. His other passion is steam engines. 'I own a steamroller I keep up in Woodcote.'

Peter, with Mungo Jerry sidies, also learnt to shoot as a boy. 'Everyone had a bow and arrow and an air gun.' He is a brickie by trade and lives on Bowsdown Farm on the river. Colin, sandwiched between the two, is the rifle secretary and probably the youngest of the group. He's quieter, neater, and is growing increasingly irritated with running his business because foreign clients keep quoting prices in euros. As if by divine design, to my disbelieving ears, Vera Lynn's 'White Cliffs of Dover' starts playing over the pub speakers.

The euro is the red flag to the bulls and they're soon off and charging. A stream of anti-European bile stretching all the way back to Ted Heath – 'Judas' – spews forth. The ugliest comments belong to the fourth member of the group (I actually quite like the other three) who I realise afterwards never introduced himself but instead launched straight into a vitriolic attack on newspaper journalists.

The biggest gripe of all is reserved for the restrictions placed on gun owners since the Hungerford and Dunblane massacres. Although the group still hold monthly pistol shoots, any member not owning a muzzle-loader has had to surrender his handgun. 'Since the restrictions were introduced, armed crime has risen forty per cent, so that's worked hasn't it?' Anonymous sarcastically comments. 'And candle-snuffing's banned too and that was just a pub game and the women were always best at that anyway because they only see one candle when we can usually see three by the time we get round to playing.'

Timidly I ask how the lot of muzzle-loaders might be improved. 'Bring Victoria back to the throne,' Bob suggests. I roar with laughter but it's not a joke. Guns of course are the giveaway. I doubt there are many *Guardian* readers down at the gun club. Anonymous starts taking the micky out of blacks. Out of the corner of my eye I notice Catherine courteously serving another round to Colin at the bar. She is black, and as English as they are. My heart sinks like a stone. I thought the men were mere eccentrics but they have veered so far right they're off the road and crashing through the wold of Rule Britannia, and the world-was-a-better-place-under-the-empire.

'For who?' I ask.

'For everybody!' Small-Weedy-Man-With-An-Ugly-Heart-And-A-Big-Gun replies. 'It was a civilising influence on India. Indians should be grateful we came and ruled them.' I decide it's time to bail out fast or risk an unattractive muzzle-loader hole in my forehead.

It is hugely depressing to again find Englishmen filled with hatred and perverted loathing. I try to imagine the muzzle-loader's parents or uncle dripping the first spoonful of projected malcontent into the child the man becomes. I see the liquid spreading, soaking the sugar and soon the boat sinks just like Olga's sodden Barlow boat did outside the Blisworth Tunnel.

Under a silver moon, I recognise Pam on the bridge with a large man I don't recognise. She is yelling at him, and walking away at the same time. The man follows like a puppy. Then she catches sight of me out the corner of her eyes.

'Your book will be worth nothing,' she yells and totters. 'You spend one night in Pangbourne and you think you know it all. You don't. You know nothing. I introduce you to the most interesting man in Pangbourne and you walk off and talk to people we don't even know!'

I cross the street and put a hand out to catch her but she pushes it away.

'I'm sorry if I offended you. I didn't mean to be rude but Mike had already started introducing me to the members of the muzzle-loader club.'

'Who are they!' She spits out the words with the same loathing Anonymous had reserved for Indians. 'They're nobody!' It's a scream more than language: the sound you hear when someone leaps out of the trenches and runs at the enemy. 'You spend one day in Pangbourne,' she retraces her earlier theme, 'and you think you know the place.'

It's hard to know where to go with this. 'You're right, I definitely don't know Pangbourne. But my book isn't about Pangbourne. It's a bit of everywhere I've been through.'

'Well, it doesn't matter. It's rubbish.' The word explodes like a bomb on the pavement and Pam totters off across the bridge. I descend to the meadow, passing a group of school kids camped out fishing, exploring, adventuring. I hope their parents aren't muzzle-loaders.

Tonight, in the mist drifting across this ancient meadow, I lost sight of the England I love and am proud of. Instead I stumbled into the dark matter that threatens to suck the universe back with a crunch and end the world. Tomorrow, I hope that the dark will roll back and I'll rediscover a caring and generous country again; an England that has the maturity to know we are no better or worse than others except in the way we conduct our lives.

Each new day offers a fresh start. Today is weatherless and I have a feeling I've seen the last of the summer sun. The featureless white of the sky occasionally stirs itself to drizzle. Sounds are muffled. Rounding a bend, I hear a crack like gunfire. Fortunately it's not Anonymous on his seek-and-destroy vigilante mission, but just a branch of the aptly named crack willow, grown old and arthritic, snapping. A narrowboat approaches, hips low, steering an arrow-straight line. A man in a canvas hat stands statuesquely at the tiller, beneath him the yellow border to the grey hull is garlanded with red roses. The river is quiet, the quietest in fact since I left Oxford.

On the sweeping lawns of Hardwick House (a Tudor mansion that Queen Elizabeth I once overnighted in) another posse of children have camped out overnight and are now sitting round an open fire at the water's edge. Across on the opposite bank, hundreds of Brent geese are waddling about on private missions. Soon the debris of last weekend's Reading Festival – paper, cartons, plastic bottles, cans, abandoned chairs and even a tent – appears like a battleground. Larne was among the thousands gathered here last weekend. I've spoken to her twice since, but am prompted to ring a third time just to tell her I'm passing it right now, holding hands across time.

Reading, the largest town on the river, has the highest concentration of high-tech firms in the country (mostly American – thank you America for colonising and civilising us). In the decade up to 1997, the town gained 28,000 jobs and 263 new firms and yet one-third of its population continues to live among the country's poorest 10 per cent of wards. It suffers the same malaise as the rest of the country – a widening gulf between the haves and have-nots.

Another mind-boggling fact about Reading is that Rimbaud taught French here in 1874 (four years after his first collection of poems appeared when aged sixteen; a year after he gave up literature and was shot by Verlaine; and a year before he became a gunrunner in the Middle East). Jane Austen was educated above Abbey Gateway, next door to Reading Gaol where Oscar Wilde was imprisoned for homosexuality and wrote *De Profundis* in 1897 (*The Ballad of Reading Gaol* was actually written in Paris a year later). The gaol is now, naturally, a tourist attraction, and memorable mostly for a Wilde quote in the commentary, 'The true mystery of the world is the visible, not the invisible.'

I slip past a line of gleaming boats moored next to an island fitted with floodlights on which there are two notices: 'Private Mooring', and 'Island Bohemian Bowling Club'. Soon I'm dodging a swarm of boats crewed by teenage members of Reading Blue Coat School Boat Club. Conkers are appearing on the horse-chestnut trees and acorns emerging on oaks. There is no sign, however, of the young female ghost – reputedly Isabella de Valois who married Richard II when aged eight – who's occasionally spotted walking the riverbank.

Every home lining the river is a mansion or palace with a fleet of staff working the garden and sweeping the stairs to the boathouse. As for the moored boats, they are ships, ocean liners, Antarctic icebreakers. Two ridiculously stylish women lean against inclined seats in a silent cherry-red wooden electric launch as it passes summer cottages sculpted like seashells and Swiss chalets. Wargrave comes and goes, as did its parish church, reputedly burnt down by the suffragettes in 1914 because the vicar refused to take the word 'obey' out of the marriage service.

Eventually I arrive at Henley-on-Thames, and am charged an outrageous £7 to moor overnight at Mill Meadows. If I'd pulled in two months ago, during Royal Regatta Week (first week of July), I would have been stung three times this. I've got too used to paying nothing on the Cut. Apart from the prices little else has changed since the first regatta was staged in 1839: heaving hampers are still landed, parasols raised, smart blazers unbuttoned (Henley is credited with the invention of both the blazer and the trouser turn-up); and women – for whom a new hat is *de rigueur* – are requested to desist from trousers or skimpy miniskirts if they wish to be admitted to the Stewards' Enclosure. Then, as suddenly as they appeared, hampers and boaters are stashed, and the most celebrated boating stretch of the Royal River returns to its idyllic Wind-in-the-Willows, Three-Men-in-a-Boat existence.

As I step out on the short waterfront walk into the town, the sky is an extraordinary angry yellow. Suddenly sheet lightning pancakes overhead, celestial cymbals clash, and the heavens open. Within minutes the town alleys are rivers. Although I have an umbrella, my feet and jeans are soaked. I escape into the Anchor on Friday Street where the walls are bare brick or wood-panelled; the ceiling is hung with bidets, potties and catheters; and the shelves are awash with Toby jugs. I sit at a round brass table. The most noticeable incumbent in the bar is a giant of a man acting as priest in the confessional to two regular customers, one of whom is wearing a rug across his lap. He gets the role of priest because he's the landlord and therefore pack leader.

The first man to speak is having problems with his daughter. The second is having problems with his son. The landlord, on the other hand, who is a tenant rather than an owner, is concerned booming property prices will force him out of Henley. 'Ten years ago there were forty-one pubs in Henley; now there are fourteen. The breweries just flogged off the properties because the land was worth so much.' The landlady appears behind the bar and joins the fray. 'I saw a one-up, one-down on the Reading Road on the market this morning for £310,000!'

Two nineteen-year-old boys crash through the door soaked to the skin. The landlady gets them two of her son's T-shirts to change into. 'He's away working in France so he won't be needing them.'

Through the windows, I watch the water rushing down the road outside. Suddenly the street and buildings are flooded with light. Simultaneously there's a mighty explosion directly overhead. Everyone jumps...it's as if a bomb has gone off...and then laughs with relief on discovering we haven't died. The black Labrador goes back to chewing its apple at my feet and the parrot at the rear resumes lordship of the silences. The landlord comments on the wisdom of selecting the church spire as the village conductor. 'Well, the dead aren't going to worry are they? Might liven them up a bit.'

When the rain eases I slip out into the prosperous High Street, a potpourri of Tudor, Georgian and Edwardian homes running from the

Victorian town hall to the Church of St Mary the Virgin which has stood on the bank for 800 years. Café Rouge, Caffe Uno, Loch Fyne Fish Restaurant and La Bodega Tapas Bar vie for the custom of a few drowned rats out and about. Off the High Street I find homes inhabited by those with more modest incomes who have been living here since before the Flood, or the last one anyway. Inside the Henley Social Club for the Over-60s, the only inhabitants without white haloes are bald. They sit round tables playing board games. It looks fun. At Silvers, round the corner, there's more evidence that summer is over: all 'half-sleeve shirts' are half price and striped blazers down from £225 to a mere £175.

As Henley – designated by Dickens 'the Mecca of the rowing man' – has more associations with the Thames than any other riverside town, the following morning I spend several hours exploring its state-of-the-art £14 million flagship River and Rowing Museum. Near the entrance is the sleek Aylings AX2 (built in Weybridge) that won Olympic gold in Sydney for Steve Redgrave, Matthew Pinsent, Tim Foster and James Cracknell. Pinsent lives here in Henley and Steve Redgrave, the greatest rower of all time, in nearby Marlow.

Upstairs an original map by James Brindley is displayed showing his proposed canalisation of the Thames intended to shorten travelling time (thankfully it never got the go-ahead). I discover an area at the back, with views over the river, where I can sit with headphones on listening to Thames-related poems, prose and song. I play an extract from Conrad's *Heart of Darkness* describing the broad waters of the estuary, then another from *The Waste Land* counterpointing the river's permanence with the fleetingness of human life. I then switch from literature to music, with the overture to Handel's *Water Music* performed for George I at a Thames pageant in 1717. I leave the best – the Kinks' 'Waterloo Sunset' – to last. The twangy guitar, the inimitable skinny voice taut with emotion can still bring a tear to a grown man. Muswell Hillbillies Ray and Dave, like me.

As I drift with the song, boats slip silently along the river and swallows flash like arrows. Suddenly I'm dragged out of my reverie by a muffled voice. A tanned middle-aged man is standing at my shoulder looking at what I'm looking at. I pull off the headphones. 'Wonderful, isn't it?' he repeats. We exchange coordinates and he immediately offers me a berth at Cookham where he lives. 'The boat we usually have moored up there has been hired out for a month so it's free. I'll give you my number and you're welcome to pop in for a drink, we live just round the corner.' I thank him, and scribble directions to reach the mooring.

Back on the river, I pass Phyllis Court Club, a two-storey cream-coloured hotel with lawns licked by the Thames. Skiffs sit low in the water, their wicker back-rests inclined for relaxation. Today, however, there appear to be no takers. The irony of the Thames Valley is that while the river and the waterside homes give an overwhelming impression of wealth and health, the reality in the hard economics of running a river business is somewhat different. In a recent English Tourist Board pecking poll, the Thames Valley didn't even figure among the ten top U.K. tourist destinations. It is as insane as it's incredible. As boatyards and related businesses with massive overheads struggle against falling visitor numbers, more and more bale out and cash in on their land value to developers.

Everyone's feeling the pinch, including the river's caretaker, the Environment Agency (E.A.). It has long been government policy to leave the wealthy Thames to survive off boat licence fees. But in terms of usage and impact, boating lags behind fishermen and walkers, and the fees raised fall way short of maintenance costs. Worse still, boat licences issued by the E.A. have dropped 30 per cent in the last ten years, and from these diminishing returns, the E.A. is expected to address an £11 million backlog of repairs.

While there has been considerable financial support for B.W. and its overhauling of the canal network, no similar funds have been made available for the Thames. Once the Royal River was the top of the tree and the canals decidedly downmarket, but things are changing. The old story is again unfolding: the one where the healthy child is ignored for the sake of the sick child and when the latter recovers, the former, ignored so long, is found to be terminally ill.

And so today, as I sail beneath a medallion of Isis on Henley Bridge, I have the Thames largely to myself. Downstream, gracing the western bank, is Fawley Court, designed by Sir Christopher Wren and with gardens laid out by Capability Brown. Not a bad team to move in to design your home. Half a mile further downstream, an Etruscan Greek temple, built by James Wyatt in 1771, floats above an island that marks the end of the Henley Regatta course. In the past it has entertained King Edward VII and Queen Alexandra. Today it is the turn of a large flock of squawking brent geese.

Next, up come the remains of Medmenham Abbey, now part of a private estate. Founded in 1145 by Cistercian monks, vacated on the orders of Henry VIII in 1538, it was given a sacrilegious pornographic makeover in the mid-eighteenth century for the orgies of the Hellfire Club.

Leader of the cell of the 'Mad Monks of Medmenham' and master of ceremonies was Sir Francis Dashwood, M.P. for West Wycombe from 1741 until 1763, Chancellor of the Exchequer 1761–62 and postmaster general 1766–81. His thirteen apostles reputedly included the Prince of Wales, the artist William Hogarth, Thomas Potter (son of the Archbishop of Canterbury), the Earl of Sandwich (inept First Lord of the Admiralty largely responsible for British failures in the American War of Independence, who invented the 'sandwich' so he could eat and play cards at the same time), John Wilkes (popular Mayor of London), the Earl of Bute (one-time Prime Minister), and even Benjamin Franklin when he visited our shores.

In a subsequent treason trial John Wilkes was warned by the Earl of Sandwich (a notoriously corrupt and opportunistic individual), 'Sir, you shall either be hung or shall die of a loathsome disease.' To which Wilkes replied, 'That depends, Sir, on whether I embrace your principles or your mistress.'

Although it was said collectively of the group, 'Orgies were their pleasure – Politics their pastime', Charles Dickens in his *Dictionary of the Thames* published in 1887 posthumously judges them more generously. 'They lived at a time when drunkenness and profanity were considered to be amongst the gentlemanly virtues.'

Hellfire Club members would float upstream from London under the veil of night with masked 'mollies' (aristocratic ladies) and 'dollymops' (prostitutes) who would act as masked nuns during their rites. Sometimes married couples attended too and transformed proceedings into prototype swinger parties.

Having landed beneath Medmenham Woods, revellers would steal up to the abbey. On one side of the entrance was a statue of Harpocrates, the god of silence, and on the other Volupian Angerone, goddess of secret passion. Inscribed above the door was the Hellfire Club motto 'Fay ce que voudras' ('Do as you will').

The network of caves in the wooded hills were used as monks' cells for private entertainment. Once pleasured, the 'monks' crossed an underground stream, known to the assembly as the River Styx, to reach the Inner Sanctum, a circular room where black masses were supposedly held over the naked bodies of female toffs.

In 1751, Dashwood paid for St Stephen's Church in West Wycombe to be restored and is said to have secretly modelled it on the solar temple at Palmyra. On the ceiling was a token depiction of the Last Supper, the Christian version of the *agape* of the classical Mysteries. A ball seven feet across surmounted the church spire (with room inside it for three!). It was a replica of the golden ball on the Custom House in Venice whose weather vane, in the shape of the goddess Fortuna, was believed to link the sun god, who was born at the winter solstice, with Christ as 'the Light of the World'.

When Paul Whitehead, a leading figure in the order, died in 1774, his final request was that his body and heart be buried separately. Dashwood's private army, the Buckinghamshire Militia, carried the heart in an urn three times around his mausoleum erected on a hill above the caves. Dashwood's biographer, Eric Towers, believes this ceremony was a re-enactment of the pagan Osirian/Dionysian dismemberment ritual.

The abbey's sacrilegious decor, unsurprisingly, didn't survive the nineteenth century. As I swing past the abbey remains on a broad bend in the river, a tanned wooden skiff passes in the opposite direction, reflected sunlight dancing on its thick layers of varnish.

A little further up, on the opposite bank, a large mum in a one-piece swimming costume bobs up and down in a truck inner tube. Her two children, swinging out together over the river on a rope, shout to her and point at *Caroline*. She looks over, smiles and shouts across to me, 'That's my name – Caroline.'

After a couple of overcast days, the sun is out again. It's 1.30 p.m. and four alpaca are grazing in a field by a caravan park a long way from home. Nearby Christians are clearing pots and pans from a trellis table at an old-fashioned tented camp.

I pull in at Hurley and head for the pub for lunch. Unfortunately, the decidedly drunk walls of Ye Olde Bell, which claims to date from 1135, are festooned with balloons indicating a private reception is in progress. I confirm this worst of all possible news with the patient chauffeur of a veteran Rolls-Royce waiting outside. A breeze blows down the street whipping the balloons into a frenzy.

A number of the village's brick-and-flint homes have been white-washed, others are overrun by ivy or hung with flower baskets. The smell of summer rises from mown lawns. Plastered inside the window of the Lilliputian post office in Church Cottages, are notices requesting gardeners, announcing the sale of an Ocean 30 Moonraker Class Cruiser (£31,995), outlining the details of a Practical Philosophy course, and informing villagers of the date of the next talk hosted by the Hurley Garden Club. There's also a reminder – amid the fungicide and light bulbs piled in the window – that the store is a member of Neighbourhood Watch.

At Temple Lock I'm asked by the lock keeper to ferry a brown envelope to his colleague at the next lock. It contains a map showing how to get to Teddington. I'm flummoxed. 'Doesn't he know the way?' I ask. 'You can't really get lost can you, there's only one road.' By road, I mean water road: I've forgotten there are alternative navigations.

'No, he's going by car. He's got a day's training up at Teddington and hasn't been before.'

Approaching Marlow, homes loosen another notch on their belts: languorous, palatial, confident. I slip under Marlow Bridge where tradition has it boatpeople would be mocked by locals asking 'Who ate puppy pie?' – a reference to a local baker who, frustrated with bargees complaining that his meat pies weren't sufficiently filled, baked them pies filled with recently drowned puppies.

A handsome church hugs the prime location at the riverbank allowing the dead to dip their toes in the water. Percy Bysshe and Mary Shelley lived in West Street, and so did T.S. Eliot for a short period after the First World War. At Marlow Lock, I exchange the envelope and chat. I casually mention that the last time I was in Marlow, back in March (just before I started out on my water journey) the river was in spate and Atlantic whitecaps were hurtling towards the weir. The lockman nods his head. 'Worst floods I remember. Trouble is, all we could do was hand boaters warning cards. We have no powers to close the river as it's a public right of way. Some idiots went out anyway of course and then we had to risk life and limb saving them. A couple died when they went over the weir and were smashed up.'

The wind whistles through the reeds and dapples the surface of the river as it spreads itself into another wide lazy bend. The Thames is more exposed now, close to 300 feet across, and trees on the bank are cowering beneath the wind. I pass Bisham Abbey where David Beckham and the boys were training a couple of days back in preparation for their World Cup qualifier in Munich.

At Cookham I pull in, as directed, beyond a small boat called *Lily Legato* that's partially covered by a tarpaulin. I phone Simon Davis whom I'd met in the Henley museum and he immediately invites me up for a drink. His simple bungalow home is just two hundred yards away, down a country lane, beyond a nursing home. Three resident dogs go potty behind the sitting-room window as I crunch the gravel up the short drive. By the time the door swings open, they've bolted into the hallway and are the first out to greet me. Simon, his wife Pat and friend Rita follow.

Looking round the walls and shelves, it's obvious I've wandered into a shrine to the Thames with photographs and books charting its various moods and history. Among the religious icons are two paintings of Boulter's Lock and Cookham Regatta. Simon points to the second one. 'That was 1995, a century after the first one was painted. Pat and I organised the regatta and recreated the original scene. Everyone made replica clothes and we had original or replica boats – canoes, wherries, punts, Thames skiffs – and then we all assembled in exactly the same places and poses as in the original and *voilà* – there you have it! Snap. The artist copied the photograph. Twenty-five thousand people turned up to watch that day!'

As we settle down to gin and tonics and bowls of nuts, I discover Simon and Pat were responsible for relaunching the Cookham Regatta in 1988 after a gap of fifty years. Vintage Dunkirk boats turned up, so did the old punts, skiffs, Honduras mahogany slipper launches (shaped like a woman's slipper with duck-tail rears), raters with huge sails resembling feluccas, and steam boats. There were craft fairs, rides, and a new Clergyman's Cup was introduced alongside the traditional regatta races.

'We encouraged Jonathan, a local rabbi from Maidenhead, to

challenge other clerics to a race in whalers. He kept entering every year until he eventually won the thing. There was also Dongola racing – an imitation of Kitchener's boys who had to paddle up a wadi to take a town called Dongola in the Sudan using their rifles to punt along the reach.'

'Do you still organise it?'

'The regatta? No. We leave it to the Rotary now. I presume you're going.'

'Going?'

'To the regatta. Tomorrow.'

I cannot quite believe what I'm hearing. Of all the nights I could have pulled into Cookham, I've done so on the eve of the biggest day in its calendar.

'If you do go, don't expect it to be like in the picture – you won't see the range that we had a decade ago,' Simon warns. 'The owners of the huge Tupperware boats that turn up today don't know and don't care about the history. They just stand about in their suits with mobiles glued to their ears in the corporate tents.'

Simon attended his first regatta, at Henley, at the age of eight. 'We lived in Bisham and the family travelled down in a camping punt because we had no money. It was such a great atmosphere – all the beautiful boats and the people dressed to the nines and the picnics. Everybody came down for a day out.'

Pat appears at the door in a pinny. 'I've put some pasta on. Is pasta all right? Pasta with meatballs?'

I hadn't expected a meal. I don't think Pat had planned it either; it was just a generous and gracious spur-of-the-minute thing. While she finishes off in the kitchen with Rita's help, Simon invites me into his study. Stacked on the shelves are 150 books on the Thames and the canal system. On the walls, seemingly, are an equal number of paintings and photographs. He points to one of the latter. 'That's my mother attending the Teddington Regatta in 1902.' The woman is dressed in a flouncy lace dress and a huge hat with a velvet ribbon. Another photograph shows a handsome highly varnished electric launch outside a hotel. 'That's one of our fleet – we've got seven now. We earn a living hiring them out to hotels like Oakley Court, Monkey Island and Chauntry House.' One picture shows Edward VII with wife Alexandra at Boulter's Lock; another shows him with his mistress, Lillie Langtry, at Monkey Island.

I tell Simon that I used to live in Lillie Langtry's old home, a delightful old rambling mansion bordering Richmond Park that I rented cheaply (a speculator was awaiting planning permission to transform the grounds into a town) with my brother and four of his friends in 1967 when I moved to London at the tender age of seventeen.

'That was when the flappers came down every weekend.' Rita has appeared in a crack in the door to call us to the meal table and has had her mission diverted by the photograph of Edward and Lillie at Monkey Island.

'Flappers?' I ask.

'Nice girls mostly, with bobs, who'd come down to dance the Charleston. But Monkey Island was pretty notorious and it wasn't somewhere you'd spend the whole night if you really were a nice girl though! There were several other venues too. Anyway what I really came to tell you was that food is ready.'

Pat and Rita, old friends, sit opposite each other at the round table in the kitchen and reminisce about the dances they themselves attended on Monkey Island and elsewhere as young girls in the fifties. 'Skindles Hotel, just after Boulter's Lock, was our favourite – it's boarded up now. They had very fashionable dinner-dances.'

Simon is wearing the same blue Levi-style shirt I'd seen him in at Henley. His white hair is cropped, he looks tanned, healthy and considerably younger than his 63 years. Pat, four years his junior, looks in fine fettle too. Unfortunately a recent stroke and subsequent sentence to heavy medication has resulted in her balance going awry. She is no longer able to sail her beloved Thames so instead walks the dogs two hours daily along its banks. Neither does she any longer work in the local surgery whose doctors somehow missed the telltale signs of an approaching stroke. 'My memory isn't what it was since the stroke, so it's too big a risk when you're dealing with people's lives.'

Pat and Simon start firing questions at me about my own waterborne odyssey. Having previously owned two narrowboats themselves, they're intrigued to know which canals I've covered. As I move into the fourth month and the Thames, Simon suddenly pushes his chair back and whips into the study. He returns, ferrying a thick brown handwritten tome to me. It's written by his Danish grandfather in French and is the record of his own four-month journey by skiff through the French waterways between June and September 1887. 'I'm planning recreating the journey soon. It'll be a bit like yours but in France.' His eyes sparkle at the prospect but I somehow doubt he'll be able to leave Pat, they seem so glued together.

I emerge from my cabin to a bright, chilly morning. From the stern of *Caroline*, I can see tents being raised, cars disgorging hampers and ducks squawking at the hullabaloo on Marsh Meadow. Those already staking places on the banks have mostly dressed down rather than up – shorts and T-shirts rather than blazers and flannels. I take a walk through the early regatta proceedings. A veteran Morris Ten-Four is parked alongside a Morris Minor as if they're out on a date. Victorian swingboats are already doing a roaring trade with boisterous Herberts. Regatta teams are preparing their boats, organisers hammer tent pegs into dry ground, dogs yap and a heady aroma of grilling bacon and sausages perfumes the air.

I watch an early race, led from start to finish by the same crew. As they cross the line, a polite cheer goes up from the thin crowd. Later it will become a well-lubricated roar but at the moment the bulk of the crowd is still carrying canvas chairs from their cars, or unloading hampers from boats.

I wander on to the adjacent National Trust's Cock Marsh where Pat has told me a pair of kingfishers are nesting. I have no luck spotting them but it's a delightful stroll anyway, the path backed by pale grasses, and fringed by ash, willow and hawthorn. Instead of kingfishers I do a little boat spotting. At the helm of a cherry-red slipper launch slipping past is a neatly bearded skipper adjusting a cravat beneath his navy blazer. From a back garden opposite, I watch a man rowing bottles and food across to the regatta bank. Once emptied, he returns to pick up the members of his family.

By the time I return to the cordillera of white-peaked tents, the iconographic symbol of summer on the Thames, the crowd is a fur covering the bank and races are coming thick and fast. Already the results of the first five have been inked up on the notice board by a tanned arm protruding from a short-sleeved canary-yellow Marlow-on-Thames Rotarian shirt.

I sit and watch a couple more races and am offered a chicken drumstick by one of my neighbours and a beer and a packet of Quavers by another. I think back to the picture of Simon's mother in all her finery at the Teddington Regatta. Where are the hats and flouncy dresses? It's all a bit of a corporate damp squib. I get bored, and do what I usually do – head for the churchyard. This one, like all the others, is special.

Beneath the castellated flint tower of Holy Trinity Church, a simple granite gravestone, among a forest of more elaborate memorials, spells out a tribute. 'To the Memory of Stanley Spencer K.T. C.B.E. R.A. 1891–1959. And His Wife Hilda.' There is also a verse from the gospel of St John – 'Everyone that loveth is born of God and knoweth God. He that loveth not knoweth not God for God is love.'

Inside the church two brooms have been parked either side of the absent congregation and two Japanese tourists are inspecting a copy of Spencer's *The Last Supper*. In the painting Jesus is breaking bread to an almost horizontal gathering of apostles whose legs, outstretched beneath the table, are crossed as in crucifixion and stretching to meet their companions across the divide.

I go in search of the original in the High Street at the Stanley Spencer Gallery and strike lucky again: the collection has just returned from a stint at the Tate Modern where there's been a major Stanley Spencer retrospective. Inside the small former chapel I pay my £1 admission, commenting appreciatively 'That's cheap.'

'D'you think so?' Pat Flint asks, rather taken aback. 'Lots complain because we've just doubled the price.'

Pat is one of a fleet of volunteers who help to keep prices down. Lining the walls are the brightly dressed moon-faced inhabitants of Cookham as well as its gardens and punts: its everyday transformed into the miraculous. In Spencer's *Christ Preaching at the Cookham Regatta*, full and luscious bodies lean stiffly like trees while Christ, in straw boater and flowing locks, cranes forward from a cane chair preaching fiercely at children cowering in the punt. In *Sarah Tubbs and the Heavenly Sister* (1933), Sarah, a local girl, is comforted by heavenly visitors after seeing Halley's Comet and fearing the end of the world.

Throughout the religious paintings, Cookham and heaven, the apostles and locals, are interchangeable; there is not a wafer separating the parochial from the celestial. The Thames and the Nile; Mary and Isis and Sarah Tubbs all reside in Cookham. Spencer was born in the village (the last of nine surviving children), and lived virtually his whole life here. He appears to have spent much of that time, rather like William Blake, in a state of revelation and sensual wonder, transfixed by the divinity of his little Jerusalem.

As I stare at the hammered old pram the artist used to transport his materials in, Pat Flint tells me she remembers seeing Spencer pushing it through the street back in the fifties. 'I was driving through town and my husband pointed him out. "See that scruffy old man," he said, "that's Stanley Spencer, the painter."' Like most visionaries and scruffs, Spencer was also a great British eccentric: when he was knighted by the Queen Mother, he took a bottle of milk and some Tate & Lyle sugar cubes in a plastic bag with him for tea.

Pat confides modestly that she attends an art class herself. 'I mostly paint roses on crockery, though we also paint live models. I'm finished usually before the others start! They stare at the model, do a couple of strokes, step back and take for ever. Don't know what there is to think about really.' She smiles warmly, lifting her carefully lacquered hair an inch, before stroking a blight, invisible to everyone but herself, from her floral shirt.

'When I first started working here ten years ago, I thought his paintings were horrible but I've grown to appreciate them. You can't help it when people turn up from all over the place and are so enthusiastic. Do you know, only one man has written "rubbish" in the visitors' book in all the time I've been here.'

Pat talks me into buying a video, *Stanley Spencer A Painter in Heaven*. I like the title. As she gives me my change, she pops a free guidebook into the brown paper bag. On my way out, I take one last look at *The Last Supper*, painted in 1920. Saint John rests against Christ as he breaks bread sitting beside the grain bin in the Cookham malt house. Spencer said, 'All the things that happened in the Bible, happened in Cookham.' He showed the same wonder at blossoming girls as he did Christ's resurrection (in one absent painting, *Resurrection in Cookham Churchyard*, Spencer's wife Hilda rises out of the ground while Charon plays the piano on the Thames).

As I wander back along the riverbank to *Caroline*, I try to pinpoint the common ground shared by William Morris in Kelmscott and Spencer in Cookham. Both saw the divine in the commonplace, the miraculous in the everyday. There is a clear line connecting them and that line and that common ground is the water road, the numinous Thames.

Back on the river, the Kinks' hymn to the Thames, 'Waterloo Sunset', continues playing in my head. Like Stanley Spencer, Ray Davies too found the sacred in the familiar. I think it was William Blake who said 'Everything that moves is holy.'

Just south of Cookham, I sail past the hanging woods of Cliveden. The stately home is now owned by the National Trust and part-leased as a hotel with rooms starting at £300 a night. Alternatively, if you're feeling really flush you could pull into one of the Venetian mooring posts and check into their riverside cottage from around £20,000 for a fortnight – considerably more than you'll pay to own a home outside the Middleport Pottery in Burslem.

Cliveden has maintained a strong whiff of scandal and intrigue throughout its history. Its first owner George Villiers, the drunkard Duke of Buckingham, killed the Earl of Shrewsbury in a duel here watched by his lover, Lady Shrewsbury. In the 1930s the Astors moved in and tried to sell us out to Germany (their plan was to make Oxford the capital when Nazi victory was secured). Then in the early sixties Lord Astor invited Stephen Ward to stay in one of the wooden cottages I'm now sailing past. Ward invited John Profumo up and the latter was both literally and metaphorically brought to his knees by Christine Keeler, buckling in turn the knees of the Conservative government.

At Maidenhead I sail under the largest single-span brick bridge in Europe, built by Isambard Kingdom Brunel and the subject of Turner's painting *Rain, Steam, Speed*. One hundred and fifty years later it is withstanding the far more considerable bulk of a twenty-first-century train hammering across it. Next I slip past Monkey Island where Pat and Rita used to attend their dances. Today the hotel is a peaceful oasis redolent of the Raj, with resident peacock, eighteenth-century fishing lodge and pavilion set in four and a half acres.

I've managed my usual paltry four miles or so before I'm seduced by the blue umbrellas and white linen tablecloths of the Riverside Restaurant. I sit out on the restaurant's wooden deck. Behind me is an open kitchen and pale wood floor. Seared tuna, a bottle of white, a twenty-yard stroll back to *Caroline*... this really is how a restaurant should be done.

I have a snooze and then potter down to my next overnight base at Windsor, around half a mile from the battlements of the world's largest inhabited castle. Skiffs slip through the water like salmon, and an orchestra of cruisers, narrowboats, punts, rowing boats and double-decker tripper boats sail by, buffeting *Caroline* against the bank. Cars snake along the water's edge where all manner of amphibious craft are moored beneath the groping fingers of willows.

Next to *Caroline*, the crew of the *Mistral* are swigging from San Miguel bottles as they stoke their barby and chat. Nearby a couple – aged around seventeen and both of them gorgeous – sit on a bench holding hands. The boy is Asian and dressed in black jeans and a black shirt. The girl has cropped retro-mod hair, is dressed in cargo pants and a belly-top, and reminds me of one of my great early loves. (Helen from Stockholm was sixteen, and also a mod – I met her when I was seventeen and she was attending a summer language course in Brighton.)

The lovebirds stare across at me as I stare at them. Eventually they wander across. 'Hi. Hope you don't mind us interrupting.' I look round to see what they're interrupting. Nothing. I'm sitting on deck doing zilch. The girl continues to take the lead. 'We do quite a bit of sailing and were just saying how great it would be to take one of these out. How much do you reckon it would cost to hire for a week?'

'In August, for a two to four-berth, probably around £800 and even in October they're around £550. They're not cheap.'

The two huddle and discuss. Again Helen takes the reins. 'If Matt and Ellie come that's £140 for a week; £20 a day each. It's doable.' Asian Paul nods his head and then turns back to me.

'Do you know of a company?'

'Not here. Mine's from Southall. The Windsor tourist office will know the local outfits though. Alternatively Waterway Holidays U.K. represents the majority of the main boat-hire companies and can direct you. They're based in Teddington but I don't know the number offhand.'

The young things thank me warmly and wander off still holding hands, the whole of the Thames in front of them. I try to decide whether I'm envious. Yes. I lie on my bunk-sofa looking through the back hatch, one door ajar, the other closed. The panels have traditional inverted painted images of a castle rising beside water. Beyond the open door, I can see the real thing, Windsor Castle soaring beside the Thames. Castles are paranoid places, a psychotherapist friend once told me, full of incest, fear of outsiders, dark remote places removed from the world. No doubt he'd say *Caroline* is my own castle and the journey incest with my past. Who cares. *Caroline* is the perfect-sized space to live and think in; a moveable home travelling through the world instead of watching it go by. I'm a snail with my home on my back making disappearing snail-trails through the country.

At 9.30 that evening I return to a home as cosy as childhood memories after a spectacularly successful night watching, in a blarey beery Windsor bar, England bury Germany in Munich 5–1 in the World Cup qualifier.

It is now just five days before I slip the padlock on *Caroline*'s fore doors and walk off, never to return. I speak to the family on the mobile and they all feel comfortingly within arm's length. This other intimacy with *Caroline*, the water and England is soon to be relegated to the memory shelves.

That night I have the most enchanting dream of my entire time afloat: in it the Thames rises and rises and becomes so light it floats, becoming immaterial, a song sung by a thousand voices passing downstream. I wake and find myself gasping for breath.

Each new day is a resurrection but, after my dream, the Thames has an added luminescence. I walk along the towpath, past the Sir Christopher Wren Hotel (built by Wren as his own retreat in 1676) and cross the bridge. On the far side I enter the Eton Tea Rooms and, over the full monty, bizarrely discuss Cyprus with the owner who moved to England from Kyrenia in the 1950s. As Julio Iglesias croons and the Thames flows, we discuss the island of love which I too had lived on between 1977 and 1981. From love, it's a natural gear change to football and England's storming win in Germany. My Cypriot friend brings me a few of today's papers and they're all full of the drubbing. It's a good day to be an Englishman.

At the top of the high street, I pay my £3 for a one-hour tour of the hallowed halls of Eton where boys have been gouging graffiti into the desks since before Columbus discovered America. In the early days the dormitory was squalid and there were three to a bed; now each boy has his own bedroom and it's probably still three to a bed. Each morning, however, students still file into chapel to atone for their sins and pray they don't go blind. Some things don't change.

When I eventually set sail, boys in the green of Eton College are going through their stretches outside the boathouses. I am the last of ten boats waved into Romney Lock. The skipper of a 50-foot narrowboat *No Problem* I'd met earlier on the Oxford Canal when Dad was on board, shouts across, 'What, no crew now?' He has another month left of his present trip. He and his wife are from Penkridge (scene of the infamous maggot invasion) and are typical boaters who, having taken early retirement, and annually spend the best four or five months of the year afloat.

I pass a bevy of fishermen and several fisherwomen in gumboots fishing at Datchet where Admiral Nelson also liked to fish. At Runnymede, I cross a field to pay my respects at the mushroom-shaped Magna Carta monument but am shocked to find the homage to the birthplace of democracy erected and paid for by the America Bar Association. They did so because their own founding fathers, fleeing British persecution, sailed to the U.S. with a copy of the first great democratic statute and this in turn formed the basis of the U.S. Constitution and later Bill of Rights. Is it not a little embarrassing that no similar body from our own shores felt moved to mark its significance for Britain? Perhaps the mead the monument sits upon is our own tribute, for its pale Yorkshire fog grass is still grazed by cattle in winter and left as a hay meadow in summer just as it was eight centuries ago.

The original Magna Carta had 63 clauses, many of which are still practised today – the right of inheritance, the rule of law, trial by jury, punishment to fit the crime, and the right of free and safe passage. The thirteen barons in open revolt against King John's incompetent rule forced the monarch to set his seal on the Magna Carta on 12 June 1215. He promptly asked the Pope to annul it but it was John that was ultimately annulled.

A mile further downstream I enter a concentrated urban blob. Staines. I look at the Nicholson guide. The blob continues, virtually unbroken through Chertsey, Weybridge, Walton-on-Thames, Sunbury, East Molesey and Kingston all the way into London. A shudder passes through my body.

It's very soon after Staines Bridge I come closest to my first full-on collision as a cruiser careers across my bow (two empty wine bottles on the dash), aiming for a riverside pub. I press the effeminate horn and, realising my adversary is oblivious, pull hard to port. There is maybe ten feet separating us as we slide past each other heading for opposite banks. The skipper, harangued by his passengers, instantly sobers up realising his close escape, and raises his hands in abject apology. Although it's not my fault, it's a timely reminder that it is always near journey's end that you relax, trip and tear your ligaments.

I pass under the M3 which means I now have the entire M set, M1 to M6, minus the southeasterly M2. At Molesey Lock I ask about overnight mooring and am advised to pull in on the right just after the bridge. I manage to do so between two trees and discover too late that the bank is too shallow and that I've run aground. Reversing is impossible because of the tree at the stern; accelerating forward is impossible for the one at the bow. I resort to poling. I push hard. I push some more. Suddenly the pole breaks through a mud crust and, as it does so, slips from my hands. The inevitable happens. Despite my reminders to be extra vigilant, in my fourth month afloat, and on my last full day on the nontidal section of the Thames, I finally fall in.

Fortunately I'm in shorts and sandals and the water's only thigh high so it's no big deal. In fact it feels like a final benediction. As I'm in the water, it's an easy matter to push *Caroline* sideways and free her, but then there's the little matter of getting back on board as *Caroline* decides to go walkabout. I hang on to the stern fender and somehow manage to yank myself up, using it first as a seat and then as a ladder. Fortunately it's early evening and no other boats are out on the water.

Sheepishly I slip across to the opposite bank and tie up to two trees just as the rain hammers down forcing me downstairs. Two minutes later the rain has moved on. I return to the deck to a sunshine fanfare and a final slice of the last Dundee cake. I ring Teddington Lock to check tomorrow's tide and am told I'll be able to leave the lock at 4 p.m., twenty minutes before high tide, enabling me to squeeze every drop out of the ebb on the ride down to Limehouse. With exit time sorted, I telephone Limehouse Basin to book in my expected arrival time. We work out it will be between 7.30 and 8 p.m. Lizzie, who'll be manning the lock, asks the name of the boat.

'*Caroline*,' I reply.

'Is that Adelaide Marine's *Caroline*?'

'Yup.' I'm a little taken aback. There must be a fair few *Caroline*s about.

'Well, I'll look forward to welcoming you into Limehouse between seven-thirty and eight tomorrow evening,' she signs off mysteriously. Tantalising. How does she know *Caroline*? I'll have to ask.

In the morning I find the heavy overnight rain has left a swimming pool around the gas rings of the cooker. I've already put sticky tape over the crack in the fore-end light vent that I smashed with the bike handlebars. I apply a little more tape on deck. As I do so, I watch the dense traffic streaming across Hampton Bridge to work. Upstream the twisting silver of the weir tumbles like salmon. I scramble up the bank and discover I've moored alongside a red brick wall that's all that separates me from Hampton Court. There is loud chattering in the branches above. I look up and catch sight of three yellow wagtails, a flash of sulphur-yellow and olive-green.

I buy a *Telegraph*. It's best for sport on Mondays and my thirst for more Germany 1–England 5 analysis, gossip and afterglow has nowhere near been satiated yet. I read it in a greasy spoon on the East Molesey side. On an inside page I also discover that the majority of the 93 detainees being held at the Campsfield centre in Oxford have gone on hunger strike following two attempted suicides (one by swallowing razor blades and the other by leaping from a building).

The river still glows hypnotically, seemingly breathing, as I set sail. The rudder creaks, resisting the river's restlessness, in its agitation to be finally free of locks and run and run to the estuary to merge with the seas that cover two-thirds of our planet.

I slip past a succession of weatherboard chalets, a moored spritsail barge, and a brightly painted two-storey houseboat. Beyond ghostly-white Kingston Bridge, men are whipping small sailing boats and dinghies round buoys, practising tacking and gybing. Three women sit in canvas chairs sketching, a man raises the front paw of his black spaniel to wave at me. Cormorants and gulls have joined the geese and swans. Transition.

By 11 a.m. I'm already moored above the lockgate to the tidal Thames at Teddington. I have five hours before locking out. I take a quick look at the tomblike skiff lock adjacent to the main lock, and notice the salmon ladder in the weir has been donated by Shell U.K. in their pursuit of a green image makeover. Salmon ladders – concrete chutes for salmon between water levels allowing salmon upriver to spawn – are just one aspect of the invisible care the Environment Agency are involved in on the river.

With time to enjoy rather than kill, I decide to take a walk up to Richmond, the start of my own adulthood and London journey. The past, as the Ancient Greeks noted, is all out in front of me.

On Richmond Hill I sit on a bench and gaze out over possibly the loveliest Thames view of all. Beneath me is the grazing meadow I've just walked across, and Petersham Woods curling up into Richmond Park. Richmond is London's country. It seems impossible that we're just eight miles from the heart of Europe's largest, most populous city.

In a broad lazy loop, the river snakes round Eel Pie Island where the Stones and the Who used to play in the 1960s. Twenty yards behind me is Jagger's Georgian home joined at the hip to his ex-wife Jerry Hall's identikit mansion. Directly to my left is the Wick, Pete Townshend's residence complete with recording studio. A couple of doors further up, just before the gateway to Richmond Park, is the Star and Garter, heaven's waiting room for ex-servicemen. Mick, Pete and the servicemen have their pasts laid out in front of them too.

Five of my own former London addresses are linked by the river. A couple of streets away is Onslow Road, where I shared a flat with my brother when I first arrived in the big city. Across the park is Dorincourt (Lillie Langtry's old home) where the French windows would be thrown wide allowing the summer in and the Incredible String Band or Cream to drift out across the lawn as our Sunday twenty-strong unisex football matches unfolded fuelled by copious amounts of booze.

Above the French windows was the double room I shared with my brother in which a full-length mirror led into the bathroom – my first adult den. Often at night, after the park's gates had all been locked, I'd climb over the wall with my best friend, Nick Edwards, into Charles I's 2,500-acre private hunting park. We'd sit among deer and long grass over a bottle of cheap Spanish plonk, aged seventeen, blathering on passionately but inanely about Leonard Cohen, Kahlil Gibran, and Albert Camus. Once back in the house, Nick would return to his garret to paint and I would return to my ivory tower to scribble bad poems. We were hugely pretentious but growing, feeling our way blindly along corridors.

As I sit there now staring out over 'The matchless Vale of Thames', I'm dragged from the past by a sense that living people have slipped in on benches either side of me. I look to my left and my right. Two women sit, staring into the distance over their own lives. The benches that we're sharing have been bequeathed by those left behind for those recently departed: 'Still in our hearts, con amore, Ricardo Potenza given to us for 18 years till 2.11.96. This seat was given by his friends and the people whose lives he touched.' 'John Wright 1899–1984. He loved it here.'

We are a blessed people to live in such a country. More than anything my journey along the water road reminds me of this. The river is the trunk of the country, the canals its branches. The fact our waterways have never been in a healthier state is good reason for optimism.

What more does one learn from a pilgrimage such as mine? There's no burning illumination like the tidy postscript at the end of the TV western. There's no message in a bottle. The experience is the whole point of the journey: to stop bashing up against your indoor furniture and get out of yourself. You may have to wait till your family have grown or left home, you may have to save an eternity or wait for early retirement. Whatever it takes, it's worth taking, for this Everyman adventure is a door into another universe. The Water Rat got it absolutely right in *The Wind in the Willows*: 'Believe me, my young friend, there is nothing – absolutely nothing – half so much worth doing as simply messing about in boats.'

Will I live differently when I get home? I doubt it. I'll clear out the cupboard and shift it to Oxfam and then try to resist replacing it. I'll try to go more slowly – as Sebastian de Grazia declared in 1962, 'Perhaps you can judge the inner health of a land by the capacity of its people to do nothing – to lie abed musing, to amble about aimlessly, to sit having a coffee – because whoever can do nothing, letting his thoughts go where they may, must be at peace with himself.' I'll use the bike more in preference to the car. I'll try to stay in the present and focus on the task in hand. And I will fail. In London I drive and am driven. But I will get away to the water as often as I can to renew myself because I know so much depends upon the red wheelbarrow glazed with rainwater beside the white chicken.

At 3.55 I sail into Teddington Lock with *No Problem*, *Inshallah* and a host of other craft who will all peel off the Thames at Brentford for the docile waters of the Grand Union Canal while I continue on the ebb down through the city and into Limehouse.

I am back now where I started, riding England's most renowned stretch of water, the $23\frac{1}{2}$ mile reach between Teddington Lock and Greenwich. Out of a dull sky with curdled clouds, jumbo jets emerge with metronomic regularity on the Heathrow Highway. Beneath them a small boat is sailing between banks covered and uncovered twice a day by salty tides.

Just before Chiswick Bridge, half concealed in the trees, I spot the ugly brick tower of Mortlake Crematorium where my brother's ashes were scattered in a rose garden in 1968. Beside the next bridge at Barnes, I recognise a mansion block from a party I attended the following year. Pat's death in a head-on car crash was still raw and I had abandoned his chosen profession, quantity surveying, to work in a pub, the Sun Inn at Barnes. When Pat had been alive, the Sun had been our second home, the gathering point for the large circle of friends that gravitated round him. Maybe working behind the bar made me feel closer.

At around midnight on the night of the party, the landlord's son, John Fisher, and I were on the top floor of the block of flats and we were as drunk as skunks, egging each other into a race across the river and back. I was very fast down the stairs, having perfected this art through endless practice in an apartment block in Hong Kong when I was ten (three steps then two is the trick with body turned obliquely). I was consequently first out the building and across the street but by the time I'd shelled my clothes, John was even and we hit the water together. Fortunately for us the tide was slack. We were neck and neck out to a moored hopper in the middle of the river. It was on the return leg that I heard John's blood-curdling scream. He'd bumped into the hard body of a dead dog bobbing its way gently out to sea.

Thirty years on, I'm sailing by where I once swam, when a loud car alarm goes off scaring the life out of me. The BMW is parked outside the Bull's Head where I used to go with Pat to listen to live jazz acts (I remember it being bohemian and cool but the cigarette smoke hurt my eyes). A little further east, on the opposite bank, stand the obscenely expensive homes of Chiswick Mall, where I twice watched the Oxford and Cambridge Boat Race just a couple of streets from my grotty bedsit off Chiswick High Street.

My history is becoming increasingly interwoven with that of the river, another voice in its song.

The twin cupolas of Harrod's Depository rise up out of my memory like some Alhambra. One spring half-night half-day in 1976 I sat on the riverbank here watching the red brick building soften as it went through subtle gradations from darkness to dawn. I'm not sure how long I stayed because Teri, my Venezuelan girlfriend, and I had dropped tabs of LSD around 3 a.m. in my bedsit and then set out tripping. We returned to the room around 6 p.m. carrying a tree trunk between us from Richmond Park. Then the tide was high, now it's ebbing.

On a long bend I recognise the point I ran aground fourteen weeks ago between Palace Wharf and Fulham F.C.'s Craven Cottage. It's obvious to anyone with eyes that running aground on silt deposits would be probable rather than possible hugging such a bend. Obvious now but not then.

Chelsea Harbour slips by, a slingshot from where I used to pay £8 a week in the late sixties for a controlled-rent flat. At the time, I lived in an area of Chelsea known as World's End. My street, Park Walk, was just a few roads from where my father grew up on Edith Grove. When his own father returned from the First World War permanently incapacitated, Dad left school to support the family. He was thirteen at the time. Each early evening, just before closing, he'd whip into the World's End Bakery to ask if they had any leftover bread (half a century later I'd be buying my own bread there). Occasionally, if flush, Dad would pop into the World's End pub where he'd sit with local women who'd brought their spuds to peel in buckets while they enjoyed a tipple of stout and sucked on their clay pipes. Before Sloane Rangers and Range Rovers, the World's End had been anything but a wealthy area.

Just a year before embarking on the water road, I booked myself, Max and Dad, three generations of Chelsea F.C. fans, into the Chelsea Village on a nostalgia weekend. In the morning, we walked the streets of Dad's childhood and my early adulthood. Max listened politely to my father's tales of playing *penny-up* against the wall, and how in those days 'neighbourhood watch' meant looking out for police and offering sanctuary to anyone fleeing them. Max was less patient with my own tales of inhabiting a block of flats opposite a house owned by John Lennon, and watching the Soft Machine at Chelsea College of Art.

Beyond Chelsea's blue-plaque country, A.K.A. Cheyne Walk (a chichi riverside village of Queen Anne terraces that once housed Rossetti, George Eliot, Swinburne, Hilaire Belloc) are the playing fields and courtyards of Wren's Royal Hospital – founded for war veterans in 1682 – that now serve as 'barracks' to the Chelsea Pensioners. As an ex-soldier with 25 years service, Dad is eligible for the Pensioners' 'barracks' and every year he gets a Christmas card from them reminding him of this fact. He says if Mum dies first, he'll probably move in. I'm sure the major incentive is the free admission to home games at Stamford Bridge. Living in a council flat in Poynton, Cheshire, he's already used to a lack of personal space. He is, however, concerned about whether he'll be able to squeeze in his two-foot-tall speakers, CDs, tapes and vinyl.

Sailing beneath Vauxhall Bridge, where huge frost fairs were held on the iced-over Thames in the seventeenth century, I'm reminded of my feeling of vulnerability on the outward journey as the rain lashed down. Now I feel as comfortable as I did pottering along the Oxford Canal.

On my left are the Houses of Parliament outside which police frogmen escorted *Caroline* during the May Day disturbances, and beside which I'd stood with my family on Millennium Eve watching fireworks and a twenty-year marriage of a close friend splutter and burn out. Gulls are squawking argumentatively and dancing on the 150 yards of mud-swirling water that separates North London from South. Just as Greenwich Observatory is the divider of east and west, so the river separates North Londoners from their southern counterparts. To those living in Camden, Islington and Highgate, the south is an ugly and unfathomable concrete jungle; to those in Crystal Palace and Bermondsey, Kentish Town is as foreign as Newcastle.

When William Blake wrote about 'dark satanic mills', he was not in the Potteries, Manchester or Birmingham. He was on that northern bank staring across the river, for at the turn of the nineteenth century, the South Bank was a centre for heavy industry. Nowadays its satanic architecture is pretty much confined to the Royal Festival Hall, the legacy of the 1951 Festival of Britain.

I can think of no other capital that has made more of its waterfront location than London over the past two decades. New life has been breathed into moribund warehouses and wharves, and out of the abandoned and derelict seemingly every day a new restaurant, art gallery, theatre, or attraction rises. Meanwhile amid this frenzy of renovation and innovation, new cycleways and pedestrian paths have made the riverfront ever more accessible.

I slip under the sublime Millennium Bridge, the first span erected across the Thames in 100 years. A blade of light at night and a work of art by day, it is a perfect symbol of the easy coexistence between the old and the new, connecting the Tate Modern to venerable St Paul's. Beside the Globe, a hoarding is emblazoned with the only bit of Shakespeare the entire nation knows – 'All the world's a stage'. As slippery as a Japanese koan, I try to fathom its meaning. Is it the kernel of a nihilistic vision that nothing really matters because we are mere part-time players and doomed? (Think of Macbeth: 'Life's but a walking shadow, a poor player that struts and frets his time upon the stage and then is heard no more; it is a tale told by an idiot, full of sound and fury, signifying nothing.') But Shakespeare gave up ownership of his play's meaning once he put his quill down. For me the existential position is a liberating starting point: if there are no dress rehearsals or curtain calls, this is the only life and all the parts are up for grabs (but it helps if you've been to Eton).

A train lumbering across London Bridge shakes the world. *Caroline* and I shudder, and in adjacent Southwark Cathedral, John of Gaunt, England's first poet, turns in his tomb. There is a low roar as if inside a cave, the familiar sound of London. A chill has returned to the air. Spring has turned to summer and just round the corner is autumn.

Just 100 yards from London Bridge the river's first span, built by the Romans, once stood. On it numerous miracles were performed and visions seen. Close by stood a temple devoted to Isis and another to Mithras. In the seventh century St Peter appeared to a fisherman and was ferried across the river; Edward the Confessor and William Blake are just two of the hundreds of others who have had visions beside the Thames.

Light rain starts to fall and now it's the turn of a darker dance. To my left is the bricked-up Traitor's Gate where the condemned entered the Tower of London after trial upstream. Once decapitated, their heads were left on spikes on the bridge as warning to others. I pass the Angel, haunt of pirates, smugglers, bawds and press gangs. But the most infamous resident to squeeze his bulk through its narrow passageway was undoubtedly Hanging Judge Jeffreys who is said to still haunt the balcony. I strain my eyes. There's no trace, spectral or otherwise. But then it may be a little early for him. Give it a few hours and no doubt he'll be seen stumbling around, looking for another drink to help him silence the persecutory voices of those he executed on the opposite bank.

The river has dropped 370 feet and flowed 175 miles. It still has 40 more to go before Southend and open sea. But I am nearly done. The lights of Gotham City's skyscrapers have closed in. Up ahead Canary Wharf, Europe's tallest building, blinks as if blinded by the city's glow. Beneath it, on the opposite bank, lies Greenwich where everything from tin to slaves was sold, Romans and Danes landed, time started and so did my journey.

For over a million years this river has heaved mud and gravel, switched tack, and separated from the Rhine with a final kiss as Britain sailed off from mainland Europe with a desire in many hearts never to return. Its banks were settled half a million years and people were back living here 8,000 years ago. And still it keeps rolling, rolling into the night. 'Sweete Thames run softly til I end my song'.

10

Gathering the Parts

The London Ring: the Regent's Canal and the Grand Union Paddington Arm

Getting into Limehouse Basin is trickier than I anticipated. Probably because I didn't anticipate – I'd grown too relaxed and complacent. As I nonchalantly push the tiller to starboard and aim for the 50-foot-wide concrete goalposts on the northern bank, the boat barely turns. Reality hits: I may be done with the Thames but the Thames is not done with me. It has no intention of releasing its plaything until it's run us to open sea.

Caroline claws sideways like a crab. For every ten feet we manage to move to port, the ebb tide carries us ten feet further downstream. The likelihood of us missing the opening and smashing into the embankment wall is looking a banker unless I take decisive action now and turn back the way I came for a second run at the entrance. There is, however, a catch-22 – I'm almost certain the tide is stronger than the engine. It's now or never.

I hold my nerve and keep the tiller hard to starboard. The bow just squeezes into the protected waters of the outer lock. The fore-end is safe. The stern, however, is not. It's being pushed downstream by the tide and swinging the prow round so that I'm now being propelled in the direction of the upstream wall. If I decelerate to reduce impact, I'll forfeit control of the boat and a crash will be inevitable.

I push the tiller hard to port and again hold my nerve and speed until *Caroline* comes under control and is turning into a slightly more oblique collision. I then slam the boat into reverse, keeping the tiller in the same position. *Caroline* brushes, rather than demolishes, the wall.

The muscles on my neck are mountains. My shoulders sag. It's over. I'm able to moor on a half-pontoon beneath two vast cylindrical metal lockgate chambers. After a few deep breaths, I phone lock keeper Lizzie on the mobile and tell her I'm earlier than expected and waiting outside.

Within minutes what looks like a flash flood is rushing out of the cracked gates. When the water finally equalises, I sail into the belly of the lock and slip the mid-line through a holding bar. I'm fifteen dank feet below ground level in a tomb thirty yards by eight. The gate closes, bringing night's blanket down on dusk. As Lizzie raises the top paddles, something like 60,000 gallons of white water barrels towards me and slowly we rise like Grendel from the deep.

What I meet at the surface is not trolls but a futuristic marina. Lights dazzle from yachts and slither like electric eels across the basin from cruise liner-shaped apartment blocks. The slick aerial Docklands Light Railway (D.L.R.) streaks across a concertina of arches above waters that appear impossibly calm after the maelstrom of the Thames.

Lizzie is standing on the dockside to welcome me to the bright new future. 'You can moor up round the corner and use our bar if you fancy a pint.' At this particular moment I can think of no greater gift. I moor, stash my gear below deck and am heading for the bar before you can say 'windlass'.

Mainstays on yachts are being strummed by the breeze as I ring the bell at the door of the Cruise Association. Beneath the bar's painful strip lighting is an armada of pennants from sailing clubs round the globe. I order a pint and Lizzie waves me over to a table she's sharing with two friends.

There are several dead glasses on the table. Robyn, the tallest of the party, lives in one of the basin's new two-bedroom flats which currently sell for around £250,000. Tony, dressed all in black and a theatre nurse, lives on a 37-foot Birchwood cruiser in the basin in the week and spends his weekends either cruising or holed up in his Ramsgate flat.

'Nah,' Lizzie is answering Robyn. 'Gotta get going. That's my last. You'll have to get your entertainment elsewhere.'

'But you're the only entertainment round here, Lizzie. Everyone knows that. Look at last week when you tried to help that drunk up and he attacked you!'

'Get your entertainment elsewhere,' Lizzie repeats. 'Go watch the Kosovans fishing.'

'That's not entertainment, that's watching paint dry.'

'Kosovans fishing?' I interject having polished off three-quarters of my pint during the opening salvos.

'They come out when it gets dark, catch perch, roach and eels and sell them to local restaurants.'

Robyn rubs his nose vigorously. 'More fun watching the boats come into the outer lock than watching that lot.'

I anxiously look up from my beer, terrified Robyn, or worse Lizzie, has witnessed my incompetence. There's no knowing smile. I can't be certain but I think I'm safe.

'Yesterday someone on a narrowboat about your length forgot he was moored to a short pontoon and stepped off the stern straight into the drink.'

'Was he all right?'

''Course. Then last week we had a seventy-foot Dutch barge slam into the outer wall sending a wave over the people drinking on the terrace of the Barley Mow.' Lizzie finishes off her own pint. 'The funniest incident though happened last winter when Gobby Kev, as opposed to Pretty Kev, was trying to get out onto the Regent's Canal but couldn't push the lock-gate back.'

'Why?'

'There was a dead body on the other side.' Lizzie laughs.

'We had an attempted suicide too on New Year's Eve,' Tony launches into the next dark tale. 'A young man came down to the basin and stuffed his jacket full of rocks and then leapt into the water but the jacket just puffed up with air and he floated to the surface.' I start to laugh but Tony cuts me short. 'That isn't the end of it. Two black kids come by and decide to save him. Not only is he pissed off with them but a woman in a flat that overlooks the basin phones the police and tells them two black boys are mugging a white and they get arrested!' The three friends crack up with laughter.

'Make no mistake, it may be modern but it isn't soulless,' Tony sums up. 'There's a real sense of community in the basin, even if it doesn't spread up to the flats. Take me for an example. Since my partner died, the basin and the Cruise Association have been my family.'

'See? All sorts live here, even Tony!' Lizzie underscores the social mix of their waterborne neighbourhood. She then catches me sceptically staring out the window at the oceanic bows of the cruisers and £1.6

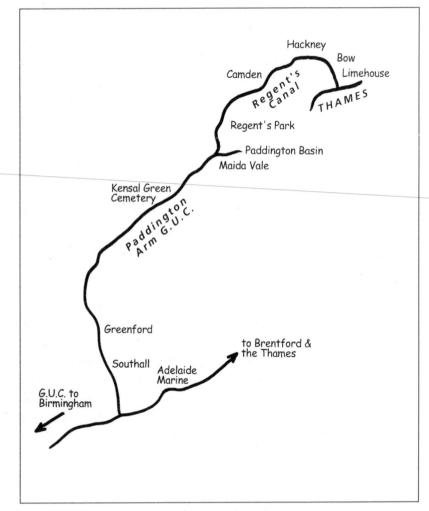

The London Ring

million penthouses. 'All right, there's people with money but they're the ones just passing through. Living here on the permanent moorings we've got a fitter, a carpenter, an I.T. engineer, a retired publican, a parole officer... What else, Tony?'

'A policeman, students, an American journalist... you know, the one who compiles those Caribbean guides?'

'And a magistrate,' Robyn chips in. 'Most live on board here in the week like Tony and then disappear either with their boats, or without, at the weekend.'

I start wondering what it would be like to live amid this Blade Runner world and ask how much it would cost to stable *Caroline* in Limehouse for the year.

'How long is the boat?' Lizzie asks.

'Fifty feet.'

She closes her eyes, the lids flicker like a fruit machine. While she computes silently, I listen to the Lovin' Spoonful playing to the virtually empty bar, 'Every night will be like tonight in the summer in the city, in the summer in the city...' Eventually Lizzie stops whirring. 'About £2,800 including water and electricity.' I run with the ball... maybe I could sell the house, pay off the mortgage, buy a boat, invest what's left, give up work, and live frugally off the rest. The great escape. Having experienced life on board when *Caroline* was packed to the gills with family, I doubt we would survive more than a week without imploding.

The conversation inevitably turns to what I'm up to and how long I've been out. When Lizzie learns I'm writing a book, I stumble on another of those wonderful English secrets I've happened on the length and breadth of the country. 'If you're a writer, you should take a gander upstairs. We've got over ten thousand maritime books in our library. It's second in size only to Greenwich.'

I look up at the strip-neon ceiling. It's impossible to imagine maritime history living upstairs but Lizzie leads the way and I discover the place is indeed packed to the rafters with books and maps. On the shelves the venerable spines, like weathered hulls, spell out the great expeditions of Vasco da Gama, Magellan, Captain Cook and the rest. Explorers don't die, they gather here – the scattered parts of the totality of our understanding – and continue chattering through the night.

'Our members spend days here planning their next trip, poring over maps and researching. In the early days of the Cruise Association you had to donate a book to join and that's how the collection got built up. We sold three hundred of the rarest ones to Cambridge University a few years back for a million pounds and bought this place with the money. A very good deal considering we can still go and see "The Cruise Association Collection" at Cambridge if we want, and we've got our place here which is probably worth around five million now.'

Downstairs, I bid goodnight to Lizzie, Tony and Robyn and head out into the night, walking to the far side of the basin to see if any Kosovans are fishing. There's just one Scottish couple hunkered down in a corner like refugees with sleeping bags pulled up to their waists. They tell me they've caught 'a couple of decent bream and a poxy perch'. The D.L.R. streaks through, bathed in its own light. Oval expressionless faces stare out from the carriages. Down on the dockside, mainstays continue to tunelessly strum and the B.W. office rotunda sprays tracer tracks like a lighthouse.

My sleep that night is filled with vivid dreams. In the first I'm unable to find anywhere close to home to park my car. A meter maid kindly grants me permission to leave it in the gap between a pond and the path leading to our front door. That's it. Or that's all I can remember of it. It may be short but it's straightforward, a kernel easy to prise open – yes, I want to return home but how do I keep the water, symbol of renewal, and the sense of adventure, with me?

The second dream is more of a challenge. In it I've murdered someone and need to hide the engine and sword I committed the act with. As my father is outside my bedroom, I temporarily slip the two incriminating objects under my bedding. When Dad leaves, I take them down through the lobby of the apartment block I'm living in. As I pass, the Asian landlord behind his desk warns me only to tune one of my televisions, rather than all of them, to the Sky network. If I don't, he says, everyone in the entire block will be forced into watching what I'm watching.

I lie in bed the next morning trying to make sense of it. Is it a warning not to bore the arse off everyone I know by being Canal Man with his canal stories when I get back? Is it a fear that by absconding for four months, I may have killed something in the family? I stare at the roses and castles design on the hatch doors. The stream running under the bridge to the castle on the lake. I don't want to live in a castle, cut off from the world. Escape is not escape if you do it all the while.

I pull out the final Nicholson guide, Book 1, from my mini-library. It's the same one I started my journey with. I'm now halfway round the London Ring travelling anticlockwise (the first leg was the Thames run from Brentford into Limehouse). There are thirteen miles, twelve double locks and two tunnels to my final overnight at Little Venice. Then it's a four- or five-hour lockless cruise to Adelaide Marine in Southall.

I make some toast and tea, and sit up on deck watching a smartly dressed woman power-walk her way through the basin, pumping her two briefcases like dumbbells. A couple of joggers jog through, chatting as they go. Boats continue to slumber in their London backwater haven.

Lock 12 opens easily – no concealed bodies thankfully – though it's a shock to be winding my own paddles again. Often referred to as 'London's Canal', the Regent's Canal was completed in 1820, linking the Paddington Arm to the Thames at Limehouse thus providing the Grand Union (then the Grand Junction) with its link to the sea. A century ago the basin would have been ringing with the clamour of colliers disgorging coal and timber onto a fleet of canal boats that then distributed the fuel domestically along London's alimentary canal. Today it's quiet and I'm the only one moving.

The city beyond Commercial Road Lock is cramped, stacked, lacking the aesthetic order and expansiveness of the Thames Valley. Alongside the Cut, road signs point to bridges smeared in the ugly identity thumbprints of tag graffiti. Old council estates peer over wastelands awaiting resurrection, the washing hanging on their balconies a reminder, just in case there's any confusion, I'm no longer in Bray, Henley or Marlow. Trains clatter and traffic streams through Mile End. Bruce, the heron, joins in for a spot of fishing.

A world away, on *Caroline*, I watch moorhens gliding round lily pads, and coots rippling their white yashmak beaks beneath blooms of purple loosestrife. The towpath, meanwhile, is a sculpture park of fishermen who haven't gone home for thirty years and have petrified with rods in hand.

I share a few locks with the 50-foot *Tarasinha II*. On board are Michael Davies and Ann Savage. A decade ago Michael swapped London banking for lock-keeping on the Thames. Now he has retired altogether and the couple are into their first six-month spell afloat. The plan is for it to become an annual migration from their flat in Maidstone.

In City Road Lock, as I wait on the bank for the water to equalise, I exchange waves and smiles with children pressed up against the mesh fence of their primary school rooftop playground.

In the upper pound, two Adelaide Marine hire boats, green and yellow twin sisters of *Caroline*, are waiting to enter as I leave. Their crews are white Zimbabwean farmers on a week's respite from the uncertainties of home. *Caroline* brags about her four-month adventure while her sisters try to beef up their more prosaic week-long jaunt. I'm comforted that *Caroline* looks to be in no worse decorative nick than they do.

More hanging baskets, and more modern developments appear as I progress westwards. At Kentish Town Lock I ask a Glaswegian lunching on the starboard balance beam if he'd mind interrupting his munching to push the gate open for me. He tries once, says it's much too heavy and sits down seemingly exhausted. I walk round the lock to open it myself and watch a dealer handing him a small plastic bag filled with illegal substances.

At the second of the three locks through Camden, a dreadlocked rasta with eyes that went AWOL several years earlier, pushes back the gate for me incomprehendingly and then sits on the beam brushing away specks from the wood.

At the third lock, alongside Camden Market, two teenage girls are sitting amid their school bags chatting and congratulating each other on their G.C.S.E. results. Two lines of gongoozlers, squeezed onto the bridge and along the wall beside Dingwalls club, stand mesmerised, like everywhere else in the country, by the arcane ritual of lockgating.

I quit Camden's tawdriness and enter the opulent emerald world of Regent's Park, passing exotic squawking birds in Lord Snowdon's 100-foot-tall aviary at the London Zoo. Beyond it is Blow-Up-Bridge where waterborne gunpowder was ignited by a spark in 1874. On the Doric columns of the replacement bridge are deep grooves worn from fifty years' rubbing by horses' towropes.

The canal continues its mock-rural romp through a collection of recently constructed neoclassical mansions more in keeping with country estates than inner-city homes. My guess is these little town pads will sell for between five and ten million, the priciest new domestic real estate currently available on the British market.

On the far side of the park, I switch on *Caroline*'s headlight for the short Maida Hill Tunnel and emerge into Little Venice, perhaps the finest stretch of urban canal in the entire country. The place is heaving with boats but I strike lucky and squeeze into a recently vacated mooring.

Flanking one bank are three-storey blocks of flats, on the other a line of handsome Regency homes that were part of John Nash's original blueprint for the area (he was also one of the three architects employed on the Regent's Canal). It must be some consolation to tenants in the apartment blocks that they have an incomparably better view than their millionaire neighbours opposite.

Having moored up, and apologised to the owner in front for giving his boat a nudge, I take a short walk to Paddington Basin where I stumble into the biggest building site I've come across since visiting Berlin in 1998. The bombsite in front of me is the latest major waterfront regeneration scheme. Where there are holes, cranes, tarpaulins and bobbing hard hats now, in a year's time there will be 1.4 million square feet of office space, canalside apartments, shops, restaurants and possibly water taxis winging commuters along the canal from Paddington Station to their London workplace.

Traffic swirls, pedestrians weave, yellow hard hats bobble. I beat a hasty return to towpath sanity.

Moored a couple of boats ahead of me is the *Ara*, a traditional butty breasted up alongside her working partner, *Archimedes*. I remember them from Norton Junction. These were the boats Nigel and Jo were looking after while the owners were holidaying in Greece. The belly of *Ara* is bulging with bags of coal, like stacked tuna in the hold of a fishing boat. Trish is leaning against the tiller drinking tea, Alan, her husband, shimmying across a plank to the bank in his steel-capped boots. I had seen him earlier unloading coal to a moored boat at City Road Basin.

'Long day?' I commiserate.

'The usual,' comes the reply, delivered in a shallow west-country timbre.

'Twelve hours a day, seven days a week, September to May.'

'Proper work. Clearly it wasn't part of the retirement plan then?'

Alan smiles. 'In fact this last winter was so bad we were still delivering into July and back at it in August.' He attempts to open a bottle of

Budweiser and adjust his mooring ropes simultaneously before calling across to Trish to throw over another beer. The first one, now topped, he passes to me. Trish duly obliges and I awkwardly stand around on the bankside as Alan walks the plank again while opening his own beer at the same time. 'Come on board if you fancy. We're just winding down.'

Descending the stairs of *Ara* to the cabin is like going down a mineshaft. Alan and Trish live like moles in the butty. 'Such a small space to live in.' The insensitive words come out unbidden, more to myself than my hosts and I immediately wish I could reel them back in. I've seen enough butties already to know what size living space to expect.

Alan shrugs his shoulders. 'We're outside all day. Cabins on working boats weren't designed for leisure living like hire boats. After work, we have something to eat and either go to the pub or to bed because bed's the most comfortable place in a back cabin. Anyway this is *big*.' Alan emphasises the final word. 'When we got *Ara*, she had the original eight-foot-six by six-foot-six living space of the traditional working butty that had to fit in the cooking range, and all this' – Alan makes an abbreviated sweep of a hand round the cabin – 'and the two bunks that had to sleep a family of five.'

Alan and Trish extended the cabin by a few feet to squeeze in a kitchen you couldn't swing a cat in. Or dog in their case. Tallow, a brown mongrel, sits in his basket loudly gnawing on a bone at our feet. 'We trade under the name Candlebridge Carrying Company, so Tallow was an obvious choice.'

Alan is perched on the bunk beside me, his beard a dreamcatcher of coal dust. I look round the cramped interior. It is a gallery of roses-and-castles metal jugs and bowls, and suspended lace plates patterned with billowing clouds from the brass kettle hissing on the range. It's hard to imagine how one person, let alone two, could fail to feel claustrophobic shoehorned into this cabin year round.

'What happens when you have a barney? How do you get away from each other?'

'Alan goes on the motor for a bit,' Trish answers without breaking concentration on the potato she's slicing into quarters on a table that will fold up and become the crockery cupboard door when she's done. 'This is the sixth winter we've been carrying freight full time. And before that we played at it for ten years so we're used to the space. Intimacy is a way of life on the canals anyway, you should know that. There's never any waste of space on canal boats.'

From intimacy, our conversation naturally segues its way to the Towpath Telegraph. 'I love it,' Trish confides, 'so does Alan. It's part and parcel of the village feel of the Cut. Don't you think so, Paul?'

I nod agreement. Just a couple of days back telephoning Adelaide Marine to inform them *Caroline* would soon be back in her stable, I was told several boaters had called in over the past few months asking what on earth one of the Adelaide fleet had been doing up on the Leeds & Liverpool or Caldon Canals.

Alan scratches his head, dislodging more coal dust. 'Just before you came on board I was chatting with a man from the Midlands who told me someone I know really well had died. I'll now be spreading the news as I travel about delivering. It's like Chinese whispers. I've got one old fellah up in Birmingham we deliver to who regularly relays messages through us to his son down on his London houseboat. Don't ask me why he doesn't simply telephone. Maybe it makes him feel his boy's closer, just round the corner.'

Before they started carrying fuel, Alan was a printer and Trish a nurse and they kept their jobs going while they learned their new trade from other boatmen. 'There's no way we'd go back. Who'd swap watching the kingfishers, and keeping tabs on the families of swans growing up? If you're on dry land or driving a car you miss it all.'

Alan believes *Archimedes* and *Ara* are the only Little Woolwich Class originals still in working trim and carrying. 'The *Ara* took the last council load on the Ashby Canal from Croxley Mills in August 1970. She then became a camping boat and took kids out for twenty years. When we bought her, she took the first load like a dream despite the fact she'd not carried freight for thirty years. She loves what she's doing just as much as we do.'

As I walk back past the line of moored boats towards *Caroline*, I ponder the fundamental difference between those living on boats full time like Trish and Alan and those inhabiting the mainland. Living afloat you become part of the land you moor to but not of it, attached but itinerant, a water gypsy. The nonattachment to a particular bed, to having a bath, to the round of familiar shops and friends, is reflected in a more outward-going attitude. Those living on boats may share the same health service and television channels, but how things are totted up, measured, and valued, is not the same. Instead of cornerstones, there are overnight moorings and overnight friends in a fluid life. It strikes me the Cut is the Buddhist 'middle way' between attachment and nonattachment: its denizens, adrift from relatives and close friends, relating to perfect strangers as the member parts of their extended family.

Back on board *Caroline*, in bed I watch the sky darken through the marbled curtains and then lighten eight hours later for the last time. I listen to the low unintelligible words of a couple talking across on the opposite bank. The smell of bacon wafts across with their voices. Is it coming from a flat or another boat? It doesn't really matter, what does matter is the smell is the most violent torture known to man. It was this heady aroma that both broke my vegetarianism and led me into my first marriage in my early twenties (after a pub lunchtime session, I smelled bacon cooking in someone's flat and promised I'd marry my girlfriend if she'd make me a bacon butty). If I'd done what I do now – make the damned thing myself – my whole life no doubt would have been different. Since then I've swapped countries, careers and even wives. And here I am approaching yet another gut-wrenching ending.

A dog yaps, a police siren wails. There is the distant rumble of a train, and the quick noisy climax of a car passing. Light ripples its liquid musical score across the curtain. I raise myself. Up on deck two duplicate gossamer cobwebs have been spun during the night from the tiller, as fragile as history.

I ring home to ask Susanna to pick me up from Adelaide Marine later in the day. Unfortunately the mobile's packed up on me. I set off in search of a B.T. box and try six before one accepts my coins – a reminder of how primitive and unreliable a navigation the telephone is. The only phone box still working is located beneath Westway. Above me is a rash of unmoving cars and beyond them, jumbo jets wheeling aimlessly in the sky, kicking their heels, awaiting a landing slot.

I clear the weed hatch for the last time, give *Caroline* her customary thirty-second warm-up and switch on. Like every other day she bursts into life first time and purrs loudly like a big cat. We slip land.

I'm not singing or whistling and wonder if *Caroline* knows today is different. Every other day we have been sailing forward. Today we're sailing back, and I'm burrowing into our recent shared past. Together we've cruised through canal villages like Stoke Bruerne whose windows and doors salaamed as we passed; struggled with lift bridges on the Caldon; tiptoed through the crumbling mills of Lancashire; been buffeted on the Aire & Calder by twelve-in-a-line coal-laden compartment boats each carrying 170 tons; lashed ourselves to a broken-down cruiser amid the tidal pandemonium of the Trent; swept beneath Lincoln Cathedral on a Roman navigation; and moored in forever-England countryside with only birds for neighbours.

Now we're passing lines of satellite dishes fixed to balconies, all facing upwards, basking in the rays of the Murdoch Sky God. Beneath the towpath, invisibly, ghosts dance their way into homes along fibreoptic cables. We are back in the city where living is done indoors. In need of help with endings, I pull in by Kensal Green Bridge. I haven't booked an appointment but the experts I'm hoping to see are unlikely to be out.

I leave *Caroline* moored between two bobble-hatted rasta fishermen and pass through a gateway. On a notice board under the arch I check out the Programme of Events 2001. There's a lecture on 'Burial Before Undertakers – Death in Early Modern England'; another on 'Disposal of the Dead in the Hindu Kush'; and a third on 'Sir Marc Brunel – The Forgotten Genius'. There's also an upcoming 'Musical Evening with a Funereal Theme in the Dissenters Chapel'. There are, however, no events today, just unmoving tombstones to walk between, the pages of the Who's Who of the Dead.

Kensal Green Cemetery, London's first private graveyard, was licensed by Act of Parliament to open in 1832 after a cholera epidemic. At the time an estimated 50,000 corpses were annually being shovelled into 150 churchyard pits and, according to Dickens, 'rot and mildew and dead citizens formed the uppermost scent' of the capital.

Among the tombstones I find one dedicated to one of the principal players in the demise of the canals, Isambard Kingdom Brunel (1806–59), engineer of the Great Western Railway and the Clifton Suspension Bridge (he also created the first transatlantic vessel, the steamship *Great Western*). Nearby stands the tomb of his father Sir Marc Isambard Brunel (1769–1849) who soared to even dizzier heights and plumbed greater depths, serving six years as New York's chief architect and engineer and six years incarcerated in the King's Bench Debtors' Prison. In between, he created a tunnel beneath the Thames between Wapping and Rotherhithe, invented an air engine, a knitting machine, floating piers, and the mechanised blocks which revolutionised shipbuilding.

There are also tombs commemorating literary figures Anthony Trollope, Wilkie Collins and William Makepeace Thackeray. It is the latter who provides most help in the art of ending, having written his own obituary on the final page of *Vanity Fair*. 'Which of us is happy in this world? Which of us has his desire? Or, having it, is satisfied? – Come, children, let us shut up the box and the puppets, for our play is played out.'

At the other end of the cemetery is the tomb of Emile Blondin (1824–97), who crossed Niagara Falls over a hundred times on a tightrope using progressively more taxing means such as a blindfold, stilts, a wheelbarrow, and another man he carried on his back. At the age of 65, in need of a change of view, he set up a stove in the middle of a tightrope from Manhattan Island to Staten Island on which he cooked omelettes for his audience. The tombstone declares simply that 'On February 22 1897, at the age of 73, he died at Niagara House in Ealing.'

Blondin, it appears, took the waters of Niagara in spirit at least with him when he moved to England. As a memento of my own journey, I've put two B.W. keys I've used throughout the journey to access lockgates and water taps inside a roses-and-castles baccy tin I bought on the Oxford Canal. When I get home I'll create a small pilgrim's shrine ('Get a life,' I hear my daughter already admonishing): a coiled hemp rope, a windlass and my roses-and-castles handbowl.

The mementos of the dead in the cemetery come in the form of Asiatic-looking caryatids, Egyptian sphinxes, and Greek temples. There are family vaults the size of my home and there's the palatial sarcophagus of Princess Sophia, fifth daughter of King George III. But even when Kensal Green Cemetery was primarily the estate of the upper and middle classes, there was a section of 'guinea graves' for the poor. Boat people too – the most recent new tenant arriving six months back – were brought by waterborne funeral cortege to the canalside iron gate for burial.

The majority of headstones are thus simple affairs, an open book on England's true Everyman – Balachou, Gavrilovic, Jimenez, Toumazou, Costa, Maric, Chen, Messai, Singh, Czolowski. They, or their predecessors, journeyed by water to these shores from every corner of the globe to contribute to the dynamism of Englishness and Britain's success. And on their final ebb tide they gathered here at this 77-acre canalside garden in the heart of the capital.

Back on the Cut, *No Problem* sails past in the opposite direction from Brentford with a smile as wide as the canal. I pass three stationary trains outside Willesden Junction, hold-ups now as reliable as arrival times on the Japanese Shinkansen. It's then on to an aqueduct across the North Circular Road. The traffic is backed up all the way to Hanger Lane.

A little further on I wave and exchange greetings with *Inshallah*. A coconut bobbles by, a sure sign I'm close to Southall and Adelaide Marine as these are often launched into the water during Hindu funerals. Sometimes they even conceal ceremonial bracelets and chains.

Thirty years ago when I briefly taught in Southall at Featherstone and Villiers secondary schools, I remember during Diwali seeing the canal lit up with floating candles like the Ganges at Varanasi. Baptism, anointment, catharsis, resurrection, the river of life, the ferry ride of the dead: it is by water we journey into life and into death. I am at the end of the Cut's parallel universe. An Indian boy and girl are snogging enthusiastically on the towpath. Spring is long gone but not for them. Above them white gilded clouds drift across a factory chimney. Sikh kids in their topknot hankies charge about Southall park playing football.

I have reached the end of sanity, of slow time, of watching stars from my bed, of exchanging stories and assistance in lockgates, of waving to every passing boat. My wife will be waiting with the car and a boot full of empty kitbags needing filling. It's time to pack my memories and head home. Winter's approaching – time to burrow back into the city and wait for spring. By then the Anderton Lift will be operating. I can descend to the Weaver Navigation, sweep down to Runcorn, and up to Lancaster on the new link. Over the next few months my family will find me hunched over *Canal Boat* and *Waterways World* in the morning at breakfast, scouring the ads for cheap narrowboats. I take *Caroline* back to her green and yellow family and return to mine. Bruce is nowhere to be seen. The Grand Union continues on its journey to Birmingham.

Addendum

For more information on Britain's inland waterways, contact British Waterways on 01923 201120 or visit website www.britishwaterways.co.uk

To check out the major boating fleets on individual stretches, contact Waterways Holidays U.K. on 08702 415956 or visit website www.WaterwayHolidaysUK.com

Drifters (08457 626252; www.drifters.co.uk) is a consortium of boat operators which also features a late booking service (call Latelink 01905 610550).

For advice and information on sailing the Thames, contact the Environment Agency on 01189 535000.

Bibliography

David Blagrove. *Braunston…A Canal History*. (self-published) 1995

James Boughey (revised by). *Hadfield's British Canals*. Budding Books 1998

Anthony Burton. *The Canal Builders*. Eyre Methuen 1972

Joseph Campbell. *The Masks of God: Occidental Mythology*. Condor, 1974

Margaret Cornish. *Troubled Waters*. M & M Baldwin, 1994

Judi Culbertson and Tom Randall. *Permanent Londoners*. Robson Books, 2000

Jane Cumberlidge. *Inland Waterways of Great Britain*. Imray Laurie Norie & Wilson Ltd, 1998

Mark Davies and Catherine Robinson. *Our Canal in Oxford*. Towpath Press, 1999

Charles Dickens. *Dictionary of the Thames*. Old House Books, 1995

James Gleick. *FASTER*. Abacus, 1999

Charles Hadfield. *The Canal Age*. David & Charles, 1968

Jerome K. Jerome. *Three Men in a Boat*. Penguin Books, 1999

Ian Mackersey. *Tom Rolt and the Cressy Years*. M & M Baldwin, 1985

Hugh McKnight. *The Shell Book of Inland Waterways*. David & Charles, 1981

William Morris. *Selected Writings*. Nonesuch Press, 1946

George Orwell. *The Road to Wigan Pier*. Penguin Books, 1989

David Owen. *Exploring England by Canal*. David & Charles

Edward Paget-Tomlinson. *The Illustrated History of Canal and River Navigations*. Sheffield Academic Press, 1994

Jeremy Paxman. *The English*. Penguin Books, 1999

John Payne. *Journey up the Thames*. Five Leaves, 2000

Joseph Priestley. *Historical Account of the Navigable Rivers, Canals and Railways throughout Great Britain*. Frank Cass & Co, 1967.

Brian Roberts. *Britain's Waterways: A Unique Insight*. Geo Projects, 1999

L.T.C. Rolt. *Narrow Boat*. Budding Books, 1994

Mary Shelley. *Frankenstein*. Penguin Classics, 1992

Samuel Smiles. *James Brindley and the Early Engineers*. Tee Publishing, 1999

R.L. Stevenson. *An Inland Voyage*. Chatto & Windus, 1925

Sally Varlow. *A Reader's Guide to Writer's Britain*. English Tourist Board, 1996

Derek Williams. *Romans and Barbarians*. St Martin's Press, 1996

Susan Woolfitt. *Idle Women*. M & M Baldwin, 1995

Maps

Geo Projects maps – B.C.N., G.U.C. maps 1–4, Oxford Canal, Staffs and Worcs Canal; Trent & Mersey Canal; Birmingham Canal Navigations; Inland Waterways of Britain route planner; Thames Ring and London Ring Atlas. *Nicholson Guide to the Waterways* Books 1–7;

Waterways World Inland Waterways of Great Britain: A Complete Route Planning and Restoration Map.

Navigator Series, Belmont Press. Route planner.

Good sources for inland waterways related books and maps are: M & M Baldwin, 01299 270110; canal@mbaldwin.free-online.co.uk and the Inland Waterways Association (IWA) Mail Order Service telephone 01923 711 114; email iwa@waterwaysworld.org.uk

Index

Bradford 177, 238

Braunston 49, 52, 242, 244–54, 257

Braunston Marina 49, 52

Braunston Tunnel 252

Brayford Pool 207, 211

Brent River Park 21

Brentford 12, 328, 339

Brentford Gauging Lock 19

Bridgewater Boats 32

Bridgewater Canal 114–20

Brindley, James 7, 13, 70, 80–3, 91, 94, 103, 105–7, 118

Brindleyplace 62, 64

British National Party (B.N.P.) 153, 169, 232, 239

British Waterways (B.W.) 14, 33, 71, 104, 108, 155, 188, 193, 202, 235, 236, 291, 301

Brixton 237

Broad Street, Birmingham 67

Brown, Alice 247–8

Brown, Capability 310

Brunel, Isambard Kingdom 320, 345

Brunel, Sir Marc Isambard 320, 345

Bulbourne Workshops 33

Bulger, James 170

Bull's Bridge 24, 248, 250

Bull's Head, Barnes 328

Burgess, Dorling & Leigh 97–103

Burnley 151–3, 210

Burnley Straight Mile 151, 153

Buscot Lock 287

Burslem 82–5, 99

Burton, Anthony 103

Butcher's Arms, Kings Sutton 260

Byrne, David 281

Byron, Lord 218

Caine, Michael 292, 296, 297

Calf Heath Wood 76

Campbell, Joseph 278

Campsfield 292, 325

Camp Hill 61

Canal Mania 144–5

Canal & Riverboat magazine 299

Canal Boat magazine 299, 348

Canal du Midi 120

Canary Wharf 332

Candlebridge Carrying Company 342

Cannock Chase 81

Cape of Good Hope, West Bromwich 54, 247

Caroline – her history 18–19

Carpenter's Arms, Slapton 36

Cassidy, Mick 215

Cassiobury Park 31

Castleford 186, 187, 188

Castle Gardens, Leicester 227, 230

Cathedral of the Boatpeople, Braunston 244

Catherine de Barnes 59

Catherine Wheel, Goring 297

Celebrity Balti House, Birmingham 65

Chamber, Craig 85–6

Chapman, Chris 201–6, 213, 214–18